THE SECRET OF LIFE

THE SECRET
OF LIFE

*Rosalind Franklin, James Watson,
Francis Crick, and the Discovery
of DNA's Double Helix*

Howard Markel

W. W. NORTON & COMPANY
Independent Publishers Since 1923

For information about permission to reproduce selections from this book, write to
Permissions, W. W. Norton & Company, Inc., 500 Fifth Avenue, New York, NY 10110

For information about special discounts for bulk purchases, please contact
W. W. Norton Special Sales at specialsales@wwnorton.com or 800-233-4830

Manufacturing by LSC Communications, Harrisonburg
Book design by Chris Welch
Production manager: Anna Oler

Library of Congress Cataloging-in-Publication Data

Names: Markel, Howard, author.
Title: The secret of life : Rosalind Franklin, James Watson, Francis Crick,
and the discovery of DNA's double helix / Howard Markel.
Description: First edition. | New York, N.Y. : W. W. Norton & Company, 2021. |
Includes bibliographical references and index.
Identifiers: LCCN 2021013018 | ISBN 9781324002239 (hardcover) |
ISBN 9781324002246 (epub)
Subjects: LCSH: Franklin, Rosalind, 1920–1958. | Watson, James D., 1928– |
Crick, Francis, 1916–2004. | Wilkins, Maurice, 1916–2004. | Pauling, Linus, 1901–1994.
| DNA—Structure—Research—History—20th century. | Genetic code—Research—
History—20th century. | Scientists—Biography.
Classification: LCC QP624.5.S78 M37 2021 | DDC 572.8/60922—dc23
LC record available at https://lccn.loc.gov/2021013018

W. W. Norton & Company, Inc., 500 Fifth Avenue, New York, N.Y. 10110
www.wwnorton.com

W. W. Norton & Company Ltd., 15 Carlisle Street, London W1D 3BS

1 2 3 4 5 6 7 8 9 0

In memory of
Dr. M. Deborah Gordin Markel
August 1, 1958—October 16, 1988
A dedicated scientist whose life was heartbreakingly cut short by cancer
ת' נ' צ' ב' ה'

Contents

PART IV: MORATORIUM, 1952

PART V: THE HOME STRETCH, NOVEMBER 1952–APRIL 1953

PART VI: THE NOBEL PRIZE

THE SECRET
OF LIFE

PART I

PROLOGUE

Toutes les histoires anciennes, comme le disait un de nos beaux esprits, ne sont que des fables convenues.
(All the ancient histories, as one of our wits has said, are but fables that have been agreed upon.)

—VOLTAIRE[1]

For my part, I consider that it will be found much better by all Parties to leave the past to history, especially as I propose to write that history myself.

—WINSTON CHURCHILL[2]

[1]

Opening Credits

Every schoolboy knows that DNA is a very long chemical message written in a four-letter language . . . Of course, now that we know the answer, it all seems so completely obvious that no one nowadays remembers just how puzzling the problem seemed then . . . In research the front line is almost always in a fog.

—FRANCIS CRICK[1]

On February 28, 1953, shortly after the chapel bells struck noon, two men hurtled down a stairwell of Cambridge University's Cavendish Physics Laboratory. Bursting with exhilaration, they had just made the scientific discovery of a lifetime and wanted to tell their colleagues about it. Reaching the ground floor landing first, with a loud thud, was James D. Watson, a twenty-five-year-old, wooly-haired biologist from Chicago, Illinois. One step behind him, and more careful in his descent down the steps, was Francis H. C. Crick, a thirty-seven-year-old British physicist from the hamlet of Weston Favell in Northampton.[2]

If this moment was made into a Hollywood motion picture, we might begin with an aerial view of the University of Cambridge and move in to the lovely English gardens of Clare College—where Watson once took his rooms. The camera then pans over the banks of the shallow Cam River, focusing briefly on a punter negotiating his narrow, square-bowed boat downstream. Next, we move across "the backs," the magnificent lawns of Trinity and King's colleges, before gazing upward at too many stone spires to count.

Our two scientists, now running breathlessly, ties askew and jacket

St. Bene't's Church, (Exterior), Cambridge.

E 32180

St. Bene't's Church, Cambridge.

tails flapping behind them, exit through the double Gothic-arched doors of the Cavendish Laboratory. They dash down Free School Lane, a short, winding path made up of irregular slabs of weathered flagstone. Rushing past a thicket of ancient trees obscuring St. Bene't's parish church, with its squat Saxon tower dating back to 1033, the two men dart around a wrought-iron fence cluttered with bicycles, the chief mode of transportation for many Cambridge students, fellows, and professors.

The duo's destination that breezy but unseasonably sunny afternoon was the Eagle pub.[3] The Eagle—a mere one hundred and one steps from the Cavendish—first opened its doors on the north side of Bene't Street in 1667. Known then as the Eagle and Child, the public house's major attraction was "beer for three gallons a penny." It has been a favorite watering hole for Cambridge dons and students ever since. During the Second World War, the Eagle was the unofficial headquarters of the Royal Air Force units stationed nearby. In one of its rooms, a megillah of graffiti, doodles, and squadron numbers is drawn, burned, and scratched onto the walls. A now-forgotten pilot even managed to adorn the ceiling with a painting of a voluptuous, scantily-clad woman.

Six days a week, Watson and Crick lunched in a cozy salon sand-wiched between the RAF Room and an oaken bar studded with the colorful pulls of stouts, ales, and lagers. On February 28, when they approached its doorway, panting and perspiring, the Eagle was already packed with academics tucking into steaming plates of bangers and mash, fish and chips, steak and kidney pie, and other noonday offerings. Between bites and sips, these brilliant men of Cambridge loudly debated nearly every aspect of the human condition.

The biologist and the physicist were there to make even more noise. They had just worked out the double helix structure of deoxyribonucleic acid, better known as DNA. Francis "winged" into the pub, shouting at the top of his voice, "We have discovered the secret of life!"[4] That is, Jim Watson *claimed* this was how it happened—even though, for the rest of his life, Crick politely but firmly denied ever making such a statement that fateful afternoon.[5]

The Eagle pub.

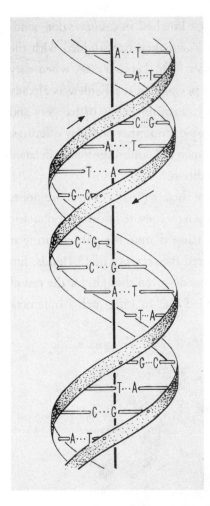

Schematic illustration of the double helix. The two sugar–phosphate backbones twist about on the outside with the flat hydrogen-bonded base pairs forming the core. Seen this way, the structure resembles a spiral staircase with the base pairs forming the steps. (Artist: Odile Crick, 1953).

Boasting of this sort was looked down upon among the scholarly tribes of Cambridge, a code to which Crick only occasionally subscribed. Yet it is indisputable that on this particular day Watson and Crick *did* discover the secret of life, or, at least, its central biological secret. Their elucidation of DNA's structure followed a centuries-old dictum, which continues to this very day: one must determine the *structure*, or *anatomy*, of a biological unit before being able to fully understand (and manipulate) its *function*. When it comes to DNA, virtually every advance in our modern understanding of how genetic information is carried is predicated on this foundational discovery. We can say, without fear of contradiction, that February 28, 1953, represents a light-switch moment in the history of science—for that matter, in human history, period. Once that switch was turned on, nothing about our understanding of heredity, the life sciences, and the body was ever the same again. Everything changed, as if emerging from an age of darkness into one of total illumination.[6]

The elucidation of the double helix explained the central role DNA plays as a single, living cell divides into two new cells, each containing and manifesting a copy of the parental DNA. The building blocks of DNA are called nucleotides; each one consists of a sugar, or carbohydrate, attached to a phosphate group (a phosphorus atom bound to four oxygen atoms) and a nitrogen base. The nitrogen bases in DNA are classified chemically as purines and pyrimidines. We now know that the purines (guanine and adenine) in one helical chain are hydrogen-bonded with complement pyrimidines (cytosine and thymine) in the other, creating the stairs to a spiral staircase. Its dual railings, or backbones, are the connected sugar and phosphate groups. This molecule, joined with billions more DNA molecules, contains within its double helical structure a precise order of purine and pyrimidine bases.

The long double helix of nucleotide molecules comprises the so-called secret of life—what we now refer to as the genetic code. Watson and Crick's insight ultimately led to the elaboration of the $E = mc^2$ formula of genetics, which Crick later termed the "central dogma of molecular biology": *DNA→RNA→protein.*

DURING THE FIRST HALF OF the twentieth century, physicists ruled the scientific roost.[7] They thrilled the world with the important discoveries of the atom, X-rays and radioactivity, the photoelectric effect, the special and general theories of relativity, and, for those taking measurements of such fundamental physical phenomena, the "uncertainty" of it all. These achievements drastically altered our view of nature and elevated science to a level of social prominence that its practitioners could hardly have imagined in 1900.[8]

A signal triumph in modern physics was the theory of quantum mechanics—developed (and reworked to accommodate other theories) by "the great Dane" Niels Bohr, the Austrian Erwin Schrödinger, the German trio of Max Planck, Albert Einstein, and Werner von Heisenberg, the Budapest-born Leo Szilard, and many others. These scientists sought to explain the physical world by gazing deeply, where no human eye could possibly see, far into the workings of the atom and its tiniest components—the electron, neutron, and proton, and, more recently, sub-atomic quarks and the Higgs boson. They dreamed up a series of breathtaking mathematical abstractions to explain and even predict natural science. Hence, it was the theoretical physicists who became world famous, as opposed to the drones who ground out the experimental data needed to prove their brilliant theories.[9]

During the Second World War, Allied physicists worked with mathematicians, chemists, and engineers to develop radar, sonar, the jet engine, and plastics, advance the fields of electronics and electromagnets, and break the Nazis' Enigma code using the newly emerging technology of computers.[10] Other physicists, based in Los Alamos, New Mexico; Oak Ridge, Tennessee; and Hanford, Washington, harnessed the power of uranium and plutonium atoms to develop the atomic bombs that devastated Hiroshima and Nagasaki.

Once they realized the horror of what their work had wrought, many of these scientists pledged never to engage again in making munitions. Instead, the loudest conversation on new scientific vistas to explore shifted to the mechanics of life at the smallest, or quantum, level: the molecules that make up blood, muscles, neurons, organs, and cells (hence the terms "molecular biology" and "biophysics"). As James Watson recounted, "in post–World War II academia, the only subject that there was universal excitement about was physics. The chemistry revolution came out of physics. The biology revolution, which also had its origins in physics, really didn't start until the DNA structure."[11]

IN 1950, NO ONE—not even the best scientific minds on the planet—knew how genes passed an organism's critical data and characteristics on to its progeny. How did genes function? Were there intermediary messengers in a cell's cytoplasm or residing within its nucleus? How did these two different parts of a cell—the cytoplasm and the nucleus—interact? Was there a genetic code and, if so, how did it communicate such diverse information? Given their long, tortuous chains of amino acids, with the potential for an infinite number of permutations, were proteins the key player in programming how cells replicate? Or was it the poorly understood DNA? And, if so, how could DNA transmit such complex genetic information? Since it contained only four different nitrogen bases (adenine, guanine, thymine, and cytosine), wasn't DNA's chemical lexicon too limited, too simple, to act as the Rosetta Stone of life?

Perhaps the most influential guide in the long slog from physics to biology was Erwin Schrödinger. He was best known for creating an equation that allowed physicists to calculate the wave function of a system and, to address his increasing discomfort with quantum theory, a thought experiment known as "Schrödinger's cat."[12] He shared the Nobel Prize in Physics in 1933 for "the discovery of new productive forms of atomic energy."[13] Schrödinger entered the annals of biology in 1944, after publishing a slim book, *What Is Life?: The Physical Aspect of the Living Cell*, based on a series of lectures he gave in 1943 at the Institute for Advanced Studies of Trinity College, Dublin.[14] No other single publication can claim such enormous influence on the conception of molecular biology. In independent interviews, James Watson, Francis Crick, and Maurice Wilkins all recalled being gobsmacked by Schrödinger's book, and the outsized impact it had on their scientific lives.

What Is Life? described the work of a German-born physicist named Max Delbrück and posed four biophysical queries: What is the gene? Is the gene the smallest unit of heredity? What are the molecules and atoms that make up genes? How are parental traits passed on to successive generations? To this end, Schrödinger hypothesized the existence of an aperiodic crystal or solid, "a gene—or perhaps the whole chromosome fiber," which is made up of molecules that repeat or arrange themselves in a specific, regular sequence.[15] He further suggested that within the chemical bonds of these genes resided the genetic information guiding life, disease, and reproduction. This focus convinced the young James Watson—and many others—of the critical importance of determining the exact positions of the gene's constituent atoms, not only the many chemical bonds but also their precise arrangement in space.

Beginning in 1947, the Medical Research Council of Great Britain granted the department of physics at King's College, University of London, £22,000 to conduct "an experiment in biophysics . . . by study of cells, especially living cells, their components and products." Determining the structure of DNA, and the role it played in the lives of cells, was one of the several objectives of this grant.[16] The King's group had the best experimental equipment, the best DNA samples, and, at least on paper, the right people to figure it out the old-school, scientific way—with the slow and steady accretion of data. Unfortunately, their work was plagued by a dysfunctional relationship between two key investigators: the high-strung, bumbling Maurice Wilkins and the sharp-tongued, chary Rosalind Franklin. Their every interaction was battered and bent by a chain reaction of fights, slights, gender and cultural differences, patriarchy, and power dynamics, and the resultant delays in their research.

Cambridge's Cavendish Laboratory housed the coincidental pairing of James Watson and Francis Crick, each of whom could finish the other's sentences before the words were spoken. They had been assigned to the same small office by their supervisors, who were tired of being interrupted by both men's nonstop, boisterous banter. The Cavendish, too, had received a hefty grant from the Medical Research Council, but its Biophysics Research Unit was charged with determining the structure of hemoglobin, the molecule in red blood cells that binds and carries oxygen. Watson was so uninspired by this work that he proceeded to rupture Britain's scientific rules of conduct. The brash Midwestern American cared not a fig about following the English gentlemen's dictum never to poach another unit's designated topic of research. He was hell-bent on figuring out DNA, no matter what the cost or offense to his colleagues at King's. Perhaps the most egregious example of this behavior was Watson's "borrowing" of Rosalind Franklin's data, without her knowledge, to complete the puzzle.

An ocean and a continent away, in Pasadena—was Linus Pauling of the California Institute of Technology, uniformly praised as the world's greatest living chemist. In 1951, working with the full faith and credit of the Rockefeller Foundation, Pauling humiliated the Cavendish Laboratory by beating them to the discovery of the helical configuration of proteins.[17] These roles reversed in 1953, when Pauling postulated a structure for DNA that proved disastrously wrong.

FIFTEEN YEARS AFTER THE FACT, Watson commandeered the historical record with boundless guile and cunning. The sledgehammer he used was a mean-spirited yet irresistible memoir that appears to have been written while he was still a young man. In reality, Watson carefully composed this book over several summer vacations as a Harvard biology professor in his late thirties. The result was the iconic 1968 best-seller, *The Double Helix: A Personal Account of the Discovery of the Structure of DNA*.[18] To return to the idiom of Hollywood, Watson's *The Double Helix* might be best pitched as a story of Boys Meet Girl—Girl Humiliates Boys—Boys Conspire to Win, and Do. A masterpiece of scientific sleuthing, *The Double Helix* all but guarantees Watson's version is the loudest voice in any subsequent DNA narrative.

Forty-eight years later, on May 16, 2016, a Who's Who of molecular biology gathered at the Cold Spring Harbor Laboratory, a leafy campus on the north shore of Long Island devoted to exploring the genetic keys to life and disease. The tallest building there is a terra cotta and red-brick clock tower featuring a double helical staircase. On each of the tower's four sides are green Connemara marble plaques marked with the abbreviations for the bases that make up DNA: *a, t, g, c,* for adenine, thymine, guanine, and cytosine. It is Watson's monument to himself even if he grumbled that the architect got it wrong by using lower-case letters instead of upper-case ones, as every scientist has done since Watson and Crick's remarkable discovery was published in the April 25, 1953, issue of *Nature*.

The occasion was titled "Celebrating Francis," in honor of the one-hundredth anniversary of Francis Crick's birth (he died at the age of eighty-eight, on July 28, 2004). Opening the symposium was another eighty-eight-year-old man, his shoulders stooped from years spent at a desk, and with wisps of white hair taking flight off from his pink and mottled scalp. He stood at the podium in the well of a beautiful, new auditorium he had recently raised the funds to build, reveling in the attention of an audience he had commanded to appear in Cold Spring Harbor. He was "King James" Watson and this was his undisputed scientific kingdom.

Watson began his remarks by repeating the Eagle pub story he told so famously in *The Double Helix*. This time, however, he finally admitted to inventing Francis Crick's exclamation of having discovered the secret of life—"for dramatic effect."[19] Two years later, in the summer of 2018, sitting in the shadow of the Cold Spring Harbor double-helical bell tower, he explained his word choice more emphatically: "Francis *should* have said it and *would* have said it. So, it was totally in character when I wrote that, and everyone would think it."[20]

It never happened. The most famous scientific announcement of the twentieth century was not made in precisely the way most of us were taught in high school. This apocryphal moment, like so many others constituting the epic search for DNA's structure, has long been exaggerated, altered, shaped, and embellished. An unwieldy tower of memoirs, journalistic accounts, and biographies have each told the DNA story from the viewpoint of one participant or another, so that by now the history has taken on the contours of a *Rashomon*-like legend. Much of what the layperson concludes depends on whose version he read last.

James Watson often dismissed his detractors with the quip, "There are only molecules. Everything else is sociology."[21] Yet history teaches us that the course of human affairs rarely moves along such narrow binary or reductionist paths. The lives of these young, driven, brilliant scientists consisted of a multitude of installments, some of which loomed large in real time and others that appeared fleeting or were easily dismissed, only to be recognized as important years later. Coincidental but critical events occurred alongside those that were long plotted yet ultimately irrelevant. There were random pairings of the right people at the right time, leading to a joyful noise, and, when the wrong people were joined at the wrong time, to a miasma of despair. There were flashes of victory and droughts of failure, acts of camaraderie and petty infighting. The story also represents a string of episodes populated and propelled by flawed human beings, whose deportment often reflected the competitive race for priority that characterizes so many great discoveries.[22] Buried under layers of interpretation, explanation, and obfuscation, the discovery of DNA's molecular structure is one of the most misunderstood whodunnits in the history of science.

And now it is time to tell how it really happened.

The Monk and the Biochemist

The laws governing inheritance are quite unknown; no one can say why the same peculiarity in different individuals of the same species, and in individuals of different species, is sometimes inherited and sometimes not so; why the child often reverts in certain characters to its grandfather or grandmother or other much more remote ancestor; why a peculiarity is often transmitted from one sex to both sexes or to one sex alone, more commonly but not exclusively to the like sex.

—CHARLES DARWIN, 1859[1]

In the beginning, high on a hill in Brünn, Moravia (now Brno in the Czech Republic), there was an abbey. In 1352, the Augustinian friars built an L-shaped, two-story, stucco and stone monastery topped by a gabled roof of orange clay shingles. The ground floor was arranged around a refectory and a library; directly above was a long, open dormitory for the friars. These rooms overlooked the confluence of the Svitava and Svratka rivers on one side and, on the other, the Gothic, red-bricked Basilica of the Assumption of Our Lady. The powers that were named it the Abbey of St. Thomas, after the apostle who initially doubted the resurrection of Jesus Christ (hence the moniker "Doubting Thomas").

The abbey's halls and arcades were overwhelmingly silent except for the chirping of birds, kept on the grounds within mesh-wired cages that the friars built to keep away predators. Wafting about were the smells of boiling hops, yeast, and grains, courtesy of the abbey's immediate neighbor, the Starobrno brewery, which had quenched the thirst of Brünn's villagers since 1325. Tucked into a corner of the abbey's central courtyard was a well-tended garden surrounded by a manicured stretch of grass. There, a monk named Gregor Mendel grew tomatoes and beans and cucumbers.[2] He was proudest of his pea plants, which sprouted in a veritable Punnett square of all shapes, sizes, and colors.[3]

Born in 1822, Johann Mendel (he took the name Gregor upon joining the Augustinian order) hailed from a family of farmers who tended a plot of land near the Moravian–Silesian border. As a boy, Mendel loved gardening and beekeeping. He sailed through the local schools and matric-

Gregor Mendel in the days of the experiment.

ulated into the nearby university in Olemac in 1840. Three years later, in 1843, he was forced to disenroll before taking his degree because money was short and tuition was high.

Determined to continue his studies, Mendel gave up his few earthly possessions and entered monastic life at St. Thomas in 1843. He thanked God in his nightly prayers for no longer having to worry about eking out a living or repaying his family's debts. His cot was comfortable and the meals bountiful. And because the abbey was the intellectual hub of Brünn, in 1851 Mendel convinced the abbot to find the discretionary funds to cover the expense of sending him to the University of Vienna.[4] There, Mendel excelled in physics, agriculture, biology, and research on the hereditary traits of plants and sheep. Intellectually gifted, Mendel was less the doubting St. Thomas than he was a finder of things and ideas, like St. Anthony.

When Friar Gregor returned to Brünn in 1853, the abbot assigned him the task of teaching physics in the local high school, even though he twice failed the oral examinations to become a certified teacher. He much preferred tending his garden to his parish duties. On this tiny patch of land, Mendel cultivated the modern study of heredity. Each day, he carefully recorded his observations of seven variations in the successive, self-fertilizing generations of his pea plants: height, pod shape and color, seed shape and color, and flower position and color.

Soon after Mendel began cross-breeding the tall plants with the "dwarf," or short, plants, he noted that all plants in the successive generation were tall. He thus called the characteristic of "tallness" a *dominant* trait and the trait of "dwarfism" *recessive*. In the next generation, bred from the hybrid plants, he observed both traits, tallness and dwarfism, expressed in a three-to-one, dominant-to-recessive ratio. Mendel found fixed ratios in the pea plants' other dominant and recessive traits as well. Eventually, he developed mathematical formulae to predict how these traits would express themselves in successive generations and fertilizations.[5] He believed these phenomena to be caused by "invisible factors"—what we now know as genes.

THE FRIAR PRESENTED HIS STUDIES at two consecutive meetings of the Brünn Natural Science Society on the evenings of February 8 and March 8, 1865. Today, it might seem odd to attend a scientific seminar and find a monk dressed in his ankle-length black woolen habit with a long, black pointed hood, or capuche, adorning his back. But the Brünn Natural Science Society was well attended by friars of the abbey, intellectual townsmen, and even curious farmers from the adjoining countryside. With only a chalkboard to present his complex formulae and in a voice almost whispery from years of quiet solitude, Mendel simultaneously impressed and bewildered the forty-odd members in the room.

Later than year, Mendel published his lectures in the society's proceedings. Sadly, *Verhandlungen des naturforschenden Vereines in Brünn* (Transactions of the Natural Science Society of Brünn) did not enjoy a wide circulation, and Mendel's discoveries failed to set the world afire. Armchair historians have often cited the obscure venue where he published his work as the cause of this delayed recognition, but it was more complicated than that. Mendel's description of heredity as occurring in discrete, predictable units ran counter to his era's explanations of the workings of the body and reproduction. The conventional wisdom of the day held that the balance of four bodily humors (blood, phlegm, yellow bile, and black bile) controlled the functioning of our organs and even the personalities of the children one produced.[6] This centuries-old theory was just plain wrong, but disproving it required several more decades of scientific inquiry. Further, the mathematics Mendel used to interpret his data was foreign to the ways biologists and natural historians thought about science, many of whom were still struggling to comprehend, if not accept, Darwinian theory. The natural historians of Mendel's day were far more comfortable collecting, naming, and classifying different species based upon morphological characteristics.[7]

Alas, Mendel spent the last seventeen years of his life as the abbot of St. Thomas and became embroiled in a dispute with the imperial Austro-Hungarian bureaucracy over the abbey's tax bills. He died at sixty-two of chronic kidney disease, in 1884. Sixteen more years elapsed before, in 1900, a Dutch botanist (Hugo de Vries), an Austrian agronomist (Erich von Schermack–Seysenegg), a German botanist (Karl Correns), and an American agricultural economist (William Jasper Spillman) independently applied some intrepid librarianship, ferreted out Mendel's paper from the dusty stacks, and verified his results.[8] Only the most obsessive recall these four scientists today because they so graciously (and honestly) gave the credit of primacy to Gregor Mendel. In recent years, a small band of post hoc detractors have suggested that Mendel fudged his data, because the mathematical ratios reported in his paper were too perfect to be statistically possible. Many more biologists and biostatisticians, however, have fervently come to Mendel's defense.[9] Most historians now agree that Mendel was certainly correct and probably honest in his reporting.

The rediscovery of Mendel's "laws" governing the transmission of simple recessive and dominant traits became the foundation of modern genetics. He has since earned his posthumous immortality as the father—or at least the friar—of what came to be known as classical or Mendelian genetics. The major problem of this system is that most inherited traits are not simple and arise from the interaction of several genes, which can also change their expression under environmental, social, and other influences.

DURING THE AUTUMN OF 1868, three years after Mendel published his paper, Johannes Friedrich Miescher was wringing out pus from bandages he had just collected from a surgical ward in Tübingen. A newly-minted Swiss physician (MD, Basel, 1868), Miescher came from good stock and station. His father, Johann F. Miescher, was a professor of physiology at the University of Basel; his uncle, Wilhelm His, a professor of anatomy at Basel, revolutionized the fields of neurobiology, embryology, and microanatomy.[10]

Since childhood, Miescher had contended with significant hearing loss, the result of an ear infection that smoldered into his mastoid sinus. This presented a problem for him as he segued from the classroom into the hospital and clinic, making the verbal give-and-take between physician and patient difficult. His father and uncle agreed that it might do him some good to take time off before embarking upon a clinical practice. They used their connections to arrange a plum research position for him in the laboratory of Professor Felix Hoppe-Seyler at the University of Tübingen. Hoppe-Seyler was the founder of modern biochemistry. Among his many discoveries was the oxygen-carrying function of red blood cells—the role played by the protein hemoglobin and its key ingredient, iron.

Hoppe-Seyler's laboratory was situated in what was once the basement vault of Hohentübingen Castle. It consisted of a suite of narrow rooms with deep-set, arched windows overlooking the river Neckar and the Ammar Valley. Miescher fell in love with the place and, under Hoppe-Seyler's guidance, studied the contents of neutrophils and leukocytes, or white blood cells, which course through the bloodstream in search of foreign invaders and attempt to ward off infection. He chose white blood cells because they are not embedded in tissue and are thus more easily isolated and purified; also, they have especially large nuclei, which serve as the cell's command center, that can be visualized when placed under the magnifying objective of a light microscope.

As it turned out, there were few better ways to collect white blood cells than from greenish-gray, pus-saturated bandages that had wrapped surgical patients. Surgeons of the mid-nineteenth century espoused a now discarded concept known as "laudable pus." Deeming pus to be the by-product of healing after a gruesome operation, they considered that the more pus a wound produced—often the result of the surgeon's dirty knife and hands—the more likely it was to heal. In most cases, we now know, the overproduction of pus translates into an ongoing postoperative infection. The all too common result of "laudable pus" was that the infection spread through the bloodstream and sent the patient into the death spiral known as sepsis.

AS OFTEN HAPPENS IN SCIENTIFIC INQUIRY, Miescher benefited from a temporal coincidence of technology created by another investigator. His benefactor was Dr. Viktor von Bruns, director of the University of Tübingen surgical clinic, who had just created a woven, highly absorbent cotton material he called "woolen cotton." We know it today as gauze. Postoperative infections aside, this new, sponge-like bandage was instrumental in Miescher's daily collection of pus.[11]

In time, Miescher learned how to better free the delicate white blood cells from the liquid portion of the pus collected in these bandages without damaging or destroying them entirely—no easy task. Fortunately, he had what surgeons call "good hands" and developed a series of chemical techniques, precipitating out a heretofore-undescribed substance rich in phosphorus and acid. Miescher determined that this substance was found only in a cell's nucleus and named the new entity *nuclein*. Today, we call Miescher's substance deoxyribonucleic acid, or DNA.[12] In casual conversation, people often mistakenly state that Watson and Crick *discovered* DNA, when in fact they discovered *the molecular structure* of what Friedrich Miescher chemically identified eighty-four years earlier in 1869.

Miescher left Tübingen in
1871 for Leipzig, where he worked
under the renowned physiologist
Carl Ludwig.[13] During this year,
he prepared a paper on his studies
of nuclein and, after a scrupulous
review of his highly reproducible
data, Dr. Hoppe-Seyler agreed
to publish his findings in an 1871
issue of the prestigious journal
he edited, *Medicinisch–chemische
Untersuchungen* (Studies in
Medicinal Chemistry). In an editorial accompanying Miescher's
paper, Hoppe-Seyler added his
powerful endorsement of nuclein's scientific novelty.[14]

Fredrich Miescher at time he discovered
DNA.

The following year, Miescher returned to his hometown of Basel to
serve his Habilitation, a postdoctoral lectureship and entry-level academic position for young physicians in Germany, Austria, and Switzerland during the nineteenth century.[15] In 1872, at age twenty-eight, he
was offered the position of chair and professor of physiology at the University of Basel. Because both his father and uncle had held prestigious
professorships there, jealous colleagues made unfounded complaints of
nepotism. Miescher proved them wrong and thrived in the role of scientific investigator.

The University of Basel was situated on the banks of the river Rhine.
Its location allowed for another wonderful coincidence. Salmon fishing
was a major industry in Basel. Salmon sperm cells, too, were easily isolated and purified using the chemical techniques of Miescher's era. They
also happen to contain especially large nuclei and, hence, more nuclein
to extract and study. Thus, Miescher enjoyed fishing for an unending
river of salmon gonads. In the laboratory, he determined that nuclein
consisted of carbon, phosphorus, hydrogen, oxygen, and nitrogen.

Miescher's earlier attempts at studying nuclein, incidentally, were often contaminated by stray proteins and their constituent, sulfur.

In 1874, Miescher reported many similarities (and some subtle differences) of nuclein across different vertebrate species. At one point in his paper, Miescher hovered near the scientific jackpot with a somewhat tepid sentence: "if one . . . wants to assume that a single substance . . . is the specific cause of fertilization, then one should undoubtedly first and foremost consider nuclein."[16] After a great deal of hemming and hawing, however, he ultimately could not fathom how so complex a process as reproduction could be guided by a simple chemical entity with such "limited diversity." A few sentences later, he concluded, "there is no specific molecule that could explain fertilization."[17]

Like Gregor Mendel, poor Miescher sank into the quagmire of administrative squabbles at the expense of time better spent in contemplative thought. He died of tuberculosis, in 1895, at the age of fifty-one. An institute for biomedical research is named for him at the University of Basel. Outside of his hometown, however, few recall his name and work. It took a little more than another half-century before anyone figured out what DNA actually did. Before that happened, unfortunately, the academy's understanding of heredity careened off the rails.

[3]

Before the Double Helix

The demand that it should be made impossible for defective people to propagate defective offspring is a demand that is based on most reasonable grounds, and its proper fulfilment is the most humane task that mankind has to face. Unhappy and undeserved suffering will be prevented in millions of cases, with the result that there will be gradual improvement in national health . . . a blessing for the present generation and posterity. The temporary pain thus experienced in this century can, and will, save thousands of future generations from suffering.

—ADOLF HITLER, 1925[1]

Beginning in the late 1880s, and reaching its zenith during the first three decades of the twentieth century, many White Anglo-Saxon Protestant upper-class men (as well as their wives and children) obsessed over the future of their nation's gene pool.[2] Supporting their fears was a pseudoscientific framework proposed in 1883 by the British naturalist—and Charles Darwin's cousin—Sir Francis Galton. He invented a new word to characterize his theories: *eugenics*, from the Greek root εὐγενής or *eugenes*, "good in stock or hereditarily endowed with noble qualities." Galton proposed a plan to improve the public's health by "giv[ing] to the more suitable races . . . a better chance of prevailing speedily over the less suitable."[3] Before long, Sir Francis's eugenics spread like wildfire among white intellectuals throughout Britain and Europe and to America.

In the United States, during what historians have named the Progressive Era (1900–20), a generation of reformers sought to confront the leading social problems of the day—including urban poverty, education, assimilation of the huge number of immigrants coming to American shores, public health crises ranging from epidemics to staggeringly high infant mortality rates, and an explosive growth in population. These reformers often applied inappropriate eugenic explanations to the people they deemed to be undesirable: the so-called "mental defectives" (whom doctors and psychologists labeled with newly created clinical terms such as "imbeciles," "idiots," and "morons"), the blind, the deaf, the mentally ill, the "halt, lame, and crippled," epileptics, orphans, unwed mothers, Native Americans, African Americans, immigrants, the poor living in city slums and in the mountains and hollows of Appalachia, and many other "outsider" groups. All these "inferior races" represented, the progressives claimed, an existential threat to the economic, political, and moral health of American society.

Eugenics offered Americans in positions of power an authoritative scientific language to substantiate their racial biases against those they feared as dangerous. The solution of the day was to quarantine, cordon off, and prevent the undesirables from contaminating the "superior," dominant, white, native-born Americans.[4] Those deemed "eugenically superior," specifically White Anglo-Saxon Protestants, were encouraged to reproduce at greater rates, a concept referred to as *positive eugenics*. Those adjudged to have inferior genes—virtually everyone else—were actively discouraged from reproducing through *negative eugenics* programs, such as state-mandated sterilization laws for mental defectives, restrictions on who could marry whom in the form of racial, or miscegenation, laws, mandatory marriage license blood tests for sexually transmitted diseases, birth control methods, and harsh adoption laws. A more ominous social policy emerged out of nativist calls for restricting the entry of those whom they deemed "inassimilable" immigrants. Using eugenics propaganda as a so-called evidence base, the U.S. Congress passed the Immigration Act of 1924, which shut the gates for over forty years. This policy signed the death warrant of millions of Jews in Germany and Eastern Europe because they could not escape Hitler's insanity by emigrating to the United States.[5]

The epicenter of the American eugenics movement was the Station for Experimental Evolution and the Eugenics Record Office in Cold Spring Harbor, Long Island, directed by Charles Benedict Davenport, an indefatigable, Harvard-trained biologist and member of the prestigious National Academy of Sciences.[6] The ERO was founded in 1910, thanks to a huge bequest from Mary Harriman, the wife of railroad tycoon E. H. Harriman, along with beneficent donations from the Carnegie Institution of Washington, DC, John D. Rockefeller, Jr., and Dr. John Harvey Kellogg, the creator of corn flakes and medical director of the Battle Creek Sanitarium. It is now the home of the Cold Spring Harbor Laboratory, long directed, expanded, and promoted by James D. Watson until his racist rants led to his dismissal.[7] To this day, graduate students at the CSHL's School of Biological Studies reside in a gloomy Victorian dormitory that was once the home of Charles B. Davenport.

In the years following the rediscovery of Mendel's work, his theories generated a maelstrom of public conversation. Nowhere did this elaboration become more fruitful and multiply than at the ERO. The fruit fly in the proverbial ointment was the eugenicists' incorrect application of Mendel's observations of pea plants to a number of complex social problems. Davenport declared war on those he considered a threat to the purity of the nation's gene pool.[8] At a 1910 meeting of the Committee on Eugenics of the American Breeders Association, he bellowed, "Society must protect itself; as it claims the right to deprive the murderer of his life so also it may annihilate the hideous serpent of the hopelessly vicious protoplasm."[9]

To this end, Davenport directed an army of social workers, field workers, sociologists, and biologists, who collated long, faulty, and highly

Charles B. Davenport, director of the Eugenics Record Office, 1914.

influential pedigree analyses asserting the hereditary basis for all sorts of behaviors, including lust and criminality, which Davenport claimed were common among Italians; the Jewish heritable traits of neurasthenia, tuberculosis, and craftiness in business dealings; feeblemindedness, which was endemic to those living in poverty-stricken Appalachia; nomadism, a trait of gypsies and hoboes; and even a love of the sea, or thalassophilia, seen among sailors.

In Davenport's mind, East European Jews posed an especially grave threat to American society. On April 7, 1925, Davenport fulminated to his friend Madison Grant, "Our ancestors drove Baptists from Massachusetts Bay into Rhode Island but we have no place to drive the Jews to. Also, they burned witches but it seems to be against the mores to burn any considerable part of our population."[10] Grant, a conservationist, lawyer, and trustee of the American Museum of Natural History, was an equally prominent proponent of eugenics. In 1916, he wrote *The Passing of the Great Race*, which promoted anti-immigrant policies, the segregation of "unfavorable" races, and, because he deemed so many Americans to come from "inferior stock," mandatory sterilization laws. The book's most sinister impact occurred in Nazi Germany. Adolf Hitler referred to Grant's magnum opus as "my bible" when designing the notorious "racial hygiene" programs that exterminated six million Jews and millions more homosexuals, gypsies, disabled, political or religious prisoners, and others whom Der Führer deemed unfit for Germany's Third Reich.[11]

THE INDELIBLE STAIN OF eugenics aside, there were several scientists hard at work during this era creating the beginnings of modern genetics. The most significant work was conducted by a cohort of geneticists who demonstrated that threadlike structures in a cell's nucleus known as chromosomes carried part or all of an organism's genetic material—what we would today refer to as genes. In several laboratories, biochemists developed the methods to determine that chromosomes are composed of proteins and deoxyribonucleic acid (DNA). Still other scientists created the field of population genetics to study genetic variation within and between different groups of people.[12]

What eluded them all, however, was the biological mechanism of how living beings reproduced. Before they could get to that critical question, a whole new science called molecular biology had to be developed. The form, or structure, of the genes had to be delineated at its smallest level, in terms of molecules and the atoms comprising them, before their function could be fully understood. Obstructing progress along this path was an intense debate over whether the genetic material resided in the DNA or the protein or both. For much of the first half of the twentieth century, the safe (and, as it turned out, wrong) bet was on the far more chemically complex proteins. DNA was thought by many to act passively, as a molecular scaffolding upon which genes rested.[13]

This scientific debate was especially loud within the walls of the Rockefeller Institute for Medical Research in New York City. Endowed in 1901 by the foundation of the same name and the bounty of the Standard Oil monopoly, the Rockefeller was the first independent, fully-funded medical research enterprise in the United States. The Rockefellers *père et fils* made certain that their institute was a gleaming beacon of modern medical research.[14] To begin, they understood that substantive scientific research requires substantial plots of real estate. So, in 1903, John D. Rockefeller, Sr. and Jr., paid the enormous sum of $650,000 for thirteen acres on a bluff overlooking Manhattan's East River between Sixty-Fourth and Sixty-Eighth streets. The institute assumed its permanent home on this site in May 1906, and, four years later, opened an adjacent sixty-bed hospital which treated, free of charge, anyone afflicted with one of the five priority diseases under study: poliomyelitis, heart disease, syphilis, "intestinal infantilism" (celiac disease), and one of the leading stalkers of humankind, lobar pneumonia. As time progressed, so, too, did the hospital's clinical mission. The Rockefellers prided themselves on rewarding "their" scientists with all the resources they desired, predicting that such generosity would produce the discovery of many new and great things. "Junior," the aging oil baron told his son, "we have money but it will have value for mankind only if we can find able men with ideas, imagination, and courage to put it into productive use."[15]

ONE OF THE ROCKEFELLERS' MOST productive workers was a physician named Oswald T. Avery. Born in Nova Scotia, Halifax, Canada, the son of a clergyman who moved his family to New York City in 1887, Avery spent the rest of his life there. Even as a young man, Oswald displayed a stiff bearing, stern visage, and impressive command of medical arcana. Bald, with an egg-shaped head, Avery wore pince-nez spectacles clipped

Oswald Avery of the Rockefeller Institute for Medical Research and Hospital, 1922.

to the bridge of his long nose. He was small in stature, soft in voice, mild in manner, and always impeccably dressed. His students referred to him with an admixture of reverence and sophomoric sarcasm as "the Professor."[16]

Dr. Avery's practice and research centered on the organism *Streptococcus pneumoniae*, or pneumococcus, the bacterial cause of most cases of "community-acquired pneumonia." Before the advent of antibiotics, pneumonia killed more than one hundred per one hundred thousand Americans every year.[17] Once it had been proven that pneumococcus was the cause of most cases of community-acquired pneumonia, many researchers attempted to develop a serum created from pneumonia victims' white blood cells and other immune constituents of the blood. The goal was to inject this serum into the bloodstreams of newly diagnosed pneumonia patients and thus passively bolster their immune systems. The scientific focus began to change after 1928, when a British bacteriologist and public health officer named Frederick Griffith observed that heat-killed virulent pneumococci converted a nonvirulent strain into a virulent one.[18] Additional studies conducted in the early 1930s by researchers at the Rockefeller Institute and Columbia University established that mixing cultures of virulent Type III or S-pneumococcus (which has a smooth surface because it is encapsulated by polysaccharides) with cultures of nonvirulent Type II or R-pneumococcus (which has a rough surface because it is not encapsulated), transformed the nonvirulent pneumococcus into the virulent variety.[19]

No one knew what the active transformative (also known as transformation) factor in this microbial process was, how it transmitted virulence to another strain of bacteria, or what its chemical constituents might be. Some postulated that the polysaccharide capsule of the pneumococcus was acting as the template for self-replication. Others posited that the factor was a protein–polysaccharide antigen found within the cell itself. Beginning in 1935, Professor Avery set out to answer these questions. Working with two younger colleagues, Colin Macleod and Maclyn McCarty, he did so with painstaking exactitude and glacial speed. Many scientists believed that his work merited the Nobel Prize;

alas, a baker's dozen of nominations between 1932 and 1948 were pitched to deaf ears in Stockholm.[20]

MONKLIKE, AVERY TENDED HIS microbial garden. He spent years developing biochemical techniques for growing, manipulating, and centrifuging huge volumes of pneumococcal cultures. More times than not, his efforts failed. "Disappointment is my daily bread; but I thrive on it," the Professor often said. On especially daunting days, he expressed his frustration more openly: "many are the times we were ready to throw the whole thing out the window."[21] Eventually, he created a set of reliable and reproducible procedures to isolate and analyze the "transformative substance."

In addition to the many technical difficulties that had to be overcome in the laboratory, Avery developed Graves' disease, an autoimmune disorder of the thyroid gland leading to debilitating hyperthyroidism and "moods of depression and irritation that he did not always manage to conceal despite valiant efforts." He underwent a thyroidectomy sometime in 1933 or 1934 (the hospital records have since been destroyed). While he regained much of his former health, he often used the illness as an excuse to minimize his social commitments, avoid academic meetings, and "devote himself more completely to his work."[22]

By early 1943, Avery had determined that the transforming substance was deoxyribonucleic acid. That May, long into the night, he wrote to his brother Roy, a physician at Vanderbilt University, about his discovery. This fourteen-page letter remains one of the pivotal documents in the history of DNA:

> Who could have guessed it? This type of nucleic acid has not to my knowledge been recognized in pneumococcus before—though it has been found in other bacteria . . . Sounds like a virus—may be a gene . . . It touches genetics, enzyme chemistry, cell metabolism, and carbohydrate synthesis, etc. [T]oday it takes a lot of documented evidence to convince anyone that the sodium salt of

desoxyribosecnucleic acid, protein-free, could possibly be endowed with such biologically active and specific properties and this evidence we are now trying to get. It's lots of fun to blow bubbles—but it's wiser to prick them yourself before someone else tries to . . . It's hazardous to go off half cocked—and embarrassing to have to retract later.[23]

The paper Avery published in 1944 was based on an extensive palette of chemical, serological, electrophoretic, ultra-centrifugal, purification, and inactivation techniques. He found that the transformative substance was composed of carbon, hydrogen, nitrogen, oxygen, and phosphorus (the elemental components of nucleic acids). It was active at one part per 100,000,000 and disabled by enzymes that attacked DNA but *not* by the enzymes that degraded ribonucleic acid (RNA) or those that digested proteins or polysaccharides. Moreover, when the transformative substance was exposed to UV light, a technique that provides a "fingerprint" of a molecule, it absorbed exactly the same wavelength as nucleic acids. Using the physician's diagnostic process of elimination, Avery concluded, in understated prose, that "the evidence presented supports the belief that a nucleic acid of the desoxyribose type is the fundamental unit of the transforming principle of *Pneumococcus* Type III."[24] Avery and McCarty published two follow-up papers in 1946 documenting improvements in isolating the transformative substance and even stronger claims that genes were made of DNA.[25] What Avery could not do, however, was determine how DNA actually worked or what it looked like, in terms of its precise atomic structure. And as we saw with both Mendel and Miescher, the Professor's work did not immediately alter the scientific landscape.

This was because the powerful "protein men" stubbornly held onto the notion of protein's hereditary primacy. They bellowed loudly against Avery's word at several academic symposia between 1945 and 1950. Perhaps Avery's most formidable foe was his Rockefeller Institute colleague, the world-class biochemist Phoebus A. Levene, who elaborated the "tetranucleotide hypothesis." This theory held that because nucleotides DNA contain only four bases (adenine, guanine, cytosine, and thymine), it was neither complex nor diverse enough to carry the genetic code. Instead, Levene insisted that the protein component of the chromosomes, and the many amino acids that comprised them, must serve as the basis of heredity. Levene's coup de grâce was the equivalent of a poison pill: how could Professor Avery be certain there was absolutely no trace of protein in his preparations, which, in turn, might *really* have been the cause of the transformation?[26]

SEVERAL AFTER-THE-FACT historiographical studies have claimed that Avery's work was premature and ignored by most scientists, especially geneticists. One frequently offered explanation is that he published his work in a periodical read more by doctors than scientists, the *Journal of Experimental Medicine*.[27] This claim reeks of nonsense. The *Journal* enjoyed a long and distinguished reputation, having been founded in 1896 by William Henry Welch of the Johns Hopkins Hospital and published by the Rockefeller Institute. It was easily accessible in every American university, as well as those abroad, on the shelves of their medical libraries. If geneticists could dig out Mendel's entirely obscure but landmark paper surely they could have walked across their campuses to look up and read Avery's work.

In fact, from the mid-1940s well into the 1950s the Avery paper was widely discussed at academic conferences attended by physicists, molecular biologists, and bacterial geneticists. In 1944 the British physicist William Astbury, who in the 1930s became the first X-ray crystallographer to turn his apparatus to imaging the structure of DNA, exalted Avery's work as "one of the most remarkable discoveries of our time."[28] Herman Kalckar of Copenhagen, who would later supervise Watson's postdoctoral fellowship, claimed to have been aware of Avery's work since 1945.[29] And in 1946, Avery spoke at the premier gathering of geneticists of the day—the Cold Spring Harbor summer meeting on "Heredity and Variation in Microorganisms."

Joshua Lederberg—who would go on to win a Nobel Prize in Physiol-

ogy or Medicine in 1958 and, in 1978, become the president of the Rockefeller University—read Avery's paper when it first appeared. He deemed
it to be "the most exciting key to uncovering the chemical nature of the
gene."[30] Lederberg often cited Avery's work in the papers he published
during the 1940s and 1950s. For the rest of his distinguished career,
Lederberg politely but firmly contested the subsequent claims of Avery's
so-called obscurity. In a 1973 letter to the editor of *Nature*, he declared
that the notion that "Avery's work on pneumococcal transformation was
not well recognized by geneticists in the decade following his 1944 report
is somewhat at odds with my own recollection and experience."[31]

The influential geneticist and Nobel laureate Max Delbrück heartily
agreed with Lederberg. He first visited Avery's lab at the Rockefeller
Institute sometime in 1941 or 1942 and followed the Professor's work
when it was published in the *Journal of Experimental Medicine*.[32] Three
decades later, in 1972, Delbrück recalled how difficult it was to fight
Levene's prevailing tetranucleotide theory during the 1940s: "Everybody
who looked at it, and who thought about it, was confronted with this
paradox, that on the one hand you seemed to obtain a specific effect with
DNA, and on the other hand, at that time it was believed that DNA was
a *stupid* substance, a tetranucleotide which couldn't do anything specific.
So, one of these premises had to be wrong."[33]

PART II

THE PLAYERS' CLUB

Most people are other people. Their thoughts are someone else's opinions, their lives a mimicry, their passions a quotation.

—OSCAR WILDE[1]

[4]

Take Me to the
Cavendish Laboratory

I have never seen Francis Crick in a modest mood.

—JAMES D. WATSON[1]

The opening sentence of James Watson's *The Double Helix* simultaneously rankled Francis Harry Compton Crick and described him perfectly. Unlike his harsh descriptions of Rosalind Franklin, Watson meant no disrespect towards Crick. He simply wanted to express that Crick was so brilliant, he had no need to be modest. For quite a few years, however, Crick took the line rather poorly. Shortly after reading the manuscript of Watson's book, Crick rallied together Maurice Wilkins and several other scientists offended by its glib prose. They petitioned Nathan Pusey, then president of Harvard University, to command its independent university press not to publish it. Crick won the battle but lost the war. Although Harvard University Press dropped the book in 1967, Watson's editor, Thomas Wilson, left Cambridge, Massachusetts, for the New York publishing house of Atheneum with Watson's manuscript under his arm.[2] The following year, *The Double Helix* became an international best-seller and has sold more than a million copies since.[3]

Francis Crick was born on June 8, 1916, in the village of Weston Favell, near Northampton, in the East Midlands region of Great Britain. His parents, Harry and Annie Crick, were well-to-do, thanks to Harry's profitable boot and shoe works and a chain of family-controlled retail shops. Young Francis devoured science books and encyclopedias with complete recall. He once told his mother that he was worried that by the time he grew up everything would already have been discovered.

After grammar school in Northampton, Crick boarded at the Mill Hill School in London where he excelled in mathematics, physics, chemistry, and prank pulling. On one occasion, he rigged a radio—a forbidden appliance during evening study time—so that it automatically turned on whenever his housemaster patrolled the halls and turned off when he entered Crick's room in search of the noise. He further alienated his teachers by stuffing glassware with various explosives to make "bottle bombs."[4]

In 1934, Crick matriculated into University College, London, after failing the entrance examinations for both Oxford and Cambridge. He read physics and graduated at twenty-one, with a second-class honors degree. Interestingly, both Maurice Wilkins and Rosalind Franklin also gained second-class degrees, a distinction that should have relegated them to second-class careers in British science but in their remarkable cases did not.[5] Crick took the path of least resistance by accepting a student research position under Professor Edward Neville da Costa Andrade at UCL while living in cosmopolitan London on the dime of his uncle Arthur Crick, who taught him the arts of glassblowing and photography. At UCL, Francis worked on the "dullest problem imaginable, the determination of the viscosity of water, under pressure, between 100 and 150 degrees C."[6]

Fortunately for the future of biology, in 1939 a German bomb blew up Crick's laboratory and his carefully wrought research apparatus, effectively ending this line of inquiry. The following year, 1940, Crick began a six-year wartime stint in the offices of the Admiralty working on magnetic and acoustic mines, which exploded without direct contact by an invading vessel, thus making them many times more effective than the older mines. After the close of the war, weapons experts concluded that these new British mines had sunk or ruined a thousand or more of the enemy's seagoing vessels.[7]

Francis Crick at University College, London, in 1938.

Crick's personal life at this time was somewhat complicated. His first wife, Ruth Doreen Dodd, was a fellow student at University College, London, who read English literature and was especially drawn to the picaresque novels of Tobias Smollett. When war broke out and all hands were needed, she boxed up her books and began clerking at the Ministry of Labour.[8] The pair married in 1940; nine months later, almost to the day, they had a son, Michael. In 1946, Crick fell in love with Odile

Francis Crick and his son, Michael, c. 1943.

Speed, a French woman who had come to Britain in the 1930s to learn English and study art; during the war, she joined the Women's Royal Naval Service. Ruth and Francis divorced in 1947 and he had little to do with the rearing of his little boy. In 1949, Francis and Odile embarked on a very happy marriage, blessed by two daughters, which lasted until his death.

At the war's close, Crick wondered if the best way to leverage his "not-very-good degree," incomplete doctoral thesis, and advancing age was to take a job as a civil servant in the British government. The Admiralty brass, too, were uncertain that they wanted to commit to many more years of employing this voluble young man. After a second interview, chaired by the physical chemist and novelist C. P. Snow, Crick was offered a job. By this time, however, Crick was certain that he "didn't want to spend the rest of [his] life designing weapons" and he turned Snow down.[9]

Odile and Francis Crick's wedding portrait, 1949.

CRICK NEXT CONSIDERED a career as a scientific journalist and inquired about a position in the editorial offices of the journal *Nature*. He withdrew his application soon after realizing that he wanted to conduct his own scientific inquiries rather than edit and report on the work of others. In his spare time, he kept abreast of the chemical literature by reading a superb book on the nature of chemical bonds in organic molecules— as described in his own words—by an "author [who] had an unusual name—Linus Pauling." He was also reading "Lord Adrian's little book on the brain," *The Mechanism of Nervous Action: Electrical Studies of the Neurone*, and Sir Cyril Hinshelwood's *The Chemical Kinetics of the Bacterial Cell*.[10]

As the journalist Matt Ridley noted, Crick was "determined, not just to break into science, but to do something heroic in science and, above all, to explode a mystery."[11] The explosions he most wanted to set off were to be found either by solving how the human brain works, creates, and dreams, or by divining the molecular mechanisms of heredity. But how was he to accomplish such lofty aspirations?[12]

Fortunately, Crick's primary mentor at the Admiralty was a mathematical physicist from Australia named Harrie Stewart Wilson Massey, who in 1945 moved to chair the physics department at University College, London. During one of their many talks, Massey lent Crick a copy of Schrödinger's book *What Is Life?* Massey lent the same book to Maurice Wilkins, second-in-command to John Randall, head of the Medical Research Council Biophysics Unit at King's College, London.[13] On Massey's advice, Crick approached Wilkins and the two became friends. They were the same age (they would also die in the same year, 2004), had both divorced their first wives and surrendered the parenting of their first-born sons, and were fascinated by the structure and function of genes. They often had din-

ner together and, at one point, Crick inquired about working in Randall's laboratory, but Randall rejected his application out of hand. He received a similar response from the X-ray crystallographer J. D. Bernal after applying for a position in Bernal's lab at Birkbeck College, London—the same laboratory which would reject Rosalind Franklin in 1949, then eventually accept her in spring 1953.[14]

Crick next competed for a scholarship to complete his doctorate under the aegis of the Medical Research Council. His application began with some

Erwin Schrödinger, winner of the 1933 Nobel Prize in Physics and author of the 1944 book *What Is Life?*, which inspired Crick, Wilkins, and Watson to study the gene and DNA.

bold and brilliant intentions. "The particular field [that most excited his attention] was the division between the living and the non-living, as typified by, say, proteins, viruses, bacteria, and the structure of chromosomes." Crick went on to explain his "eventual goal" in describing these biological entities in terms of "the spatial distribution of their constituent atoms, in so far as this may prove possible." The clinching line to his application was a most prescient conclusion: "This might be called the chemical physics of biology."[15]

For the scholarship, Crick interviewed with A. V. Hill, the Cambridge muscle physiologist and 1922 Nobel laureate in Physiology or Medicine. Hill wrote a glowing letter of endorsement and arranged for Crick to meet Sir Edward Mellanby, the powerful secretary of the Medical Research Council.[16] Mellanby, who discovered vitamin D and its role in preventing rickets, was similarly struck by the young man's energy and breadth of knowledge. After less than an hour of discussion, he advised Crick, "You should go to Cambridge. You'll find your own level there."[17] After the interview, Mellanby scribbled on Crick's MRC application, "I was very much attracted to this man."[18]

For his first two years at Cambridge, 1947–49, Crick worked at the Strangeways Laboratory. The lab was founded in 1905, by Dr. Thomas Strangeways at the Cambridge Research Hospital, to study rheumatoid arthritis. When Crick arrived there, the Strangeways Laboratory was focused on tissue culture, organ culture, and cell biology under the direction of a superb zoologist named Honor Bridget Fell, one of the few British women scientists in a leadership position.[19] Crick recalled spending his time there trying to "deduce something about the physical properties of the cytoplasm, the inside of the cell. I was not deeply interested in the problem but I realized that in a superficial way it was ideal for me, since the only subjects I was fairly familiar with were magnetism and hydrodynamics." The work was substantive enough to facilitate Crick's earliest scientific publications, one experimental and the other theoretical, in the journal *Experimental Cell Research*.[20]

During his second year at the Strangeways, Fell asked Crick to deliver a short talk on "important problems in molecular biology" for a group of guest researchers visiting Cambridge. He recalled the guests waiting "expectantly, with pen and pencils poised, but as I continued they put them down. Clearly, they thought, this was not serious stuff, just useless speculation. At only one point did they make any notes, and that was when I told them something factual—that irradiation with X-rays dramatically reduced the viscosity of a solution of DNA." When he retold this tale at the age of seventy-two, Crick yearned to know exactly what he had spoken about nearly four decades earlier, but his "memory [was] so overlaid with the ideas and developments of later years" he felt he could "hardly trust it." There are no surviving notes from this lecture and all Crick could do, later, was speculate that he spoke on the important role of genes in reproduction, the need to "discover their molecular structure, how they might be made of DNA (at least in part), and that the most useful thing a gene could do would be to direct the synthesis of a protein, probably by means of an RNA intermediate."[21]

A diploma from Cambridge University represented Crick's last, best hope of becoming a great man of science, and he was determined to make the most of it. Realizing that he had no real future at the Strangeways, Crick convinced Sir Edward Mellanby of the need to change venues. After a few telephone calls, he was reassigned to the Cavendish Laboratory's Biophysics Unit, under the tutelage of Max Perutz and his assistant, John Kendrew.[22] Crick's primary assignment was to help them discern the molecular structures of hemoglobin and myoglobin; in turn, Perutz would help Crick earn his Doctor of Philosophy degree.[23]

Crick's first visit to the Cavendish Laboratory had a less than auspicious beginning. Bounding off the platform at the tiny Cambridge train station after an extended trip to London, he treated himself to a cab ride. With the excitement of a serious student about to begin his career in earnest, Crick could feel his pulse bounding on the occasion of visiting the premier institution of its kind in the world. Placing his valise inside the cab and settling back into his seat, he told the cabbie, "Take me to the Cavendish Laboratory." The driver turned back and looked through the dividing glass to ask, "Where's that?" Crick was puzzled, until he realized "that not everyone was as deeply interested in fundamental science as I was." He fumbled through the papers in his battered briefcase, found a sheet of paper with the Cavendish's address, and informed the cabbie that it was located in the Free School Lane, "wherever that is." The driver recognized it as being not far from the Market Square, turned the car in the right direction, and drove off to their destination.[24]

☥

FROM THE LATE NINETEENTH CENTURY to well after the Second World War, there were two places to study physics: the Cavendish Laboratory at Cambridge University and everywhere else.[25] Many might right-fully claim that the modern study of physics began at Cambridge. In 1687, Isaac Newton of Trinity College wrote his famed *Principia*, a sci-entific masterpiece describing gravity, the law of universal gravitation, and many of the principles now known as classical physics. Nearly two centuries later, in 1874, the Cavendish was founded and named for the reclusive Henry Cavendish, an eighteenth-century English genius who discovered "inflammable air," or hydrogen, and successfully measured the force of gravity between masses, yielding accurate values for the gravitational constant.

The laboratory's first professor was the Scotsman James Clerk Max-well (1831–79), who sported impressive salt-and-pepper muttonchop whiskers and a wiry beard that split into two, making his appearance positively Dickensian. As a Cambridge undergraduate, Maxwell had aspired to spend the rest of his days figuring out the physical world, even if that meant upending the Biblical scriptures that shaped his world-view.[26] During his scholarly career, he described how electric charges and currents create electrical and magnetic fields, using a set of mathematics still known as Maxwell's equations. Maxwell also reintroduced physi-cists to the Aristotelian "thought experiment," a method Einstein, Bohr, Heisenberg, Schrödinger, and others elevated to the scientific art form now known as theoretical physics.[27]

Cambridge students called the Cavendish "the center of all things physical." Constructed of limestone, brick, and slate, the three-story structure was replete with Gothic arches and narrow stairwells. Inside was a lecture room, with "steeply raking seats for up to 180 students," the professor's office and laboratory, and a workshop; above that was a cramped apparatus room and a student laboratory; and on the top floor, or attic, was an experimental electricity room.[28]

Maxwell, who died at age forty-eight, was replaced in 1879 by John William Strutt, Lord Rayleigh. Rayleigh won the Nobel Prize in Physics in 1904 for his work in describing the densities of several of the most important gases and discovering argon. He devoted his prize money to improving the wretched state of the Cavendish. And in 1882, Rayleigh instituted the novel custom of allowing women to take classes there, a decision in equity that had great import, five decades later, for Rosalind Franklin.

In 1884, J. J. Thomson, a slight, balding man with a scraggly attempt at a walrus mustache, was tapped to become the next Cavendish professor. Looking more like a banker than a physicist, he was only twenty-eight when he took the helm. Thomson discovered the electron, measuring both its mass and charge. He was so clumsy, however, that his lab assistants constantly contrived new ways to keep him from breaking the delicate apparatus they developed to define the electron. No small feat, this discovery was the basis of the understanding of chemical bonds at the molecular and atomic levels. Thomson's work laid the groundwork for a plethora of power sources, artificial lighting, radio, television, telephones, computers, and the Internet.

In 1919, a burly physicist named Ernest Rutherford returned to Cambridge from his native New Zealand to succeed Thomson. Rutherford became the father of nuclear physics by successfully splitting the atom, discovering the proton, elucidating the concept of radioactivity, and defining the concept of radioactive half-life—all the while whistling (or if especially ebullient, singing) Sir Arthur Sullivan's "Onward Christian Soldiers."[29] The element rutherfordium, number 104 in the periodic table, is named for him. Both Thomson and Rutherford won the Nobel Prize in Physics, in 1906 and 1908 respectively. During this same period, James Chadwick, the Master of Gonville and Caius College and a Cavendish physicist, discovered the neutron. He won the Nobel Prize in Physics in 1932.

WHEN IT COMES TO DNA, however, the Cavendish's most important professor was Sir William Lawrence Bragg, who held the post from 1938 to 1953. He was thirty-nine when he was called to Cambridge University even though he had never taught physics or directed a large department.[30] With his father, William Henry Bragg, he had developed the science of X-ray crystallography, for which they won the 1915 Nobel Prize

The Cavendish Physics Laboratory.

in Physics, the only father–son shared award in the Prize's history.[31] The theorem they developed, known as Bragg's equation, explains how crystals reflect X-ray beams at specific angles. At Cambridge, Bragg was tasked with modernizing the Cavendish, something Rutherford had failed to do during his tenure. Playing to his strengths, Bragg moved the research spectrum away from nuclear physics and toward the field of X-ray crystallography. He evolved into a superb administrator, known for both his tact and his leadership skills. Despite the consequences of the Great Depression and two world wars, Bragg rebuilt the Cavendish into a world-class operation.[32]

Bragg's first undertaking was to do something about the Cavendish's cramped, outmoded facilities. By the late 1930s, there were too many physicists and not enough space for experimentation. In 1936, Bragg successfully courted the automobile manufacturer Sir Herbert (later Lord) Austin to donate £250,000 for a new wing to be named after him. The Austin Wing was a nondescript, four-story, utilitarian box of a building clad in light brown-gray brick. Aesthetics aside, it added ninety more rooms—thirty-one of them for research, another thirteen for offices—as well as a glassblower's room, an apparatus or machinist's workshop, a library, a tearoom, and a special techniques workshop where "delicate operations requiring the highest technical skills are carried out."[33] This was the building in which Watson and Crick conducted their DNA work from 1951 to 1953.

OUTSPOKEN AND ENDLESSLY UPBEAT, Crick lacked a brake between his incredible brain and his rapid-fire mouth. He embodied the buoyant wit of Oscar Wilde and the authoritative arrogance of George Bernard Shaw's Professor Henry Higgins—with quite a few dollops of Albert Einstein's genius to complete the picture.[34] Rosalind Franklin's biographer Anne Sayre found Crick's conceit to be "superhuman."[35] Easily bored and prone to drift from one project to the next without making any real advance toward completing his doctorate, he was all but destined to earn Bragg's enmity. Crick dominated most conversations with an endless, Joycean free association of ideas and theories. His grasp of biophysics, down to the molecular level, was dazzling. He was often so precise in his attack (and resolution) of other researchers' projects that many feared discussing their work with him, lest he appropriate their intellectual property as his own. Francis most closely identified with the theoreticians, who came up with grand ideas, rather than the experimentalists, whom he considered drudges—existing only to prove the great ideas of geniuses like himself. Yet, it was the rare colleague who could listen carefully enough to silently edit and mine Crick's nonstop scientific soliloquies for the best thoughts. As the novelist Angus Wilson noted in 1963, "all the false hares and nutty suggestions, all the hours of exhausting listening, and strained disagreement, are finally, miraculously, made infinitely worthwhile when a man like Dr. Crick eventually talks himself into one of the great revolutionary scientific theories of the century."[36]

In July 1951, Crick delivered a departmental seminar for his fellow Cavendish physicists. John Kendrew suggested that Crick title his talk "What Mad Pursuit," a phrase from the first stanza of John Keats's poem "Ode on a Grecian Urn." In the lecture, Crick went through each method of interpreting X-ray crystallography pictures, from Patterson analyses and Fourier transforms to Perutz's protein work and Bragg's optical method, known as the "fly's eye." Generating a cloud of chalk dust as he scrawled mathematical formulae on a blackboard, Crick demonstrated the futility of each and boldly concluded that "most of the assumptions that had been made in those papers were not substantiated by the facts." The one exception, he argued, and Perutz agreed, was a method known as the isomorphous replacement of atoms, wherein one replaced the atoms of the molecule in question with foreign atoms that strongly scattered X-rays, without changing the structure.[37] Crick recalled in his memoirs—also titled *What Mad Pursuit*—how livid Bragg was after the presentation. Here was Crick, new to the Cavendish, already telling the founder of the field, his staff, and his students "that what they were doing was most unlikely to lead to any useful result. The fact that I clearly understood the theory of the subject and indeed was apt to be unduly loquacious about it did not help."[38]

During a subsequent departmental seminar, Crick took the even more foolhardy step of suggesting that Bragg had appropriated one of his ideas. This was nearly the final straw for the professor, who turned his reddened face and bulky body around to face his accuser and angrily hissed, "Don't rock the boat, Crick! We were getting on quite well before you got here. And by the way, when are you going to do some work for your Ph.D.?"[39] After this event, the great man took to slamming his office door whenever Crick entered the laboratory so as not to have to listen to his "daily prattling." More than sixty years later, James Watson recalled that "Crick's voice . . . was just too loud to put up with . . . [his laugh] . . . was really powerful."[40]

The Third Man[1]

DNA, you know, is Midas' gold. Everybody who touches it goes mad.

—MAURICE WILKINS[2]

Maurice Hugh Frederick Wilkins, PhD, CBE, FRS, was a bag of neuroses packed into a tall, wiry body. The self-perceived horrors of his life made almost every human interaction awkward, if not distressing. A habitué of far too many Harley Street psychoanalysts, Wilkins was a Freudian analysand until he became a certified Jungian pile of fears, phobias, and complexes—a process of introspection that must have seemed without end.[3] When speaking to others, Wilkins rarely made eye contact. He preferred to contort his body so that his back faced the other person.[4] Soft-spoken and slow in cadence, his sentence structure was so tangential and meandering that he often tried the patience of his listeners. It took him forever to arrive at a definite point.[5]

Anne Sayre described Wilkins as "a sufferer who has genuine problems in human contact; it gives him pain; it cripples him." He was tortured by resentments and anger turned inward, especially when it came to Rosalind Franklin.[6] Francis Crick offered a briefer assessment of Wilkins's mental health: "You don't know Maurice. In those days he was very, very emotionally tied up."[7] Yet he was also generous to a fault and especially well-liked by his male colleagues, assistants, and students, if not pitied for his emotional difficulties.

Wilkins was born on December 15, 1916, in a house made of rough-hewn timber situated in the mountain country of Pongoroa, New Zealand.[8] His father, Edgar, was a pediatrician who trained at Trinity

Medical College in Dublin. His mother, Eveline Whittacker, was the daughter of a Dublin police chief. Wilkins described her as "an affectionate woman with long, blonde hair and a great deal of common sense."[9] The couple left Ireland in 1913 for a better life in distant New Zealand. Unable to rise to the level of success he had hoped, Edgar moved the family to London in 1923 and took up studies for a doctorate in public health at King's College, London.

In 1922, Maurice's older sister by two years, Eithne, developed an infection of the blood, bones, and joints that required a painful series of hospitalizations and orthopedic surgeries at Great Ormond Street Children's Hospital. Maurice's recollections of Eithne in those pre-antibiotic days still have the power to disturb. On visiting days, the little boy tightly clutched his parents' hands as he "climbed up big institutional staircases." Together, they passed through dozens of open wards, with beds lined up like regimental soldiers, each holding an ill, lonely, crying child. It seemed like forever before they reached Eithne, who "was not only swallowed up by the great hospitals: she had become physically almost unrecognizable." The nurses had shaved off every lock of "her beautiful blonde hair" and her face was so swollen with infection that the little boy hardly recognized his own sister. What he could not forget was how "she was a victim of some nightmarish scheme, trapped in her bed by a great framework of ropes and pulleys that held her legs up in the air." It reminded him of a medieval torture chamber in the Tower of London.[10]

Imagine the six-year-old Maurice's horror when Eithne confessed to him that she wished to die. Their father was helpless to use his medical acumen to cure his own child and he mournfully wandered the streets of London, filled with the horrible angst any parent feels on the occasion of possibly burying his child.[11] This crisis affected the psyches of the entire Wilkins family in different but equally damaging ways. In Maurice's case, it may have been the genesis of his difficulties in trusting and communicating with women. When Eithne finally came home, he felt betrayed by his once-closest playmate's refusal to engage in the childish things they had so loved. From that point on, he reported, "there was very little communication between us."[12]

In 1929, the family moved to Birmingham, where Edgar Wilkins practiced as a school pediatrician and, later, authored a prominent text on the subject, *Medical Inspection of School Children*.[13] From 1929 to 1935, young Maurice attended one of the best day schools in England, King Edward's School, where he became fascinated by astronomy and geology. In 1935 he won admission to the prestigious St. John's College, Cambridge, and a scholarship from the Worshipful Company of Carpenters.

At Cambridge, Wilkins pursued "science that was directly related to the problems of human life." One of his first gurus in this quest was Mark L. E. Oliphant, deputy to Ernest Rutherford, then director of the Cavendish Physics Laboratory. Oliphant served as Wilkins's tutor at St. John's and taught him that "a physicist should make his own apparatus," a concept that reminded Wilkins of the wonderful hours he had spent tinkering in his father's workshop.[14]

Another mentor was John Desmond Bernal, who worked at the Cavendish until 1937, when Ernest Rutherford refused to grant him tenure and he moved to Birkbeck College, London. A brilliant scientist who hailed from County Tipperary, Ireland, and was Sephardic Jew on his father's side, Bernal used X-ray crystallography to study the structure of viruses and proteins.[15] Wilkins was mesmerized by both Bernal's science and his Communist politics. During the 1930s, Bernal, who was reverentially known as "Sage," attracted many like-minded faculty and students and, in 1932, founded the pacifist Cambridge Scientists Anti-War Group.

As did many students of his era, Wilkins read the work of Karl Marx and "formed a very high opinion of his materialistic theories of history" [even if Marx failed] to find a way toward a humane non-dictatorial Communist society."[16] Wilkins joined CSAWG and several other leftist student organizations concerned about the rise of Nazism, the Spanish Civil War, and the "acute problems of Indian Independence."

At St. John's, the young man sought out fun by visiting modern art museums and galleries and attending meetings of the Natural Science Club. He often went to the local cinema, where he was exposed to a different set of Marxist theorists as wisecracked in the madcap motion picture comedies of Groucho, Chico, Harpo, and sometimes Zeppo Marx.

Maurice especially enjoyed European art films and, in keeping with his political beliefs, adored the 1925 Soviet-produced, Sergei Eisenstein silent epic, *Battleship Potemkin*.[17] For a brief period, he engaged in fencing, but gave it up because he "simply wasn't fast enough."[18]

Despite his intellectual interests, Wilkins was plagued by insecurity and feelings of inferiority, especially when he compared himself to the wealthier, self-assured undergraduates at his college. In 1990, he confessed that it was while at Cambridge he "came down a peg in [his] own estimation . . . it was just that some of these other people seemed so damned clever."[19]

He was especially stunted when interacting with the opposite sex. In 1937, he fell in love with a fellow student and CSAWG member named Margaret Ramsey. Unfortunately, he was so shy that he could not figure out how to tell her about his feelings. One evening, while they were seated at opposite ends of his room in St. John's, he simply blurted out the words, "I love you." Margaret was baffled by this abrupt behavior and, after a pregnant pause, got up, bade him farewell, and left. The young man's only physical encounter with a woman during his college years seems to have occurred when he accidentally collided with a "shop girl" at a London department store. Fifty years later, he could still summon up the erotic experience of that young woman's "remarkable softness, warmth and perfume." And while his inexperience in matters of love was hardly rare for young men of his generation, these two episodes describe how helpless and befuddled Wilkins felt in so many of his romantic and platonic interactions with women.[20]

During his final year at Cambridge, in the fall of 1938, Wilkins spiraled down into a serious depression, a mental health problem he contended with for the rest of his days. The depression did little to help his academic standing. Upon taking his final examinations in 1939, he was awarded a lower-second-class degree in physics, a mark that for Wilkins was more scarlet (and damning) than any worn by Hester Prynne. A second-class degree obliterated a student's chances of continuing on to advanced degree studies at the University of Cambridge. This disappointment—well understood by any student who has had the misfortune of failing at a particular goal—was absolutely devastating for Wilkins. Indeed, he felt as if his "world had ended."[21]

Wilkins's failure to win a first turned out to be a very good thing because it forced Wilkins to leave the comfortable cocoon of Cambridge.[22] He had by then developed an interest in thermoluminescence—the ways electrons move through crystals and emit light. After rejection by the graduate programs at Cambridge and Oxford, he shifted his sights to the University of Birmingham, where, in 1937, his former Cambridge mentor Mark Oliphant had assumed the helm of the physics department to construct the "biggest atom-smasher [cyclotron] in Great Britain."[23]

Helping Oliphant in this work was John Turton Randall, a physicist who was remarkably entrepreneurial in building scientific empires. Bespectacled, bald, and plain, Randall was the son of a gardener at a local nursery. He made up for his humble origins and physical deficits by wearing bespoke suits of finely-loomed wool, snappy silk bow ties, and—like the floorwalkers at Harrod's department store—a freshly cut carnation in his lapel. From 1926 to 1937, Randall worked at the General Electric Company's research laboratory at Wembley, where he commanded a cadre of physicists, chemists, and engineers in creating profitable lines of luminescent lamps. In 1937, Oliphant hired Randall as a Royal Society fellow.

When Wilkins approached Oliphant with the idea of doing his doctoral research at Birmingham, Oliphant heartily accepted and assigned him to work in Randall's lab.[24] Wilkins was impressed by the "religious enthusiasm" Randall brought to his scientific pursuits. For Randall, science was better than religion because "it had the seductive aspect of leading to personal recognition and fame."[25] "Unusually knowledgeable about top-rate scientists, and a good judge of form," Randall gave his staff the freedom to pursue their scientific inquiries wherever the data took them and, unlike many of his peers, hired both men and women to toil away his laboratory.[26] Yet working for John Randall was not all merriment and inspiration. He could be a petty boss who demanded quick results, absolute loyalty, and a strict chain of command. Assuming a Napoleonic air, Randall entered the laboratory and made everyone, in Wilkins's words, "go hop."[27]

Wilkins worked in Randall's luminescence laboratory from 1938 to 1940 and completed his doctorate, in 1940, on solid state physics, phosphorescence, and electron traps.[28] He also developed tools to advance his research and made important contacts with senior physicists who would help him progress in his academic career, all of which enabled him to "eclipse" his second-class degree from Cambridge.[29]

What Wilkins could not eclipse was his utterly inept relations with women. During the Blitz (the German bombing raids on Britain in 1940–41), he met a young violinist he refers to, in his memoir, only as Brita. They cycled in the countryside, shared meals, and enjoyed each other's company, or so Wilkins thought. The relationship remained "very prim" because of his inability to express his feelings, let alone his ignorance of how to act on them. As he had done with Margaret Ramsey, he took a seat "at the far end of Brita's room" and, in the manner of "making a philosophical statement," he pronounced his love for her. This odd approach worked about as well as it had in Cambridge. Many years later, he confessed, "I imagine she was discouraged by my unromantic approach, for she did nothing to help me out of my difficulty. I retreated frozen with horror and despair."[30] In the depths of the dumped, Wilkins "decided to give up [his] Love and immerse myself in higher things." Seeking inspiration in how "Spinoza gave up his Love and began grinding telescope lenses," he immersed himself in quantum mechanics. As he gained mastery over the subject and moved beyond his emotional pain, he gained "control of [his] life" and concluded that the break-up with Brita "made a crucial contribution to [his] career."[31]

IMPRESSED BY WILKINS'S WORK, Randall arranged a postdoctoral position for him at Birmingham University, set to begin in January 1940. During this period, Randall was embroiled in a bitter rivalry with Mark Oliphant over primacy in what each physicist considered to be *his* greatest invention, the giant cavity magnetron. When working well, the device made microwave radar possible and could detect objects in the air, such as the German planes that were destroying Britain's landscape and morale. In its earliest days of development, the machine was unreliable and kept "switching from one frequency to another." It was Randall's job to diminish that unreliability and he soon succeeded in creating what some considered to be "the most important invention that came out of the Second World War."[32] To ensure his share of the credit for the magnetron, however, Oliphant blocked Randall's funding requests. In turn, the physicists working in Randall's lab locked their doors whenever Oliphant's team visited, and, in the evening, their desks. Such internecine competition deeply troubled Wilkins, especially when Britain seemed to be losing an existential war with Nazi Germany.[33]

Much to John Randall's chagrin, Wilkins left his lab in 1944 and accepted an invitation to join Oliphant's new wartime physics unit, the Birmingham Bomb Lab. After two senior members of the team—Rudolph Peierls and Otto Frisch, both Jewish émigrés from Nazi Germany—figured out that less uranium would be needed to build an atomic bomb than initially thought, the Oliphant unit was dispatched to the United States to help with the Manhattan Project, the largest scientific endeavor of its era. Wilkins was assigned to work at the University of California, Berkeley, where the legendary nuclear physicist and 1939 Nobel Prize winner Ernest Lawrence had devised a giant cyclotron particle accelerator. Wilkins's task was to determine the means of vaporizing uranium metal, which eluded him until Lawrence figured it out himself. Although working at a lower level of security clearance than Lawrence or Oliphant, Wilkins knew he was helping to build a weapon of mass destruction. As did many pacifist scientists enlisted into the war effort, he rationalized this work because the threat of conquest by the Nazi–Japanese Axis was so grave. That did not mean he had to feel good about it.

At the war's end, Wilkins eagerly left the bomb-building business.[34] He not only loathed the unchecked violence of nuclear weapons, he also objected to how the ethos of secrecy—necessary during wartime—had seeped into the standard practices of scientists he admired, to negative effect. He was still naïve enough to believe that the best science only proceeded in "an atmosphere of openness and cooperation."[35] Wilkins's antinuclear sentiment did not go unnoticed. Both the American FBI and the British secret intelligence service MI5 suspected that one of the nine New Zealand or Australian scientists working on the Manhattan Project had leaked top-secret information. In fact, Wilkins was under MI5 surveillance from 1945 until, at least, 1953. The spies failed to find anything incriminating, although "an informant" described Wilkins as "a very queer fish," "a caricature of a scientist," "incapable of dealing with ordinary human situations," and more likely "a socialist rather than a communist."[36]

While still in California, Wilkins fell in love with a Californian art student named Ruth Abbott. Like so many of Wilkins's romances, this relationship quickly hit shoals rockier than the Pacific coastline. Abbott became pregnant and Wilkins proposed marriage. He later admitted that he had mistakenly assumed Abbott had the same views on marriage that he had, with respect to the male being the dominant partner in the relationship, especially regarding career and major life decisions. In fact, she did not agree with his old-fashioned views, a discovery which surprised him and caused no end of discord during their brief period of matrimony.[37] They lived, fought, and fumed in a big house in the Berkeley hills. After a few months, Abbott told Wilkins that she had made an appointment for him to meet with an attorney, who informed him that she wanted a divorce. Understandably shocked, he had little to do with her or his son thereafter and, as he later recalled, "I returned to England alone."[38]

Wilkins received only one academic job offer: an assistant lectureship in the department of natural philosophy at St. Andrews University. The offer was made by John Randall, who had by then forgiven his prodigal student and moved to Scotland, far away from the academic politics of Birmingham. On August 2, 1945, his final day of a solitary vacation spent hiking in the Sierra Nevada mountains, Wilkins wrote to Randall accepting the position and telling him that his wife would not be coming along. Politely referring to her as "a very fine girl," he told Randall about the divorce and how it had cost him a much needed two hundred dollars. Unfortunately, it would not be final in Britain for three years unless Abbott remarried, and his lawyers advised him to keep the legal proceedings a secret until well after he returned to Britain. Indeed, for a long while, Wilkins did not even tell his parents about it.[39]

Wilkins spent the academic year of 1945–46 puttering aimlessly in the lab, cursing his humiliating personal situation, and desperate for contact with his infant son. It was then that the British physicist Harrie Massey, who had worked on minesweepers at the Admiralty Research Laboratory with Francis Crick, gave him a copy of Schrödinger's book *What Is Life?* (just as he had done with Crick). Massey sensed that Wilkins was at a career crossroads, uncertain about what to study next, and suggested, "You might be interested in reading this." His implicit message was that Wilkins should think about molecular or quantum biology.[40] As a student at Cambridge, Wilkins had been a great admirer of Schrödinger's work on quantum physics and his explanations of the complex ideas of wave mechanics, "which had a down-to-earth character, like Einstein's thoughts about how a boy would see the universe if he were sitting on a light wave."[41] As he bolted down Schrödinger's paragraphs, pages, and chapters, Wilkins was first inspired to segue from physics to biology. Schrödinger's description of the structure of the gene as an aperiodic crystal resonated deeply with Wilkins, because his research interests were focused on solid state physics and crystalline structures.[42]

That same year, Randall was offered the prestigious Wheatstone Chair of Physics at King's College, London. Soon after his appointment, in 1946, Randall won £22,000, the Medical Research Council's equivalent of a golden ring—a huge, multiyear grant to create an elite Biophysics Unit within the department of physics, an elite cadre of biologists and physicists chosen to study the structure of biological systems, or, as Randall mesmerized his reviewers, "to bring the *logi* of physics to the *graphi* of biology."[43] The field even took on a new name, now a familiar part of the scientific lexicon: "molecular biology."[44] Randall brought his entire team from St. Andrews to London and made Wilkins assistant

Maurice Wilkins at work,
1950s.

director of the Biophysics Unit. Unbeknownst to the MRC, Randall had also won a huge grant from the bountiful Rockefeller Foundation to purchase equipment for his molecular biology research. When the college administrators began asking questions about the obvious double-dipping, Randall glibly replied that the Rockefeller grant funded the King's physics department and the Medical Research Council award was for the Biophysics Unit.[45]

Predictably, Randall's grants drew envious taunts from his less well-funded colleagues at King's and beyond. Randall studiously avoided the controversy, concentrating on organizing the huge research group under his supervision. The menu of projects was long and varied; it aimed to apply physical methods to the study of cells, cellular membranes and nuclei, chromosomes, spermatozoa, muscle tissue, nucleic acids, and the structure of DNA.[46] These lanes, in accordance with British standards of research etiquette, were sacrosanct; in other words, from 1947 forward, the King's unit "owned" DNA research, just as the Cavendish unit would soon be secure in its claim (and MRC funding) to discovering the structure of hemoglobin and myoglobin. Competition existed within, but not between, MRC units.

※

FOUNDED IN 1829, King's College was the Anglican alternative to the University of London's nonsectarian University College (itself founded as a response to the Church of England colleges at Cambridge and Oxford). King's was a modern university committed to providing an educational experience that prepared its students for work in the rapidly evolving world. As late as 1952, the most prominent feature of the King's College campus was a giant bomb hole—more than 18 meters long and 8 meters deep—in the center of its quadrangle, courtesy of a Luftwaffe raid during the Blitz.[47] Overlooking the river Thames and Waterloo Bridge and stately Somerset House to the south and opening up to the busy Strand to the north, the college long suffered the scars of bombs, bullets, and privation. The Biophysics Unit was housed in the basement of the ornately pillared main building. In distinct contrast to the finer digs of the Cavendish, every day the King's biophysicists descended from the hustle and bustle of the Strand down a narrow flight of stairs into their underground laboratory.

Wilkins was more than happy to leave the glens, lochs, and loneliness of Scotland. Unable to secure his own apartment, he moved into a spare room of his now-married sister Eithne's home in Hampstead, an area of London filled with "artists, intellectuals and refugees from Nazism." In his spare time, he took to frequenting the art galleries of the West End. At one of them, he met a Viennese-born artist named Anna, and began seeing her romantically. In his old age, Wilkins chuckled over the facts that Anna was at least a decade older than him and that their relationship was hardly monogamous. The affair ended abruptly after he confessed to also seeing one of her close friends. Tortured by the emotional maelstrom brought on by the exit of still another woman from his life, he sought solace in Freudian psychotherapy. A year of structured introspection with a female therapist, assigned to him by the "official Freudian organization," did little to calm his frayed nerves. Eventually his therapist fired him after he complained to her supervisor, "that woman will never get anything out of me." He heard no more from the Freudians and sank even lower into depression. Despite thoughts of suicide, he held off because he did not want to upset his mother, who was actively grieving the recent death of her husband.[48]

Fortunately for his mental health, Wilkins's days were kept busy conducting his own research and serving as Randall's "right-hand man." He dove head-first into the pool of DNA. He began by studying Oswald Avery's work and, like Watson and Crick, grew ever more convinced that it was DNA, and not proteins, that carried the gene's most important information. At last, Wilkins found love in the marriage of biology and physics. Intellectually, everything seemed to be finally coming together for this shy man. His path was now clear. Biophysics and the structure and function of living molecules fascinated him.[49]

Wilkins was not entirely free as long he worked in a subservient role to his professor. As is often the case in laboratory hierarchy, one man governed all of its endeavors. Every morning, Wilkins endured Randall's griping about running the MRC project. He tried his best to be unflappable, but soon tired of Randall's slapdash administrative practices and the senior professor's insistence on taking a byline on all his papers—often in the lead author's slot. Increasingly, Wilkins engaged in "stand-up rows" with his boss over research questions and resources. At the root of the conflict was Randall's desire to be in control of the laboratory's DNA work, even though his administrative chores prevented him from conducting any meaningful research. The never-ending constraints on Randall's time irritated him and, as a result, he took to irritating Wilkins, who christened the laboratory "Randall's Circus." More bluntly, Wilkins dismissed his relationship with Randall in one succinct sentence: "I admired and respected him, but I cannot really say that I found him very likeable."[50]

For more than a year, Wilkins applied ultrasonic waves to induce mutations in DNA, and ultraviolet and infrared light to make it visible under a microscope. After making only meager progress, he consulted

Maurice Wilkins with John Randall, chief of the King's
College Medical Research Council Biophysics Unit.

John Kendrew at Cambridge and some of the biologists at King's and
the Plymouth Marine Station on next steps. Sometime in early 1950,
Wilkins embarked upon a series of attempts at using X-ray diffraction
techniques on DNA samples extracted from the nuclei of calf thymus
cells, harvested from batches of sweetbreads purchased at slaughter-
houses. Rudolf Signer, a Swiss organic chemist at the University of Bern,
generously donated this cache to Wilkins, preserved in just the right mix
of isopropyl alcohol and salt and stored in a jam jar. Wilkins described
Signer's 15 grams of viscous gold as looking "just like snot." With each
passing day, he became more adept at spinning long, 10-to-30-μm

(1 μm = 10^{-6} meters) fibers, or chains, of remarkable structural uniformity, perfect for crystallographic analysis. The coveted Signer samples were an essential, if now forgotten, contribution to the discovery of the molecular structure of DNA.[51]

Yet, even though Wilkins had obtained such terrific research material, Randall harangued and humiliated him for taking too long to deliver results. Randall could not accept that the dexterity required for X-ray crystallography took almost as long to develop as mastering the violin. By the spring of 1950, Randall lost patience and began searching for a professional crystallographer to take over his lab's X-ray diffraction work. The candidate he ultimately chose was a thirty-year-old physical chemist named Rosalind Franklin. Having just completing a four-year postdoctoral fellowship in Paris perfecting the crystallographic analysis of coal, she was ready to return home to London and point her X-ray camera at biological structures.

Like Touching the Fronds
of a Sea Anemone

When Rosalind wanted proof, it was proof she wanted, and no approximation would do . . . a rough description of Rosalind's character sounds almost like contradictions—that she combined honesty and tact, logic and warmth, an enormous abstract intellect with vivacious humanity and sensitivity; but this is all true, and it was not contradictory but harmonious. I would give anything to be able to get all this, with the delicate shadings, down on paper accurately.

—ANNE SAYRE, LETTER TO MURIEL FRANKLIN
(ROSALIND FRANKLIN'S MOTHER), FEBRUARY 5, 1970[1]

I suspect you won't agree with my Asperger's diagnosis of Rosalind, but I am not the first to make the possible connection.

—JAMES D. WATSON, LETTER TO JENIFER GLYNN
(ROSALIND FRANKLIN'S SISTER), JUNE 11, 2008[2]

From an early age, Rosalind Elsie Franklin intuited she was different: different from her pampered siblings; different from her well-heeled family of Anglo-German Jewish financiers and do-gooders; different from the other girls at her school, who were preoccupied by early-twentieth-century strictures of domesticity; and wholeheartedly different from the strange looking and sounding *Ostjuden*—the East European Jewish refugees who settled in the East End, London's equivalent of New York City's teeming Lower East Side.[3] James Watson often speculated that Rosalind's distinctive nature came from her "upper-class" social station.[4] This point is not without merit, but the monied circles the Franklins inhabited were primarily Jewish ones.

Rosalind's mother, Muriel, came from a distinguished Anglo-Jewish family, the Waleys, who numbered several prominent lawyers, financiers, poets, and politicians in their ranks. In 1835, the Waley relative David Solomons became the first Jewish elected sheriff of London, though he was not allowed to take office because the mandatory oath required him to swear allegiance to the Christian faith. Solomons was more successful in serving as the first Jewish member of the House of Commons (1851) and London's first Jewish Lord Mayor (1855).

Although the Franklins had long been bankers, their history was char-acterized by far more than the mere accumulation of pounds, shillings, and pence. A family history compiled by Rosalind's paternal grandfather, Arthur, claimed direct descendance from King David.[5] Royal assertions aside, there were several distinguished rabbis in the Franklin family tree, including Rabbi Judah Loew ben Bazelel of Prague (1512?–1609), the Tal-mud and Kabbalah scholar who, as folklore has it, created the Golem, an artificial man made of clay, as a means of protecting the Prague ghetto from anti-Semites. The Golem legend, a staple of Yiddish literature, was likely the source for Mary Shelley's 1818 novel *Frankenstein*.[6]

In 1763, the Fraenkel (soon to be anglicized to Franklin) family emi-grated from Breslau, Germany, to London, when fewer than 8,000 Jews resided in all of England. Arthur Franklin liked to boast that three out of his four grandparents were born in Britain, as a means of displaying his family's long-standing residence. Although not as established as the Sephardic Jews who fled the Spanish Inquisition of 1478 nor as wealthy as the Rothschild clan, the Franklin family were members of Britain's Jewish elite—"a compact union of exclusive brethren with blood and money flowing in a small circle which opened from time to time to admit a Beddington, a Montagu, a Franklin, a Sassoon, or anyone else who attained rank or fortune, and then snapped shut again."[7]

The history of modern Israel, too, contains many Franklin family members on its pages. Rosalind's paternal aunt Helen Caroline "Mamie"

Franklin Bentwich (1892–1972) was a feminist activist who developed nursery schools, arts centers, and other social programs in the 1920s. Her husband, Norman, was Attorney General for Mandatory Palestine from 1920 to 1931; his father, Herbert, a copyright lawyer, was one of Theodore Herzl's first followers in England and a major figure in the early Zionist movement. Most prominent was Rosalind's great-uncle Herbert Louis Samuel (1870–1963). In 1915, Viscount Samuel wrote the "secret" memorandum to the British cabinet that resulted in the Balfour Declaration of 1917, establishing the concept of "a national home for Jewish people" in Palestine. Samuel was appointed the first High Commissioner of Palestine in 1920, three weeks before Rosalind was born.[8]

AS A LITTLE GIRL, Rosalind distinguished herself from her siblings (one older brother, David; two younger brothers, Colin and Roland; and a younger sister, Jenifer) by being quiet of voice, observant of those around her, and perceptive in her judgments. Overly sensitive, especially if she felt slighted or wronged, her response as a youngster was to retreat and ruminate. Her mother, Muriel, the very model of the traditional Jewish

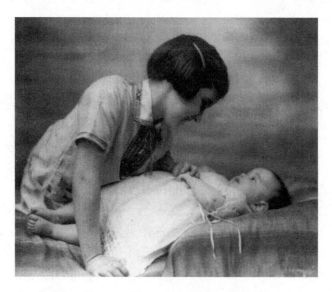

Rosalind Franklin, age nine, with her baby sister, Jenifer.

wife, wrote more than a decade after her second child's death, "When Rosalind was upset she would figuratively curl up—like touching the fronds of a sea anemone. She hid her wounds and trouble made her withdrawn and upset. As a schoolgirl I always knew when something had gone wrong in school by her silences when she got home."[9]

Such sensitivity often obscured her deeper talents. In 1926, Rosalind's aunt Mamie described to her husband a visit with her brother and his family on the Cornwall coast. She provided a wonderful characterization of the six-year-old Rosalind: "[she] is alarmingly clever—she spends all her time doing arithmetic for pleasure, & invariably gets her sums right."[10] Equally apt was Muriel's recollection of young Rosalind's "immensely glowing character, very strong and brilliant—not just intellectual brilliance, but a brilliance of spirit."[11] Perhaps we ought to give the last word to the eleven-year-old Rosalind, soon after her mother introduced her to the science behind developing photographs: "It makes me feel all squidgy inside."[12]

"All her life Rosalind knew exactly where she was going," Muriel insisted; "her views were determined and clear cut."[13] As a teenager, Rosalind had already developed a sharp tongue and pointier elbows. She was unafraid of expressing her distaste or critique of others, especially in the cause of science. To those she loved, she was an ideal companion, funny, mischievous, and incisive of thought. Such was not the case for those who disappointed her in some manner or whom she found to be not up to the mark. Muriel knew all too well how her daughter could be devastatingly blunt and, to less grateful sets of ears, humiliating: "Rosalind's hates, as well as her friendships, tended to be enduring."[14]

Like many gifted young people, Rosalind Franklin erroneously assumed that her intense intellectual focus and quick, logical mind were universal and common. Throughout her life, she had a difficult time tolerating the mediocrity of others, often at the expense of her professional development. "Absurdities exasperated her," Anne Sayre observed. She responded to such people and situations with "fierce and stubborn indignation."[15] According to Muriel, people whom Rosalind deemed not to be very bright irritated her to distraction because of her "natural efficiency

in whatever she was doing was characteristic, and she could never understand why everyone could not work as methodically, and with equal
competence. She had little patience with well-intentioned bungling and
could never suffer fools gladly."

ROSALIND GREW UP IN the London neighborhood of Notting Hill, when
it was the heart of the affluent Anglo-Jewish community. The Franklins were wealthy but careful never to appear ostentatious. Household
accounts were meticulously kept by Muriel, on a strict allowance her
husband meted out each Monday. Ellis Franklin refused the easily
afforded luxuries of either a second home or a driver, preferring, instead,
to take the Underground to his office in the City, at the family's private merchant bank. Weekends were spent away from peering eyes at
his parents' estate in the Buckinghamshire village of Chartridge, a home
modernized by the same architect who designed the familiar façade of
Buckingham Palace.[16]

The Franklins' life was bounded by family ties, propriety, and a love of
all things English. Ellis and Muriel Franklin inculcated in their children
the importance of education and devoting their time to those less fortunate. Ellis Franklin's favorite charitable organization was the Working
Men's College, on Crowndale Road in the borough of St. Pancras, London. Founded in 1854 to introduce laborers to their university-educated
counterparts, the college offered courses ranging from economics and
geology to music and cricket. Ellis Franklin was a vice-principal there
and long taught courses on electricity.[17]

Equally important was the Franklins' deep-rooted Jewish faith. The
family regularly attended services at the New West End Synagogue in
Bayswater—which Ellis helped fund, and later reorganize, under the
assimilative principle that "the whole idea is that Judaism is a religion
not a race . . . the English Jews are as much English as other English."[18]
Yet in an era when Jews were considered to be a distinct and not necessarily admired race of their own, no matter how well Ellis Franklin's
family assimilated into British life, they remained foreign and apart

from those they most wanted to be like. Stereotypes die hard, and this was a time when most English people were introduced to Jews by watching Shakespeare's Shylock overcharge his Christian customers or reading how Dickens's Fagin plotted to subvert runaway boys into pickpockets. As George Orwell noted in 1945, there were only 400,000 Jews living in Great Britain, about 0.8 percent of the total population, and they were "almost entirely concentrated in half a dozen big towns." Worse, he concluded, "there is more anti-Semitism in England than we care to admit, and the war has accentuated it . . . It is at bottom quite irrational and will not yield to argument."[19] When Rosalind Franklin became an adult, she, too, had to contend with not being "as much English as other English." Not only was she one of few women in British physical science, she was a Jew trespassing the ivory tower inhabited and controlled by Christian, white men. Saddled as she was with the baggage of gender and strong personality, the addition of a subtle but omnipresent anti-Semitism in British academic circles all but doomed her chances for success—if it were not for her brilliance.

At the age of nine, in early 1930, Rosalind was sent away to the Lindores School for Young Ladies at Bexhill on the Sussex coast, overlooking the choppy English Channel. Although she focused on her studies and demonstrated superb hand-eye coordination in her handicrafts course, she was often homesick and wrote to her parents how much she missed them and her baby sister, Jenifer. It was clear that Lindores was not an ideal place for her, so in January 1932 her parents enrolled her in the middle fourth form of St. Paul's School for Girls, in West Kensington, London. St. Paul's had the double advantage of being a short bus ride from the family home in Notting Hill and, because it had no church allegiance despite its saintly namesake, being well-populated by the daughters of educated Jewish men.[20] Rosalind was happy to be transferred closer to home, though she objected to her parents' characterization of her as being too "delicate" for boarding school, an objection that festered into a resentment at being sent away from home at such a young age.[21]

St. Paul's was of its time in that its science courses emphasized the

feminine qualities of maintaining a neat appearance, meticulousness in
one's studies, and reviewing one's answers over and over again. Grand-
standing or bold measures were not encouraged. Ahead of its time for
a girls' school, St. Paul's was well equipped with brand-new laboratories
staffed by three "highly-qualified mistresses" who taught biology, physics, and chemistry. After her first four years at St. Paul's (the equivalent
of American middle school years), Franklin entered the sixth form and
declared her intention to work toward the Higher School certificate with
a focus on chemistry, physics, and mathematics. She bypassed the biology and botany courses, which tended to be taken by those interested
in attending medical school. According to Anne Crawford Piper, one of
her closest friends at St. Paul's, Franklin did a great deal of independent
scientific study in order to advance ahead of her peers.[22]

At seventeen, Franklin sat for the Cambridge University entrance
examination in mathematics and physics. Like many students, she experienced test and interview anxiety but soldiered on. In October 1938, she
was offered a place by both women's colleges at Cambridge, Girton and
Newnham. The two colleges, while beautiful by any architectural standards, were not nearly as grand as the many men's colleges. Newnham

Newnham College, Cambridge.

College, which she ultimately chose, was a red-brick, Queen Anne-style suite of buildings, with white-painted window frames and prominent chimneys, grouped around a lush green garden.

Cambridge, though an "ancient university," did not accept women students until 1869, nor Jews before 1871. For decades thereafter, a young woman's options for matriculating into Cambridge were limited to the two women's colleges, compared to twenty-two for men; there were only 500 places for women, compared to 5,000 places for men.[23] This was merely the tip of a long list of inequalities endured by young women seeking a first-class education in pre–Second World War Britain. A dismissive comment scrawled on too many transcripts explained such sorry matters well: "Good, but female."[24]

Unlike Oxford, which began granting women degrees in 1921, women students at Cambridge were not accepted as "members of the University" until 1947. Instead, they were merely listed as students of Girton or Newnham. Women were not eligible to receive the degree of B.A. Cantabrigia (the Latin form of Cambridge); their diplomas read "degrees titular," or "degrees tit.," a shorthand that inspired many a snicker among the male students. Beyond the sexist wisecracks, the women of Cambridge were forced to accept their secondary status each day by taking seats in a segregated section (the front row) of the classroom.[25] If a female student arrived late to class and could only find a seat in the male section, the men would stamp their feet on the back of her wooden chair and throw wadded-up paper spitballs at her.

In October 1928, the novelist Virginia Woolf delivered lectures to the Arts Society at Newnham College and the ODTAA ("One Damn Thing After Another") Society of Girton College. Never one to waste her carefully wrought sentences, Woolf published the talks that same year as a book pointedly entitled *A Room of One's Own*. In it, she described the anxiety women students experienced when they accidentally stepped on the verdant lawns of the men's colleges—a serious infraction of university rules. Only college fellows could ignore signs ordering mere mortals to keep off the grass, but only men were eligible for fellowships at these colleges.[26] In consolation, the Newnham women were free to trek its seventeen acres of grass and garden during most of the year.

As a reader, Franklin was no admirer of Virginia Woolf. After putting down Woolf's novel *To the Lighthouse* without finishing it, she wrote to her parents: "I like long sentences well put together, but hers are so arranged that the beginning is meaningless until the end is reached, which I consider unjustifiable."[27] Criticisms aside, Franklin's undergraduate years coincided with the publication of Woolf's 1938 book *Three Guineas*, which discussed women's rights. In its pages, Woolf dispassionately ticked off the inequalities between British men and women with respect to education, ownership of property and capital, valuables, patronage, and access to the professions. Most important, Woolf encapsulated this plight of separate and unequal in one crystalline, perfect phrase that would have rung true to Franklin's ears: "Though we see the same world, we see it through different eyes."[28]

EVER THE DUTIFUL DAUGHTER, Franklin wrote home twice a week. Taken in full measure, these letters reveal a curious, empathic, hardworking, ambitious, and earnest young woman, one in possession of a strong sense of humor about herself and others. She displays an unquenchable thirst for scientific knowledge, even if it meant taking extra courses, attending additional lectures, or staying in the laboratory for stretches of eight or more hours. In sum, these letters are a joy to read. They allow one to experience vicariously the development of a young adult who both loved the freedom of university life and was eager to maintain her connection with her family in London.

Early in the correspondence we see that although she chose Newnham College after much deliberation and anxiety, Franklin was able to poke fun at her selection. The day after a severe Nazi bombing raid struck Cambridge, she wrote her parents, "I was right to come to Newnham . . . Girton was in flames last week. They repeatedly remind listeners that the [local Royal Air Force station] makes Cambridge University a military target."[29] In one letter, she apologizes for her poor handwriting (a lifelong issue) and focuses on girlish things.[30] In another, she asks that her mother "please send me your rumours" as well as her "evening dress (tulip one), evening shoes and evening petticoat. Shoes in the bottom drawer of the wardrobe (gold *or* silver)" for "Commemoration, a sort of old girls' dinner next week to which present student[s] have been invited . . . because not enough old girls can come."[31] After a severe bomb attack, she mixes the urgent with the mundane: "Please may I have my GAS MASK!! Also need my pyjamas and handkerchiefs from last week's laundry, and the strapping which you will find in the right-hand draw-

Rosalind Franklin mountain climbing in Norway, c. 1939.

ers of my walnut piece. It is essential for skating."[32] She later consoles her father, who was struggling with some divisive board members of his beloved Working Men's College, "It is really difficult to believe that anyone could be so destructive."[33]

In 1940, Franklin wrote to her parents about meeting Adrienne Weill, an eminent French Jewish physicist who lost her husband in the war and fled Paris for England. Weill and her daughter initially took up residence at Newnham, and later operated a hostel for French refugees. For Franklin, Weill represented a direct link to the keystone of women scientists, Marie Curie, as she had studied under the great Curie at the Institut du Radium from 1921 to 1928.[34] In Cambridge, Weill conducted physics and metallurgy research at the Cavendish Laboratory, under Bragg. After hearing Weill lecture on Curie's work, Franklin gushed to her parents, "of all the French people now in England she should be the first I met . . . She is a delightful person, full of good stories and most interesting to talk to on any scientific or political subject . . . I was really thrilled by her lecture."[35] As with so many young people called to academia but who hail from families that do not fully understand such pursuits, Franklin's connection to Weill represented a watershed moment in her life.[36]

Politics and religion were often discussed at the Franklin family dinner table. Like most young adults, Rosalind used such topics as a vehicle for teasing her parents, especially her doting and demanding father. She tended to take the liberal, if not socialist, view, in contrast to her father's more conservative beliefs. During the war, Ellis Franklin deeply insulted his daughter by suggesting that she might be more useful to the war effort if she gave up her scientific studies and took a clerical job for a government agency, and spent her evenings rolling bandages and making cups of tea for soldiers on leave. According to some friends, this was an affront she never forgot. By the same token, when Rosalind wanted to seriously discuss her work with a trusted family member, it was almost always with her father.[37]

Ellis Franklin dealt in a world of equities, bonds, mortgages, and bottom lines. When it came to Judaism, the Torah taught him the infinite splendor of God and that the Jews were His chosen people—and in Ellis's view, these teachings were definitive. Rosalind, on the other hand, refused to accept such statements unless presented with proof, facts, and reason. As a six-year-old, she asked her mother for evidence establishing the existence of God. No matter what answers the exasperated but believing mother came up with, the child quickly shot them down with precocious and penetrating follow-up questions, topped off by the rhetorical flourish, "How do you know *He* isn't a *She*?"[38]

During the summer of 1940, the bombing raids were so severe that there was talk of closing down Cambridge for the fall term. In a four-page letter written during this chaos, a twenty-year-old Franklin responded to her father's accusation that she was making a religion out of science. Although she disagreed with him over the ultimate authority—God versus "scientific truth"—she did it in a manner so bright, thoughtful, and loving, it is hard to imagine a father who would not be proud to be jousting with his daughter's rapier-sharp mind:

> You frequently state, and in your letter, you imply, that I have developed a completely one-sided outlook, and look at everything and think of everything in terms of science. I think this is a completely distorted view. Obviously, my method of thought and reasoning is influenced by a scientific training—if that were not so, my scientific training will have been a waste and a failure. But you look at science (or at least talk of it) as some sort of demoralizing invention of man, something apart from real life and which must be cautiously guarded and kept separate from everyday existence. But science and everyday life cannot and should not be separated. Science, for me, gives a partial explanation of life. In so far as it goes, it is based on fact, experience, and experiment. Your theories are those which you and many other people find easiest and pleasantest to believe, but so far as I can see, they have no foundation other than they lead to a pleasanter view of life (and an exaggerated idea of our own importance).
>
> I agree that faith is essential to success in life (success of any

sort) but I do not accept your view of faith, i.e., belief in life after death. In my view, all that is necessary for faith is the belief that by doing our best we shall come nearer to success and that success in our aims (the improvement of the lot of mankind, present and future) is worth attaining. Anyone able to believe in all that religion implies obviously must have such faith, but I maintain that faith in this world is perfectly possible without faith in another world . . . One further point, your faith rests on the future of yourself and others as individuals, mine on the future and fate of our successors. It seems to me that yours is the more selfish.

It has just occurred to me that you may raise the question of a creator. A creator of what? I cannot argue biologically, as that is not my field . . . I see no reason to believe that a creator of protoplasm or primeval matter, if such there be, has any reason to be interested in our insignificant race in a tiny corner of the universe, and still less in us, as still more insignificant individuals. Again, I see no reason why the belief that we are insignificant would lessen our faith—as I have defined it. . . . Well, now my normal letter . . .[39]

Anne Sayre conjectured that Franklin's struggle against her father and his lack of positive approval was psychologically harmful to her.[40] During the early 1970s, Sayre and Muriel Franklin argued back and forth by transatlantic post over the veracity of this claim. Muriel, to be sure, had her late husband's legacy to guard. Nevertheless, she avowed that he heartily supported Rosalind's career choices and she thought it "unfair" to state otherwise. She insisted there was never any serious rift between father and daughter.[41] In a subsequent letter, she objected to Sayre's characterization of Ellis Franklin as a "narrow, conservative Victorian papa," countering that Rosalind "did sometimes get imaginary grievances and these have been highlighted and blown up to give a cruelly distorted portrait."[42]

Sayre, who claimed to have discussed these issues with Franklin sometime in the mid-1950s, stuck to the narrative of a soul-damaging father–daughter conflict. On October 30, 1974, she explained to Muriel Franklin

how "Rosalind saw herself as someone who had had prejudices to com-
bat and opposition to overcome, and she expressed this so strongly and
to so many people, that it has to be considered as an element of her personality." Sayre was willing to concede Muriel's claim that there was no
"real opposition to her intentions or her ambitions from her father." But
for Sayre, such reality was less than relevant because "Rosalind thought
she had and this conviction—however mistaken in substance—did have
an effect upon her in various ways."[43]

Decades later, James Watson, Maurice Wilkins, and Francis Crick
each used this perceived discord to paint the father–daughter relationship as fraught, disapproving, and the source of Franklin's purported
troubles in dealing with men. In many public settings, they alleged without evidence that Franklin fatally disappointed her father by not taking
up the traditional well-heeled, young woman's role of wife, mother, and
charity patron. Such assertions suggest a gross misunderstanding of the
tangled relationships between women and their fathers—especially as
the former become adults and fulfill their developmental need to individ-
uate from the first important men in their lives.[44] Rosalind Franklin was,
by every measure, a good and devoted daughter, simultaneously close
to her parents and siblings and, by personality, independent of mind,
spirit, and aspiration. As with most family conflicts, the truth probably
lies somewhere in between. Ellis and Muriel Franklin would probably
have been more comfortable if Rosalind had simply married and raised a
family. It may have taken them time to come to terms with their daughter's ahead-of-the-curve career choice in what must have seemed to them
such a mysterious and highly specialized scientific field—especially
when compared to their sons, who followed their father into the family
banking business—but they did eventually get there. And because of
his long interest in electricity and physics, Ellis and Rosalind Franklin
enjoyed a dialogue on science. He loved his daughter deeply, and she
loved him just as profoundly. Like many fathers and daughters, they had
a complicated relationship, made more complicated by the coincident
generational and seismic shifts in gender roles, society, and the sciences.

Franklin's intensity, noted by all who knew her during her Cambridge years, had the potential to impede her success. For those who admired her, it was merely a function of her insatiable quest for truth and knowledge. Others, however, found her to be daunting and cold, if not harsh and prickly. One college acquaintance, Gertrude ("Peggy") Clark Dyche, recalled that Franklin could be "a difficult character—impatient, bossy, intransigent." There was no finesse or diplomacy to her arguments, but "this was all because she had such high standards and expected everyone else to be able to reach her ideal requirements.[45] Adrienne Weill agreed heartily: "she was always so frank in her likings and dislikings."[46]

Her student notebooks document a depth of study rarely found among undergraduates today. She was introduced to Newton, Descartes, and Doppler; she read Linus Pauling's *The Nature of the Chemical Bond* from cover to cover. She even spent time contemplating the nature of sodium thymonucleate, the salt form of DNA extracted from calf thymus. While reading up on the molecule that would soon represent her destiny, Franklin sketched a helical structure in her student notebook and wrote "Geometrical basis for inheritance?" She also studied the basics of X-ray crystallography and drew diagrams of the different crystal formations and their unit cells, including one type that later proved critical for figuring out DNA's structure: "monoclinic all face-centered."[47]

In the classroom, her professors adjudged her to have "a first-class mind" and to be fully devoted to her studies. During examinations, however, her perfectionism often led her astray. She tended to answer the first few questions in such detail that she often ran out of time and was forced to give short shrift to those remaining.[48] She also fretted away the nights before an important exam, tossing, turning, and losing sleep. She tried to make up for it by gulping down stimulating mixtures of Coca-Cola and aspirin, referred to by many an undergraduate of this era as "drinking a dope."[49]

The day before Franklin sat her Tripos, the final examination for her degree, she contracted a nasty cold. She completed the first two-thirds of the exam with distinction but froze during the final segment. A first-class degree, as we saw in Wilkins's and Crick's educational history, was essential for pursuing an academic career at Cambridge or Oxford.

Although her score was especially good in the physical chemistry portion of the Tripos, the missed questions pulled her down to the list of second-class degrees conferred in the spring of 1941. Fortunately, her ranking as an "upper second," along with a strong letter from her advisor, Sir Frederick Dainton, secured her a graduate research scholarship in physical chemistry. She was assigned to work under Ronald G. W. Norrish, who later won the 1967 Nobel Prize in Chemistry for his studies on the kinetics of chemical reactions. The plan was for Franklin to complete enough experimental work for a doctorate in physical chemistry.

THE NORRISH LABORATORY WAS situated in the University's department of physical chemistry, just down Free School Lane from the Cavendish. It was an unwelcoming place for any student, but especially so for a young woman. The professor was "ill-at-ease with himself, irascible and drinking too much—not the qualities that would endear him to Rosalind Franklin."[30] When she joined Norrish's laboratory, Franklin was only twenty-one, a fourth-year student seeking instruction in the thick of the war. To the erratic, hostile Norrish, she was a pesky female, easily dismissed and disrespected. Moreover, their personalities and approach to research were diametrically opposed. Franklin often displayed "characteristic total absorption in what she was interested in and a corresponding indifference to what she wasn't interested in—a kind of specialization that was not only intellectual but had some accompanying emotion."[31] According to Dainton, Franklin's logic-driven, focused, and rigid conduct did not mesh well with Norrish's "flair, which was ever so frequently wrong but sometimes brilliantly and inexplicably right. Temperamentally, in personal and intellectual values, in relative emphases given to the cognitive and affective, in cultural background they were chalk and cheese and Rosalind was right to get out."[32]

The scientific problem Norrish ordered her to tackle—describing the polymerization of formic acid and acetaldehyde—was a chemical process he and another student had already worked out and published in 1936.[33] The assignment was the chemical equivalent of grunt work and she knew it. To make matters worse, Norrish exiled the claustrophobic Franklin to a small, dark room for her workspace.

In December 1941, Franklin wrote home describing her tortured teacher–student relationship: "I'm only just beginning to realize how well justified his bad reputation is. Well, I've now got thoroughly on the wrong side of him, and almost reached a deadlock. He's the sort of person who likes you all right as long as you say yes to everything he says and agree with all his misstatements, and I always refuse to do that."[34] A few months later, in early 1942, she reported what transpired after she discovered a mistake in Norrish's work. When she proudly brought it to his attention, he nearly detonated. In her own fashion, Franklin told her parents how she refused to back down under his bullying: "When I stood up to him he became most offensive and we had a first-class row— in fact several. I have had to give in for the present but I think it is a good thing to have stood up to him for a time and he has made me despise him so completely that I shall be impervious to anything he may say to me in the future. He simply gave me an immense feeling of superiority in his presence."[35]

Fighting with one's dissertation advisor is never a good idea because that individual holds power over the stipend upon which you live, your position in the graduate program, the granting of your degree, and your future career opportunities. Still, Franklin found Norrish's drinking, abusive behavior, and wrongheadedness to be intolerable and she was quite vociferous about it. Some have written off her battles with Norrish as a harbinger of how she interacted with men. A more precise observation, however, would be that she did not work well with unreasonable men (or women)—at least, those she deemed to be such. Throughout her life Franklin was able to work well with many male and female colleagues who tolerated her quirks and looked beyond the surface to appreciate her virtues. It was the fools she could not abide.

Franklin's greatest comfort during this stressful period was found in taking up residence at Adrienne Weill's hostel on Mill Road, across the Cam River from Newnham College. The opportunity to practice her French, interact with international students, practice her facility with the

French language, and spend time with the inspirational Weill was a balm for the psychic wounds she experienced working in the Norrish lab.

⚕

IN JUNE OF 1942, the British Ministry of Labour decreed that all women research students be "de-reserved" and funneled into war-related work. Although Franklin preferred to stay in Cambridge, there were no solid opportunities for her there. On June 1, 1942, she wrote a feisty letter to her father, illustrating how fiercely she stood her ground when confronted with male entitlement, or what she perceived as unmerited authority:

> On one point you are quite unjust—I don't know where you got the idea that I'd "complained" about giving up a Ph.D. for war work. When I first applied to do research here a year ago, I was asked whether I wanted war work, and I said I did. I was led to believe that the first problem I had was war work. I soon found that I had been deceived, and since then have made repeated requests to Norrish for war work—it's one of the many things over which we have differed—and I have explicitly stated on several occasions against the advice of my elders and betters, that I would rather have war work now and a Ph.D. later.[56]

Fortunately, the war enabled her to escape Norrish's laboratory for research that allowed her to both complete her doctorate and contribute to the war effort. That August, she took a job as a physical chemist in the government-sponsored British Coal Utilization Research Association laboratory in Kingston-upon-Thames, a suburb of London. The laboratory was run by Donald H. Bangham, a kind, supportive chemist who filled the place with eager young male and female scientists, let loose to discover new uses for coal and charcoal to enhance energy resources during wartime. Franklin, in her role as assistant research officer, studied bituminous and anthracite coal mined in Kent, Wales, and Ireland. At night, she and her cousin Irene (with whom she was living in Putney) volunteered as air raid wardens.

That she would take on this latter role was characteristic of her strong personality and courage. She hated bomb shelters because of her claustrophobia. Anne Sayre, who witnessed Franklin's discomfort in close quarters, recalled that her fear "was not acute, but it was real. She concealed it habitually. She mastered the bus systems of several cities in order to stay out of the subways, and in the later stages of the war became an air raid warden, partly to stay out of shelters."[37] She combatted her fear of closed-in places by indulging in one of her favorite recreational activities during this period, arduous hiking through the mountains of North Wales (she later climbed mountain ranges all over Europe, including Norway, Italy, Yugoslavia, France, and across the Alps, and, later, in California). Like Watson and Wilkins, she loved a good mountain hike; sadly, these three scientists never found a way to take such a restorative outing together.

By the war's end in 1945, Franklin had completed enough original research to be awarded her doctorate in physical chemistry by Cambridge University. In 1946, she published her first paper, written with the BCURA chief Donald Bangham, "Thermal Expansion of Coals and Carbonized Coals." It appeared in the top-notch chemistry journal *Transactions of the Faraday Society* and described the porous properties (microscopic holes) of different types of coal and how such porosity contributed to the amount of energy produced.[38] Even Ellis Franklin had to admit that his daughter's talents were put to maximum use in studying the nation's chief source of energy.

In the fall of 1946, Franklin gave a brilliant talk at a carbon conference held at the Royal Institution. Despite her personality quirks, she was an excellent and confident speaker. Adrienne Weill had invited two French colleagues to attend the presentation, Marcel Mathieu and Jacques Mering, who were crystallographers at the government Laboratoire Central des Services Chimiques de L'Etat, in Paris. The two men were suitably impressed. A few weeks later, Merring offered Franklin a job as a physical chemist, applying X-ray diffraction techniques to analyzing the microstructure and porosity of coal, charcoal, and graphite.

IN EARLY 1947, Franklin moved to Paris and reported for duty at the laboratory—or, as everyone there called it, the *labo*. The facility was situated at 12 Quai Henri IV, in the 4th arrondissement, and featured large arched and leaded windows looking out onto the river Seine. She spent the next four years working alongside a cadre of French men and women and expatriates. She toiled as intensely as ever, reveling in the opportunity to apply her superb motor skills, sharp mind, and love of experimental research to becoming one of the world's finest X-ray crystallographers.[59]

This was no easy task. To begin, one must identify a suitable molecule to analyze. The crystalline structure of a particular molecule has to be somewhat uniform and relatively large in size; otherwise, multiple errors will be introduced on the X-ray pattern. Once an appropriate crystal is identified, the crystallographer aims a beam of X-rays at it. As the X-rays strike the electrons of the atoms making up that crystal, they are scattered—and the scatterplot is recorded on a piece of photographic paper placed directly behind the crystal. By painstakingly measuring the sizes, angles, and intensities of these scattered X-rays and then applying complex mathematical formulae to mapping them, the crystallographer develops a three-dimensional picture of the crystal's electron density. This, in turn, allows the positions of the atoms comprising the crystal to be determined, thus solving that molecule's structure.

Confounding matters further, a single X-ray image never provides the total answer. The crystallographer must rotate the specimen stepwise through hundreds of infinitesimally different angles over a spectrum of 180 (or more) degrees and take an X-ray picture at each one—and each presents its own set of smudges or diffraction patterns, making the process both time-intensive, mind-numbing, and physically cumbersome. Every one of these hundreds to thousands of X-ray diffraction patterns was, at that time, measured and analyzed by hand, eye, and a slide rule. If each step was not executed perfectly, artifacts or errors of measurement might be introduced, leading to wrong answers and conclusions.[60] A blurred image leads to an even blurrier assessment of how the atoms of that molecule are arranged. Fortunately, Franklin was astoundingly adept at these methods, and her results were superb.[61] Her colleague, Vittorio Luzzati, an Italian Jewish crystallographer, was amazed by the results that came out of her "golden hands."[62] Her supervisor, Jacques Mering, who was also Jewish, described Franklin as one of his best students, someone with a voracious appetite for acquiring new knowledge and remarkably skillful in both designing and executing complex experiments.[63]

In Paris, Franklin's social life took on a continental flair. Fluent in French, she loved shopping at the greengrocers and butchers, bolting down creamy pastries along the way, shopping for the perfect scarf or sweater, and getting lost exploring the byways of the City of Light. She adopted Christian Dior's "New Look" and took to wearing perfectly-cut dresses that featured tight waistlines, small shoulders, and long, full skirts.[64] She absorbed the local culture and politics, frequently attending films, plays, lectures, concerts, and art exhibitions with friends and potential suitors. Her vivacity, stylishness, and youthful beauty were not lost on the men in her life. Some have speculated that she developed a crush on the handsome, flirtatious Jacques Mering, but because he was married, albeit estranged from his wife, she quickly retreated, sensing there was no chance for a romantic future. [65]

For three of her four years in Paris, Franklin lived in a small room on the top floor of a house on rue Garancière, for the equivalent of about three pounds sterling a month. The landlady, a widow, had strict rules: no noise after 9:30 p.m., and Franklin could only use the kitchen after the maid had prepared the widow's dinner. Despite such restrictions, Franklin learned how to cook perfect soufflés and often made dinner for friends. She had access to a bathtub once a week but otherwise used a tin basin filled with tepid water. The rent was a third of what she would pay elsewhere and the location was perfect: the 6th arrondissement, home to the iconic Left Bank and the Sorbonne, situated between the gardens of the Palais du Luxembourg and the lively cafes of Saint-Germain-des-Prés.[66]

In the *labo*, men and women worked as equals in attending to their experiments, sharing meals and coffee, and debating scientific theory as if their lives depended on the outcome. Luzzati (who, in 1953, would share an office with Crick at the Brooklyn Polytechnical Institute) recalled that deep within Franklin was "a psychological knot" he could never unravel. She made many friends and some enemies, Luzzati explained, largely because "she was very strong and crushing, very demanding of herself and others, enduring not always to be liked." Although he often had to smooth over her verbal squabbles, he insisted that "she was a person of utter honesty, incapable of violating her principles. Everybody who worked directly with her enveloped her in affection and respect."[67]

Franklin's *aventure parisienne* was the antithesis of the very British conduct she later encountered at King's College, London. According to the physicist Geoffrey Brown, who worked with her both in Paris and at King's, the *labo* "resembled a traveling opera company . . . they screamed, stamped their feet, quarreled, threw minor bits of equipment at each other, burst into tears, fell into each other's arms—all this in the course of any discussion." At the end of such heated debates, however, "the storms blew over leaving no grudges behind."[68]

On more than a few occasions, much to the harm of her reputation, Franklin imported such high drama to the King's College lab. One afternoon, she asked Brown if she could borrow his Tesla coil, an electric circuit designed to produce the high voltages needed for X-ray to work. She failed to return it despite several polite entreaties from Brown, who needed the coil for his own experiments. As a result, he recalled, "I went and took it back, and screwed it onto the wall. And she came in, pulled it off and walked straight out." He was still a lowly student and she was a postdoctoral fellow. Rank, clearly, had its privilege in the lab. The matter, Brown recalled, was resolved without hard feelings; he, his wife, and Franklin became fast friends. Several of the other resentments she inspired at King's, however, would not pass so easily.[69]

FROM EARLY 1949 through most of 1950, Franklin made plans to return to England. She loved working at the *labo* but it was time to get on with her life in England. Writing to her parents in March 1950, she told them the projected move home was "much harder that leaving London to come here, because the break will be more permanent."[70] In 1949, she applied to work at Birkbeck College, London, under J. D. Bernal, one of the world's leading crystallographers. Bernal rejected both Franklin and one other applicant, a physicist from the Admiralty Research Office named Francis Crick.

In March 1950, Franklin had tea with a theoretical chemist named Charles Coulson, whom she had met during her BCURA days. He was now working at King's College. Coulson introduced her to John Randall, who was impressed by the young woman's credentials. At the time, Randall desperately needed to fill several staff vacancies. He was short of well-trained "senior people" and worried about meeting the many obligations required to keep his MRC grant running on track; this was a factor in his frustration at Wilkins's fumblings with the X-ray equipment. Randall was most charming when attracting a new performer for

Rosalind Franklin in Cabane des Evettes, taking a break from a mountain climb, c. 1950.

his circus, and he turned it on while recruiting Franklin. Unfortunately, he made the colossally wrong assumption that Franklin, whom he deemed to be a quiet and reserved woman, would be a perfect fit for the King's Biophysics Unit.

Her application for the postdoctoral fellowship at King's was written after much consultation with her prospective boss. The original research plan called for the "X-ray diffraction study of protein solutions and the changes in structure which accompany the denaturing of proteins."[71] In June 1950, she interviewed for a three-year Turner and Newall Fellowship, paying £750 a year, which she formally accepted on July 7. Although the fellowship typically began in the fall, Franklin requested to begin work at King's College on January 1, 1951 "so that she may complete some research that she is doing in Paris."[72]

ON DECEMBER 4, 1950, Randall wrote Franklin a letter dictating an entirely new course of research for the months ahead. This communication set in motion what can only be considered as one of the greatest human resource snafus in the history of science. Randall's letter merits quoting at length because it serves as an oracle of what would soon go so terribly wrong between Franklin and Maurice Wilkins:

> The real difficulty has been that the X-ray work is in a somewhat fluid state and the slant on the research has changed rather since you were last here.
>
> After very careful consideration and discussion with the senior people concerned, it now seems it would be a good deal more important for you to investigate the structure of certain biological fibers in which we are interested, both by high and low angle diffractions, rather than to continue with the original project of work on solutions as the major one.
>
> Dr. Stokes, as I have long inferred, really wishes to concern himself almost entirely with theoretical problems in the future and these will not necessarily be confined to X-ray optics. *This means*

that as far as the experimental X-ray effort is concerned there will be at the moment only yourself and Gosling, together with the temporary assistance of a graduate from Syracuse, Mrs. [Louise] Heller. Gosling, working in conjunction with Wilkins, has already found that fibers of deoxyribose nucleic acid derived from material provided by Professor Signer of Bern gives remarkable good fiber diagrams. The fibers are strongly negatively birefringent and become positive on stretching and are reversible in a moist atmosphere. As you no doubt know, nucleic acid is an extremely important constituent of cells and it seems to us that would be very valuable if this could be followed up in detail [italics added]. If you are agreeable to this change of plan it would seem that there is no necessity immediately to design a camera for work on solutions. The camera will, however, be extremely valuable for large spacings from such fibers.

I hope you will understand that I am not in this way suggesting that we should give up all thought of work on solutions, but we do feel that work on fibers would be more immediately profitable and, perhaps, fundamental.[73]

TO HIS LAST DAY ON EARTH, Maurice Wilkins claimed not to have seen Randall's December 4, 1950, letter to Franklin until years <u>after</u> her death. Clinging to this tall tale, he obsessively tried to paper over the wrongs he committed against a colleague whose martyrdom has only grown exponentially in the years since. The files of every historian and journalist who interviewed him are filled with corrective letters from Wilkins after reading their books and manuscripts.

This much we do know: Wilkins left the lab for his winter holiday on December 5, one day *after* the Randall letter was typed, signed, and dated. For nearly a week, he was hiking in the Welsh mountains with an artist named Edel Lange, wooing her with romantic walks under the "mild winter sun" and long evenings of reading Jane Austen together.[74] Just before his trip, Wilkins contended, he had developed "a clear crystalline X-ray pattern" of DNA. While out on his "short holiday," he decided he "must give up completely the microscope work and concentrate full-time on X-ray structure analysis of DNA," even though he makes no note of ever informing Randall of this decision until well after he returned from his winter vacation.[75]

Once back at King's, Wilkins unfurled his self-admitted, old-fashioned views about the subservience of women, whether in marriage or the workplace. Despite Franklin's doctorate and years of independent research experience, Wilkins assumed that she was hired to serve as his research assistant. James Watson later propagated this misunderstanding in print when he wrote that "She claimed that she had been given DNA for her own problem and would not think of herself as Maurice's assistant . . . The real problem, then, was Rosy. The thought could not be avoided that the best home for a feminist was in another person's lab."[76]

When speaking with Franklin's biographer Brenda Maddox in 2000, Wilkins admitted that his alibi of not knowing anything about Randall's "Rosalind letter" appeared dubious, given that he was the assistant director of the laboratory and should have been abreast of all hiring issues.[77] On other occasions, Wilkins went so far as to take credit for hiring Franklin, as if that should have earned him some type of gratitude in return. On February 6, 1951, one month after Franklin's arrival at King's, Wilkins wrote to Roy Markham at the Molteno Institute at Cambridge, "We now have Miss Franklin for X-ray work and hope to get something really done, as almost no progress was made since the summer."[78] In 2000, Wilkins told Brenda Maddox he "believed he was instrumental in getting Rosalind assigned to DNA. When he heard from Randall that she was coming to work on proteins in solution, he thought it a waste as they were getting such good results on nucleic acids. Considering her X-ray expertise, why not, he suggested, 'grab her and get her in on the DNA work?' To his surprise, Randall readily agreed." Such statements fail to support his adamancy about not knowing the precise terms of her appointment at the lab.[79]

Nonetheless, in his 2003 memoir, Wilkins made certain to deflect the blame entirely onto Randall. He insisted that his boss was completely in the wrong in having told Franklin that his and Stokes's X-ray work on DNA had ended without so much as consulting them. Wilkins accused Randall of wanting to take over the work himself, with Franklin reporting directly to him. Calling his former boss "ruthless," Wilkins insisted, "If Randall had not barged in, it might even have been that Rosalind could have worked happily alongside Stokes and me, and her professional X-ray approach could have combined fruitfully with our techniques and theorizing."[80] But Wilkins also had the humility, too late to be sure, to record his admiration at Franklin's ability to conduct her work even "when our Head of Department's secret letter was so clearly contradicted by the reality of me and Stokes giving no sign of moving off DNA. It must have been a great burden to her, and I continue to be impressed by her fortitude."[81]

Perhaps the employment arrangements at King's are best left discussed by the man who actually did the hiring. In 1970, Sir John Randall sat for an interview with Anne Sayre. She described him as "a tit-tuppy-type given to 'oh dears.'" On the one hand, he claimed full responsibility for the misunderstandings that developed between Franklin and Wilkins, but on the other, he immediately excused himself because he had "his hands full" administering a large laboratory. Sayre recorded that Randall "did not in any way remember Rosalind kindly, though he quite astonished me in remarking how good-looking she was. His feelings are very colored (and he admits this freely) by the fact he feels certain that if Rosalind and Wilkins had worked together, they would have brought out the DNA structure ahead of Cambridge. . . . He says this failure was a 'tragedy,' and blames it on Rosalind more than Wilkins but 'Wilkins may in some ways be a bit difficult.'" Without prompting, Randall emphatically added, "at no time was Rosalind Wilkins 'assistant', as Watson said . . . she was an independent worker and was in no way subject to Wilkins."[82] Sadly, Randall's declamations of quasi-support were made two decades after the fact. In the moment that was December 4, 1951, he failed miserably.

There Was No One Like Linus in All the World

The combination of his prodigious mind and his infectious grin was unbeatable. Several fellow professors, however, watched this performance with mixed feelings. Seeing Linus jumping up and down on the demonstration table and moving his arms like a magician about to pull a rabbit out of his shoe made them feel inadequate. If only he had shown a little humility, it would have been so much easier to take! Even if he were to say nonsense, his mesmerized students would never know because of his unquenchable self-confidence. A number of his colleagues quietly watched for the day when he would fall flat on his face by botching something important.

—JAMES D. WATSON[1]

Just as physicists were using quantum theory to reshape biology, Linus Pauling proposed doing the same for chemistry.[2] In 1936, at the age of thirty-five, he was named the chairman of chemistry and director of the division of chemistry and chemical engineering at the California Institute of Technology. With a river of dollars steered his way from the Rockefeller Foundation, Pauling had all the resources he needed to merge chemistry, biology, and physics into the new discipline of "molecular biology—which [was just] beginning to uncover many of the secrets concerning the ultimate units of the living cell."[3] It was a wise investment of titles and resources to make. A cursory review of his research during this period is breathtaking: it ranges from developing new methods to study the structure of inorganic and organic molecules to co-authoring an important textbook on the application of quantum theory to chemistry.[4] While pursuing these tasks, Pauling set his piercing, steel-blue eyes on an entirely new scientific vista: determining the structure of proteins, the building blocks of all living beings. Succeeding in such a Himalayan endeavor, he posited, would help scientists and physicians better understand the daily functions of life; it might also contain the keys to the heretofore locked box of genetics.[5] This was, to say the least, an understatement.

LINUS CARL PAULING WAS born in Condon, Oregon on February 28, 1901.[6] His father, Herman Pauling, was a druggist who long suffered from poor business sense and debilitating stomach aches. As a little boy, Linus loved watching his father mix and compound his own dyspepsia treatments. After his drugstore in Condon burned down, in 1909, Herman moved the family to Portland. He died the following year, from a perforated ulcer and peritonitis, when he was only thirty-four and Linus was nine. His mother, Lucy Isabelle Darling Pauling, had few skills outside of homemaking and parenting Linus and his two young sisters, Pauline and Frances. Their economic situation grew so dire that Mrs. Pauling began to eke out a living by running a small boardinghouse in Portland for itinerant travelers. Money remained tight, Mrs. Pauling was often ill, and Linus had to help supplement the family coffers with a variety of odd jobs. In between school and chores, he clocked marathon hours at the county public library, reading books of all kinds and topics. He routinely amazed his teachers not only with an ability to memorize whatever he read but also in applying the content to the lessons taught in school.

When Pauling was fourteen, his best friend received a toy chemistry set and the two boys played with it nonstop. He "was simply entranced by chemical phenomena, by the reactions in which substances, often with strikingly different properties, appear; and [he] hoped to learn more and more about this aspect of the world."[7] Soon after, he created his own basement laboratory with chemicals, glassware, and reagents he liberated from an abandoned smelter where his grandfather worked as a security guard. As with Francis Crick's boyhood antics, most of Pauling's chemical output was limited to making stink bombs and explosive firecrackers. To supplement his laboratory hijinks, he began borrowing chemistry texts from the library to learn how different substances changed when mixed with others and, more broadly, the composition of matter.

At sixteen, Pauling set his sights on a degree in chemical engineering from the Oregon Agricultural College in Corvallis, a practical goal that he hoped would feed his curiosity and lead to stable employment. OAC was especially attractive because it offered free tuition for in-state students. There was one significant problem with him moving 72 miles southwest to Corvallis: his mother desperately needed the wages he earned from his after-school machine shop job, so she demanded that he continue working and abandon his academic ambitions. Pauling held his ground, however; he dropped out of high school and soon afterward was accepted at OAC.

He matriculated in the fall of 1917 but temporarily left college in 1919 to help out at home by taking a job as a road paving inspector for the state of Oregon. Fortunately, the eighteen-year-old was so brilliant at chemistry and oratory that upon his return the college offered him a full-time position as an assistant instructor in quantitative analysis. He could now live and learn in Corvallis and send a healthy portion of his income to his mother in Portland.

During his senior year, Pauling met the love of his life, Ava Helen Miller, a bright, pretty, flirtatious freshman with long black hair. He later recalled the reason behind his infatuation, "she was smarter than any girl I'd ever met." Hailing from Beaver Creek, Oregon, Ava Helen was the tenth of twelve children of a German immigrant schoolteacher, whose liberal Democratic Party views tended toward socialism, and a mother active in the suffrage movement. Ava Helen had a wide array of interests, ranging from women's rights, racial equality, and social reform to chemistry. They met while she was taking his "Chemistry of Home Economics" course. At first, Pauling was hesitant to ask her on a date, because romantic relationships with students were discouraged among the faculty. Love conquered bureaucracy after he convinced himself that a more accurate description of their relationship was not that of the coed and the boy professor but rather one between two students. He courted Ava Helen during long walks together, sharing bags of sea foam candy, and at school dances. In late spring of 1922, before giving her final grade, he asked her to become his wife. After she accepted his proposal, he marked her final grade down one point lest he be accused of showing favoritism to his fiancée.[8] They married in the spring of 1923

Linus Pauling and Ava Helen Miller, 1922.

and embarked upon a sixty-year partnership of family, ideas, science, and activist politics. While Pauling would win the 1962 Nobel Peace Prize for his work against nuclear proliferation, it was his wife who had first introduced him to the peace movement.

Upon graduating from OAC, Pauling matriculated into the PhD program of the California Institute of Technology in Pasadena, a newly reconfigured school of science and engineering, rich in endowment money, pathbreaking research, and Nobel Prize winners. Caltech remained his academic home for the next forty years.[9] As a doctoral student, he gravitated toward the topics of X-ray crystallography, quantum theory, and atomic structure. In 1925, he completed his dissertation, "The Determination with X-rays of the Structure of Crystals," under the supervision of Roscoe Dickinson, who, in 1920, had earned the first PhD granted by Caltech. The following year, 1926, the department chairman Arthur Noyes pulled strings to win Pauling a John Simon Guggenheim Memorial Foundation Fellowship, a program for brilliant scholars of all stripes established in 1925.[10]

Pauling used those funds to travel to Munich with his wife, where he took a visiting post at Arnold Sommerfeld's Institute of Theoretical Physics. Sommerfeld was a pioneer in quantum physics and trained several doctoral students who went on to win the Nobel Prize in Physics or Chemistry, including Werner Heisenberg, Paul Dirac, and Wolfgang Pauli.[11] At the institute, Pauling met some of Europe's most prominent physicists and chemists and they, in turn, introduced him to their research. Although theoretical physics was never his métier, Pauling was convinced that quantum theory was the key to understanding the "structure and behavior of molecules," atoms, and the chemical bonds holding them together.[12] The Guggenheim Foundation granted additional funds so the Paulings could travel to Copenhagen, where he visited Niels Bohr's famed physics institute and briefly experienced *Der Kopenhagener Geist der Quantentheorie* (the Copenhagen spirit of quantum theory), an ethos of intellectual collaboration in the development of modern atomic physics.[13]

PAULING RETURNED TO CALTECH in the fall of 1927 as an assistant professor of theoretical chemistry. His rise was meteoric and by 1930, at age
twenty-nine, he was a full professor. In 1931, a German physicist who was
being recruited for a position at Caltech sat in on one of Pauling's lectures. The topic was the application of wave mechanics to understanding
chemical bonds. When asked by a newspaper reporter what he thought
of the lecture, the physicist demurred, saying, "It was too complicated
for me," and he promised to "brush up on the subject" before "again trying to engage the young Dr. Pauling in a conversation." The visitor was
Albert Einstein.[14] That same year, Arthur Noyes described the young
scientist as "a rising star, who may yet win the Nobel Prize."[15] By 1933, at
thirty-two, Pauling was well on his way to such an accolade; that autumn,

Linus Pauling in his Caltech laboratory, c. 1930s.

he was elected to the National Academy of Sciences, one of the highest honors bestowed upon American scientists.

IN 1937, PAULING INVITED the British X-ray crystallographer and pioneering molecular biologist William Astbury to deliver a series of lectures at Caltech. As a professor in the faculty of textile science at the University of Leeds, Astbury focused on the molecular structure of natural fibers, such as wool, cotton, and animal hair. He brought with him a large portfolio containing detailed X-ray diffraction photographs of keratin fibers—the principal protein of hair, nails, claws, horns, feathers, and the outer layers of vertebrate animal skin.[16] Few knew better than Astbury what an incredibly difficult task it was to interpret the lines, dots, smudges, and smears that resulted. Yet even after an interpretation of this complex collection of data was achieved, the results were prone to massive revision or rejection by other researchers.

Astbury suggested several possible protein structures which he "thought were compatible with his data." After reviewing the images, however, Pauling disagreed with the crystallographer's conclusions. Not only was there little extant knowledge on the structure of amino acids (the building blocks of proteins), Pauling complained, but "nobody was attacking the problem vigorously, systematically." Pauling, who knew the scientific literature on the subject backward and forward, concluded that the published X-ray studies of amino acids "were all wrong." As he recalled, "I knew that what Astbury said wasn't right because our studies of simple molecules had given us enough knowledge about bond lengths and bond angles and hydrogen-bond formation to show that what he said wasn't right. But I didn't know what *was* right."[17]

Seven years earlier, in 1930, Pauling had begun developing a new methodology for solving the molecular structure of inorganic, silicate minerals.[18] It combined quantum chemistry and theoretical physics with his brilliant intuition. Specifically, Pauling set out to learn everything he could about the sizes and shapes of the molecule's component parts. He then made a series of educated assumptions about the chemical bonds holding the atoms of that molecules together which, if correctly inferred, would delineate the specific angles, twists, and turns forming its three-dimensional shape. He applied this information to building models with precision-made balls, sticks, and shapes to reconstruct a much-enlarged version of these atoms and molecules, much as a college student uses less elaborate sets of multicolored sticks and pronged balls when cramming for an organic chemistry examination. Once completed, Pauling compared his model to the X-ray data to corroborate that the chemical bonds and molecular shapes he predicted fit correctly. He called this method "stochastic," from the Greek στόχος *(stókhos)*, meaning "to aim at a mark" or "to guess."[19]

LONG BEFORE COMPOSING THE final paragraph of his classic 1939 book *The Nature of the Chemical Bond*, Pauling was planning to redirect his research toward organic, or living, molecules. A protein's shape, he posited, was determined by hydrogen bonds, which were essentially electrostatic forces of attraction between positively charged hydrogen atoms and negatively charged atoms or groups. Since these bonds defined form, Pauling explained, they probably governed "properties of substances of biological importance," ranging from antibody–antigen reactions and the contraction of muscles to the conduction of electrical impulses and messages from the brain to the nerve cells. The path to an understanding of the molecular structure of proteins, he predicted, would take "many years . . . and it is my belief that this attack will be ultimately successful."[20] Pauling was being neither modest nor cagy. It would take him eleven years to define the general structure of proteins.

Before accomplishing this task, Pauling began musing about how genes replicated and passed on traits from generation to generation.[21] In 1940, Pauling co-authored a short article with Max Delbrück, his Caltech colleague and the protagonist of Schrödinger's *What Is Life?*— a book, incidentally, that Pauling considered to be "hogwash."[22] Their paper, published in the journal *Science*, was a rebuttal of the German theoretical physicist Pascual Jordan, who insisted that heredity was

Linus and Ava Helen Pauling and their children (left to right: Linda, Crellin, Peter, and Linus, Jr.) and assorted pet rabbits, c. 1941.

mediated by information passing between *identical* molecules. Pauling and Delbrück applied what they knew about covalent bond formation to predict "these interactions are such as to give stability to a system of two molecules with *complementary* structures in juxtaposition, rather than two molecules with necessarily identical structures."[23] The model they proposed was akin to a key and the tumblers of a lock; where there is a ridge on the "key" of one molecule, there is an equal trough in the "lock," or the complementary molecule. Pauling continued to espouse this tantalizing yet unproven theory throughout the 1940s,[24] and it was not lost on James Watson and Francis Crick. Complementarity was one of the key principles that enabled them to unlock the structure of DNA.

PAULING'S FAVORITE COLLABORATOR during this period was a shy X-ray crystallographer named Robert Corey. A boyhood bout of spinal polio-myelitis had left him with a partially paralyzed left arm and a pronounced limp, which required a cane when walking, and for the remainder of his life he suffered from what was then described as "a frail constitution." After receiving his PhD in chemistry from Cornell University in 1924, Corey stayed on as an instructor in analytical chemistry until 1928, when he won a fellowship at the Rockefeller Institute for Medical Research. In 1930, he was invited to stay there as an associate in biophysics. This was the same institution where Oswald Avery conducted his landmark work on the "transforming principle" in pneumococci. Unfortunately, Corey's laboratory was dissolved in 1937 because even the Rockefellers had to tighten their belts during the Great Depression.

Corey next moved to the National Institute of Health (as it was then called) in Washington, DC, for a one-year fellowship, before approaching Pauling for a position at Caltech. He was so desperate for an academic appointment that he offered to bring his own equipment and cover his own salary. Pauling agreed to name Corey a research fellow in his laboratory without stipend. Within a few weeks, however, he recognized the man's worth. Even though they were never close socially, Pauling carefully shepherded Corey's career at Caltech and nominated him several times, without success, for a Nobel Prize. Under Pauling's tutelage, Corey steadily climbed the academic ladder, reaching full professor in 1949. His friend Richard Marsh, another Caltech crystallographer, described Corey as "a private person [who] seemed to dislike social events of all kinds, preferring to be at home with [his wife] Dorothy listening to Gilbert and Sullivan or perhaps tending to his lawn." Corey was "the direct opposite of Pauling, who enjoyed the limelight and relished both adulation and confrontation . . . Pauling would give lectures so charming and entertaining that the audience might get a whiff of snake oil; but then a definitive paper would appear, carefully written and with supporting evidence supplied by Corey. Care and attention to details were the essence of Bob Corey."[25]

This odd couple worked in an almost effortless rhythm as they attacked the structure of amino acids. Corey began by ascertaining the structure of the simplest amino acid, glycine. After he successfully described every atom of that structure, Pauling assigned Corey a two-glycine dipeptide

Robert Corey and Linus Pauling, 1951.

known as diketopiperazine, and so they continued, in terms of molecular complexity, until they ground out the data necessary for more intricate amino acid structures. Their goal was to delineate, atom by atom, how proteins are built, using the bond lengths and angles of simpler molecules, deduction, and, after finding what worked in a model and what did not, the process of elimination.

FOR THE ACADEMIC YEAR 1948–49, Pauling served as the George Eastman Professor at Balliol College, Oxford. The Eastman chair was "one of the world's most respected visiting professorships," endowed by the founder of the Eastman Kodak Company.[26] At this time, there was great interest among British X-ray crystallographers about a remarkably clear set of diffraction photographs of keratin developed by William Astbury and his team in Leeds. In Astbury's estimation, the molecular structure of keratin "zig-zagged" every 510 picometers (one picometer is one trillionth of a meter), like a long-kinked ribbon. Others, however, interpreted the X-ray data as depicting a structure akin to a bedspring or a spiral—in other words, a helix. One of those firmly in the helical camp was a PhD student at Cambridge named Francis Crick. He criticized Astbury as a "sloppy model builder" who was "not meticulous enough about the distances and angles involved." Crick also insisted that it was common knowledge among protein men of this era that any chain featuring "identical repeating links that fold so that every link is folded in exactly the same way, and with the same relations with its close neighbors, will form a helix."[27] What bothered Pauling about both the zigzag and spiral models was that no one could explain how the constituent amino acids conformed to both the 510-picometer repeats and their stiff chemical bonds.

During the cold, damp winter term at Oxford, Pauling split open the hard shell of proteins. He later attributed his uncommon discovery to a common cold—the same pesky malady he later claimed, without much evidence, could be cured by ingesting massive doses of vitamin C.[28] The cold developed into a painful sinus infection, confining him to the "unsuitable" accommodations attached to his professorship. "The first day," Pauling recounted, "I read detective stories and just tried to keep from feeling miserable, and the second day, too, but I got bored with that, so I thought 'why don't I think about the structure of proteins?' "[29] He leapt out of bed, grabbed paper and pencil, and began sketching out a slew of possible structures. For the first time, he recognized the need to account for a molecular backbone, or scaffolding for the biologically active components of a particular protein. Soon enough, he started folding the paper into tetrahedrons, a telescopic tube, and—after much more folding and refolding—a decent, though necessarily imperfect, helical model. Decades later, Pauling recalled, "Well, I forgot all about having a cold then, I was so pleased."[30] What he still could not figure

out was how to model the distance between one twist of the chain and
the next in order to precisely fit the 510-picometer distance Astbury had
determined from his X-ray images. That chore required another three
years—a delay caused not only by the constraints of his laboratory methods but also by the distractions of managing a department, training fellows and students, designing new experiments, writing up publications,
teaching, and lecturing.

The most important scientific detour Pauling took during this period
was his demonstration that sickle cell anemia has a molecular etiology.
Using a new technique called electrophoresis, Pauling and his colleagues
demonstrated that a minor change in the electrical charge of a single
amino acid in the long protein chain comprising hemoglobin led to
the clinical picture known all too well to the disease's sufferers and the
doctors who treat them. A mutation in just one nucleotide in the gene
for the beta chain of the hemoglobin protein, found on chromosome
11 and passed down in an autosomal recessive manner, causes the sick-
ling, stretching, and stiffening of round red blood cells. These intrepid
but damaged travelers of the body jam up in the smallest blood vessels, blocking blood flow and oxygen delivery—an excruciating condition known as the vaso-occlusive crisis of sickle cell anemia.[31] Pauling's
discovery was not only a major advance in protein chemistry, it was the
overture to the long list of diseases that were soon to be explained at the
molecular level. Two decades later, in 1968, Pauling disgraced himself
by proposing an outrageous eugenicist approach to discouraging sickle
cell carriers and those with the disease from reproducing: "There should
be tattooed on the forehead of every young person, a symbol showing
possession of the sickle cell gene [so as to prevent] two young people carrying the same seriously defective gene in single dose from falling in love
with one another."[32]

PAULING WAS NOT THE ONLY scientific heavy hitter of his era hoping to
knock the protein problem out of the park. At Cambridge University's
Cavendish Laboratory, Sir William Lawrence Bragg and his associates

Max Perutz and John Kendrew had been laboring for several years with-out much success on the structure of complex proteins. Unlike Pauling—who worked his way up from the smallest atomic components to build a predictive model of the whole structure, and then compared it to the X-ray images—the Cavendish team began by analyzing the X-ray images of whole proteins, thus attacking the problem from the other end. This task was so tedious that Max Perutz complained that the "many nights of interrupted sleep and the appalling strain of measuring the intensities of thousands of little black spots by eye had brought me no nearer to the solution of hemoglobin, and I wasted some of the best years of my life trying to solve a seemingly insoluble problem."[33] Frustrations aside, by 1950, Bragg, Kendrew, and Perutz believed they had gathered enough data to beat Pauling and published a paper, "Polypeptide Chain Config-urations in Crystalline Proteins," in the October issue of the *Proceedings of the Royal Society of London*.[34] As soon as he got his hands on a copy of the journal, Pauling discovered to his joy that Bragg and company had not solved the puzzle at all. Instead, they simply reviewed all of the latest proposed polypeptide structures. Coming to the end of their molecular laundry list, Bragg et al. sided with Astbury's faulty theory that keratin fibers were configured in the shape of a folded, or kinked, ribbon.

Pauling had something far better up his sleeve. Working with Corey and Herman Branson, an African American physicist on leave from Howard University for the 1948–49 academic year, Pauling proposed a helical structure for protein chains that conformed to what was known about the lengths and angles of bonds between the amino acids com-prising them. Based on his experimental work, Pauling hypothesized that the peptide bonds joining together amino acids are planar, stable, and rigid. In terms of quantum mechanics, this means that the atoms reside within a single plane and form a partial double bond where there is no rotation around the bond. Pauling, Corey, and Branson subse-quently built a structure that allowed for as many hydrogen bonds as possible between the turns of the hypothesized helical structure. These deductions led Pauling to determine "the two main structural features of proteins: the α-helix and the β-sheet, now known to form the backbones of tens of thousands of proteins."[35] Pauling published his conclusions—often referred to as "one of the greatest achievements in structural biology"—in a series of eight papers in the April and May 1951 issues of the *Proceedings of the National Academy of Sciences*.[36] Over the next decade, a large number of X-ray crystallography studies proved his theory to be correct.

In a mirror image of the actions that had transpired a year earlier at Caltech when Bragg's protein paper was published, the Cavendish crew anxiously read Pauling's *PNAS* articles. Bragg was horrified to see that Pauling had proved him wrong in so public a manner. In 1963—twelve years later—Bragg delivered a lecture titled "How Proteins Were *Not* Solved," in which he admitted the errors of his 1950 article and confessed, "I have always regarded this paper as the most ill-planned and abortive in which I have ever been involved."[37] Bragg's widely publicized blunder released a miasma of apprehension that permeated every wall and stairwell of the Cavendish. All of Bragg's men dreaded being scooped again by Linus Pauling, who seemed always to be a step ahead of the rest of the scientific community.

The Quiz Kid

CHICAGO

Fierce as a dog with tongue lapping for action, cunning as a
savage pitted against the wilderness,
Bareheaded,
Shoveling,
Wrecking,
Planning,
Building, breaking, rebuilding . . .

Laughing the stormy, husky, brawling laughter of Youth,
half-naked, sweating, proud to be Hog Butcher, Tool Maker,
Stacker of Wheat, Player with Railroads and Freight Handler
to the Nation.

—CARL SANDBURG[1]

Like Saul Bellow's autobiographical character Augie March, James Dewey Watson was "an American, Chicago born" determined to "make the record in my own way: first to knock, first admitted; sometimes an innocent knock, sometimes a not so innocent."[2] The blessed event occurred on April 6, 1928, at the Gothic-revival St. Luke's Hospital on Chicago's South side. From the moment he came into this world, he was known as Jim. In keeping with his "DNA obsession," Watson loved to regale others with tales of his forebears, colonial citizens of the American experiment and brave pioneers who made their way inward to its great prairie. One relative, William Weldon Watson, was born in New Jersey in 1794 and became minister of the first Baptist church established west of the Appalachians, in Nashville, Tennessee. His son, William Weldon Watson II, traveled north to Springfield, Illinois, where he designed a house for a somber, strikingly tall lawyer named Abraham Lincoln. The Watsons and the Lincolns lived across the street from each other. When "Honest Abe" was called to Washington to take the office of president of the United States, William Watson II, his wife, and son Ben rode with the Lincolns on the president elect's inaugural train. Ben's son, William Watson III, became an innkeeper in the Chicago area. And one of William III's five sons, Thomas Tolson Watson—Jim's grandfather—"sought his fortune at the newly discovered Mesabi Range, the great iron-ore-bearing region located near Duluth, Minnesota on western Lake Superior."[3]

Jim's father, James Dewey Watson, Sr., was a product of the affluent La Grange, Illinois, public school system and a one-year stint at Oberlin College, cut short thanks to a bout of scarlet fever. He recovered enough to serve a year in France during the First World War, with the 33rd Division of the Illinois National Guard. Upon returning to Chicago, Jim, Sr., abandoned his hopes of becoming a schoolteacher and eventually took a job as a bill collector for the LaSalle Extension University, a correspondence school that offered long-distance business courses.[4] "Making money was never near [Jim, Sr.'s] heart," but he loved birdwatching and became so adept a birder that in 1920 he co-wrote a well-regarded guidebook on the birds of the Chicago area.[5] His collaborator, a teenager named Nathan Leopold, Jr., soon found lasting infamy with his best friend, Richard Loeb. In 1924, the two young men became morbidly fascinated by Friedrich Nietzsche's concept of the *Übermensch*, or superman, which held that intellectually gifted men were above the laws ruling the inferior masses.[6] This aberrant philosophy, they later claimed, inspired them to kidnap and brutally murder a fourteen-year-old boy named Bobby Franks. (Clarence Darrow represented Leopold and Loeb in what was once referred to in the press as "the trial of the century.")[7] Thankfully, only Jim, Sr.'s love of birding, and neither his choice of birding companions nor his habit of chain-smoking cigarettes, were passed on to his son, who revered his father as a man of "truth, reason, and decency."[8] Consequently, Jim Watson, Jr.'s first career goal was to become an ornithologist, who would identify new species in the wild and teach at a state university.

His mother, Margaret Jean Mitchell Watson, went to the University of Chicago for two years before becoming a secretary, first at LaSalle and, later, in the University of Chicago's housing office. Severe rheumatic fever as a teenager left her with congestive heart disease and for the rest of her life she became quickly short of breath and used the weekends for bedrest. She and Jim, Sr. married in 1920 and had two children, Jim, Jr. and Elizabeth.

In 1933, Watson's maternal grandmother, Elizabeth Gleason Mitchell ("Nana"), moved in with the family. Mrs. Mitchell, the daughter of Irish immigrants from County Tipperary, had been widowed on New Year's Eve, 1907, when her husband, Lauchlin Mitchell, a Glasgow-born tailor, was killed by a runaway horse upon exiting the Palmer House Hotel.[9] Margaret Jean was fourteen years old when her father died and became the caretaker for her grieving mother, in spite of her own chronic illness. They switched roles when it came to the care of Margaret Jean's children. Each afternoon, the warm, loving Nana was at home to welcome Jim and his younger sister, Elizabeth (known as Betty) when they returned from school and to make supper while their mother and father finished up their workdays.[10]

Unlike many of their neighbors, the Watsons were not particularly religious. Raised a Catholic, Margaret Jean Watson went to Mass only on Christmas Eve and Easter. Jim, Sr. never attended church. Thus, Watson would proudly declare himself "an escapee from the Catholic religion."[11] The Watsons may have been agnostic when it came to Christianity, but they made sure to instill a fervent faith in knowledge in both their children. In 1996, Watson reminisced, "My family had no money but lots of books."[12] Seven years later, in 2003, he insisted that "the luckiest thing that ever happened to me was that my father didn't believe in God, and so he had no hang-ups about souls. I see ourselves as products of evolution, which itself is a great mystery."[13]

BOOKS, BIRDS, AND IDEAS ASIDE, the most pressing concern for the Watson family was paying their bills on time, especially acute after the Great Depression tightened its years-long grip on the nation's collective pocketbook. In the early 1930s, Jim, Sr.'s yearly salary was cut in half, to $3000, a decrease he quietly accepted simply to keep his job. Margaret Jean's part-time work at the university was essential to keep the household going.[14] The Watsons were not only staunch supporters of Franklin D. Roosevelt's New Deal; they were beneficiaries of it.

The Watsons lived at 7922 Luella Avenue, a "heavily mortgaged," 1,604-square-foot brick bungalow just below Seventy-Ninth Street on Chicago's South Side. It was a great source of pride for his parents that they owned their home, which had four bedrooms under its second-floor eaves and a fenced backyard. The house was (and is) a little more than four miles from the University of Chicago and fewer than fifteen blocks from Jackson Park and the south shore of Lake Michigan. It was also conveniently close to Horace Mann Grammar School, which Jim and Betty attended.[15] Later, Watson "delighted in saying [his house] was closer to the steel mills of Gary, Indiana . . . than to the University of Chicago"—which wasn't accurate, as the "Gary Works" is about twenty miles to the south. Still, the billowing mills of U.S. Steel, the first billion-dollar corporation in American history, were part of his daily landscape. He merely had to look out his window to see the long gray plumes of industrial exhaust polluting the air.

Watson was a razor-thin, shy,

James D. Watson, age ten, 1938.

odd-looking, and unathletic boy with bulging eyes and peculiar facial mannerisms. By day, he preferred birdwatching to playing baseball. At night, he lulled himself to sleep by memorizing facts and figures drawn from the agate-font pages of the *World Almanac*.[16] Unpopular at school, he routinely disdained the "dumb" boys, who reciprocated by beating him up on a regular basis.

There was, at least, one social advantage to his bookishness. In 1940, an industrious Chicagoan named Louis Cowan created and produced *The Quiz Kids,* a wildly successful radio show that ran for thirteen years on the national NBC network, sponsored by "the makers of Alka-Seltzer."[17] Every week, from a studio deep within the mammoth Merchandise Mart Building, a gaggle of precociously bright children, aged six to twelve years old, were asked all sorts of questions in order to win a U.S. savings bond worth one hundred dollars. Watson recalled, "The only reason I was on [the show] was that the producer of the program was literally our next-door neighbor. I was bright enough so that I knew a lot of facts." Fourteen-year-old Jim lasted only three weeks, in the fall

Jim Watson as a Quiz Kid in 1942, second from left.

of 1942, before losing—on a Bible-related question—to an eight-year-old named Ruth Duskin, who became a popular regular on the show. At ninety, Watson still smarted over his short-lived radio career: "Well, yeah, she was a little Jewish girl. She was pretty and outgoing and perfect for a quiz show and, of course, she knew all about the Old Testament."[18] Watson, however, was able to convert his childhood losses and unpopularity into a strength. As does almost every bullied, bruised boy, young Jim Watson swore to someday get even; in the 1980s, he told a colleague, he had been "getting back" at those bullies ever since.[19]

⚕

AFTER COMPLETING GRAMMAR SCHOOL, Watson and his sister attended the Laboratory School of the University of Chicago, the famously progressive day school founded by the philosopher, psychologist, and educational reformer John Dewey. The Watson children both matriculated into the university at the age of fifteen, their early admission facilitated by its president, Robert Maynard Hutchins, a college chum of Watson's father from Oberlin. A boy wonder himself, Hutchins had become America's youngest college president upon taking the helm of the University of Chicago at the age of thirty, in 1929.[20] Two years later, in 1931, he created what became a nationally known, four-year generalist baccalaureate program, anchored by the then-novel "Great Books of Western Civilization" curriculum. In 1942, Hutchins announced an even more innovative plan whereby academically talented high school sophomores, like Jim and Betty Watson, were admitted to the university, thus propelling their intellectual development and helping them "escape high school's monotonous rote learning."[21] Jim Watson was the perfect candidate, and because his parents' house was a fifteen-cent bus ride from the University of Chicago campus, he could continue to live at home. Given his youth and immaturity, Watson's family support was critical to his success at the University of Chicago.[22]

The University of Chicago's faux-Gothic campus in the Hyde Park neighborhood of Chicago was built under the auspices of the American Baptist Education Society and the generous philanthropy of John

D. Rockefeller, Sr. A relatively new school—it was founded in 1890—
Chicago enjoyed one of the most distinguished faculties of the twentieth century as well as an eager to learn and devoted student body.
There, from 1943 to 1947, Watson pursued his paternally inspired love of
birds by studying ornithology. One of his teachers, Paul Weiss, a crusty
professor of embryology and invertebrate zoology, recalled that as an
undergraduate, Watson "was (or appeared to be) completely indifferent
to anything that went on in class; he never took any notes—and yet
at the end of the course he came [out] on top of the class."[23] In 2000,
Watson spun his own interpretation of what he learned at the University of Chicago. He discovered that his final examinations were rarely
based on the lecture material and, instead, concentrated on the validity
of the material presented: "it was the ideas, not the facts, that led to good
grades in Robert Hutchins's college."[24]

Although his transcript was riddled with B's, Watson incorporated
three major intellectual premises taught at Chicago that served him well
for the rest of his intellectual life: go directly to the original source rather
than parroting back the interpretations of others; develop a theory of
how a particular set of facts may fit together; and, instead of memorizing
facts, learn to think and rid your mind of unimportant things. Watson
described his college days more succinctly in 1993, on the occasion of
the fortieth anniversary of the discovery of the double helix: "You were
never held back by manners, and crap was best called crap."[25] It wasn't
that he felt "inherently brighter" than his student cohort; instead, he felt
more comfortable challenging conventional wisdom and theories that
did not make scientific sense. The pursuit of knowledge, not background
or wealth, was all that really mattered to him. Even as a teenager, he was
determined not to waste a moment of his life in the pursuit of money,
academic "triviality," or "idle learning."[26]

Before retiring each night, he read popular novels and short stories
of the day. One book that made a huge impact on his imagination was
Sinclair Lewis's 1925 Pulitzer Prize–winning novel *Arrowsmith*—the first
American novel to detail the life, career, and thoughts of a medical scientist.[27] Jim was equally enthralled by the gossamer fantasies spun by

the Hollywood film factories, including such cinematic masterpieces as *Casablanca*, *Citizen Kane*, and the comic gems of Charlie Chaplin and the Marx Brothers.

In 1945, the seventeen-year-old Watson found (and quickly read) Erwin Schrödinger's *What Is Life?* "I spotted this slim book in the Biology Library and upon reading it was never the same," he later recalled. "The gene being the essence of life was clearly a more important topic than how birds migrate, the scientific topic that previously I could not learn enough about."[28] For Watson, just as for Wilkins and Crick, Schrödinger's book raised more questions than answers about how "every complete set of chromosomes contains the [genetic] code."[29]

During the fall term of his senior year, Watson audited a course on physiological genetics taught by Sewall Wright, one of the founders of population genetics (the study of genetic variations and differences within and between populations). In Professor Wright, Watson found his first scientific idol, and he often enthused to his parents about Wright's "brilliant mind."[30] Wright introduced him to Oswald Avery's 1944 experimental work on DNA. He also proposed a series of questions that Watson would dedicate his life to answering: "What is the gene? . . . How is the gene copied? . . . How does the gene function?"[31] A few weeks after beginning Wright's class, Watson dropped ornithology and took up the study of genetics. Although he completed his bachelor of science degree and earned his Phi Beta Kappa key at the age of nineteen, in the spring of 1947, he knew he had more work to do if he was ever to unlock the mysteries of the gene. He first needed to gain admission into a graduate school, where he would master the scientific method and obtain the union card of academia, a PhD.

Watson's doctoral application was rejected by the California Institute of Technology. Harvard accepted him but failed to attract the young man, partly because its biology department was still mired in nineteenth-century principles of taxonomy rather than experimentation. More pragmatically, Harvard did not offer him a scholarship that paid tuition, room, and board, making the move to Cambridge, Massachusetts, a financial impossibility. He had more success at nearby Indiana

Jim Watson, graduation photo, 1947,
University of Chicago.

University in Bloomington, which offered him a full scholarship, including room and board, to its graduate program in biology. No mere state university, Indiana was a hotbed of genetic research. That charge was led by Hermann J. Muller, whose work exposing *Drosophila* fruit flies to various doses of radiation and assessing the mutations in their genomes won him the 1946 Nobel Prize in Physiology or Medicine.

Two other faculty members proved equally important in luring Watson to Indiana. The first was Tracy Sonneborn, a Johns Hopkins–trained PhD who had aspired to become a rabbi until he took a biology course. Sonneborn studied the genetics of the one-celled organism *Paramecia*. The second was Salvador Luria, a Jewish physician from Turin who fled Mussolini's Fascist and anti-Semitic Italy, first to France and then to the United States.[32] He became a mentor and lifelong friend to Watson. Like Sinclair Lewis's fictional Martin Arrowsmith, Luria studied bacteriophages— viruses that attack bacteria. They are, in essence, "naked genes, stripped-down versions of living organisms." Bacteriophages exist to infect and replicate in other living cells.[33] And because of the rapid replication rate of bacteriophages, Luria could track their genetics in experiments that took a matter of hours, rather than days, weeks, or longer.[34]

AS A GRADUATE STUDENT, Watson wore his robes lightly and enjoyed ridiculing the antics of the "Joe College" boys and flirtatious "homecoming queens" on the Hoosier campus. In an era when male students still wore a coat and tie to class, Watson preferred untucked shirts, battered dungarees, and untied tennis shoes. He did his best to ignore the popular crowd and affected an attitude of superiority toward his peers. He could be equally unkind in his characterization of many members of the biology department. After being passed over for a competitive fellowship, he masked his disappointment to his parents with the left-handed hope that "they use the money to bring someone with first rate intellectual curiosity to the department. Unfortunately, they are few and far between down here."[35] One of those few first-raters, who also entered Indiana University in fall 1947, was an Italian émigré physician named Renato Dulbecco.[36] Always preferring older men for advice and friendship, Watson bonded with Dulbecco on the university's tennis courts, where they often went to play a set or two between experiments.

As low man in the pecking order, Watson was assigned to a small office on the top floor of the zoology building, "whose original elevator was still operated by pulling ropes to go up and down." A few levels below, on the first floor, labored a lonely professor who had recently left the study of gall wasps and their evolution to forge a path on the still taboo topic of human sexuality. His name was Alfred Kinsey. His work bored the young Watson because Kinsey's findings were "so heavily statistical as to be more likely purgative than prurient."[37]

During his first year away from home, Watson continued to birdwatch and take solitary walks along Jordan Street—which was the "site of the most desirable sororities, where I would spot girls much prettier than most to be seen in science buildings." He also enjoyed attending the university football games on Saturday afternoons, sitting with 20,000 other shouting fans at the old Memorial Stadium.[38]

Like Rosalind Franklin, Watson faithfully (and weekly) wrote to his parents about his academic pursuits and sundry activities. These letters, all carefully preserved in the Cold Spring Harbor Laboratory archives, provide us with a front-row seat to this period of his life. In his first letter home, Watson wrote about meeting Salvador Luria and gaining permission to take his course on viruses, even though he had not taken the

Max Delbrück and Salvador
Luria at Cold Spring Harbor,
summer 1952.

prerequisite course on introductory bacteriology: "he let me in when I
told him my record and my intention to be a geneticist. He is an Italian
Jew, whom LaMont Cole told me treats his students like 'dogs.' However, he is undoubtedly one of the most brilliant men on campus. He is
young, 30–35, and has done good work on the genetics of viruses (a very
good field). I should learn considerable from him."[39] In another update,
he described Hermann Muller as "one of the great men in modern biology."[40] A week later, however, he complained that Muller's course (and
the requisite lab sessions) were "hopelessly confusing to all concerned."
The lectures were "more difficult and thus more interesting," but Muller
failed to convince Watson to follow him in "into *Drosophila* work. The
possibilities of working with microorganisms appeals to me greatly."[41]

Watson was fascinated by both Sonneborn and Luria. Sonneborn's
course on microbial genetics was quite popular, he reported, and "the
graduate student gossip [about him] reflected unqualified praise, if not
worship . . . In contrast, many students were afraid of Luria who had the
reputation of being arrogant toward people who were wrong." Reflecting his own arrogance, Watson testified that "I saw no evidence of the

rumored inconsiderateness towards dimwits." Before the end of his first term at Indiana, Watson chose Luria's bacteriophage genetics over Sonneborn's *Paramecia*.[42] Initially he experienced an insecurity common among students of all stripes: he worried that he was not smart enough to be accepted into his teacher's "inner circle." The young man somehow quelled his emotions and rose to the occasion. "The more I learned about phages," he recalled in 2007, "the more I became ensnared by the mystery of how they multiplied and even before the fall term was half over I did not want to complete my degree with Muller."[43]

Watson's assessment that studying genes through *Drosophila* was the past and bacteriophages the future represents a superb example of his forward and careerist thinking. Time and again, he demonstrated an uncanny ability to divine the path of science and focus on the next big thing. As an undergraduate at the University of Chicago, he chose genetics over ornithology and classical descriptive biology. Now, as a first-year graduate student at Indiana, he focused on experimenting in microbial genetics instead of with fruit flies. Yet he took a significant risk in choosing to work with the relatively unknown Luria over the Nobel Prize laureate Muller, whose fame alone could have helped him land an academic post in the years that followed. It was one of many occupational risks that would pay off in dividends but were difficult to imagine at the time Watson took them.

For his first research project, Luria asked Watson to "see whether phages inactivated by X-rays could still undergo genetic recombination and produce viable recombinant progeny lacking the damaged genetic determinants present in the parental phages."[44] Luria had proven this to be the case with phages inactivated by ultraviolet light and subsequently allowed to infect an *E. coli* host cell. For the next three years, Watson put colonies of bacteriophages through their paces as he exposed them to all sorts of radioactive mutagens.

As Watson soon learned, Luria treated no one "like a dog." "Lu" was a generous, collegial, well-organized teacher and, unlike many bad actors in the academic swamp, played an active role in the advancement of his students. Over the next few years, Luria opened many doors of opportunity for Watson. In 1948, he facilitated Watson's first meeting with Max Delbrück of Caltech, who proved to be a kindred spirit and a lifelong friend. Luria and Delbrück were the leaders of the "Phage Group." Small in number, they revolutionized genetics and their work led to a pile of Nobel Prizes.[45] Delbrück was blessed with a charisma and gentle nature that enticed many young scientists into his orbit; other pioneering molecular biologists have elevated him to a mythical status akin to a combination of Gandhi and Socrates.[46] Despite the differences in their age and rank (Delbrück was a forty-two-year-old, renowned scientist, Watson only twenty and still a first-year graduate student), from the moment they shook hands in Luria's flat, they always addressed each other by their given names. As Watson recalled their first meeting: "Almost from Delbrück's first sentence, I knew I was not going to be disappointed. He did not beat around the bush and the intent of his words was always clear."[47]

During the summer of 1948, Luria arranged for Watson to pursue

Max Delbrück and the Phage Group, 1949. Left to right: Jean Weigle, Ole Maaløe, Elie Wollman, Gunther Stent, Max Delbrück, and G. Soli.

his phage research at Cold Spring Harbor Laboratory, where he enjoyed swimming in Long Island Sound and gained access to a powerful X-ray machine at Memorial (now Sloan-Kettering Memorial) Hospital in nearby New York City. In a letter to his parents, Watson told of his visit to the city over the Fourth of July weekend, which included a trip to Ebbets Field, the baseball park in the Flatbush section of Brooklyn:

> Last night several of us went in to see the Dodgers play at night . . . It was a good game and the crowd was what I expected to see and hear. Brooklyn from my short visit appears to be a very crowded and poor city with the majority of the inhabitants being either Jewish or Italian. From all appearances a most horrible place to live.[48]

By 1949, after spending the summer in Pasadena, Watson had exhausted all the possibilities of mutating phages with X-rays and began writing up his thesis. Delbrück and Luria decided that Watson needed to broaden his scientific chops by learning some biochemistry, so, in fall 1949, they sat down with Herman Kalckar of the University of Copenhagen during a break in a Phage Group meeting held in Chicago and arranged for Watson to work as a postdoctoral fellow in Kalckar's laboratory.[49] Under Luria's guidance, Watson wrote his first grant application to fund his salary and living expenses in Copenhagen. As with most students who find themselves in this precarious situation, Watson's letters to his parents were filled with anxiety over the distinct possibility of rejection.[50]

On March 12, 1950, he was invited to interview for a two-year Merck Fellowship administered by the prestigious National Research Council of the National Academy of Sciences. A committee of several gray-haired, self-important scientists sat at a long table in the main ballroom of New York City's cavernous, art deco Hotel New Yorker. The candidates, all male, competitive, and eager to succeed, sat nervously in the lobby; at hourly intervals, a committee member opened the double doors of the ballroom and invited one of them inside to be grilled on the merits of his research project. Two weeks later, the committee sent Watson a registered letter informing him that he had been selected for the Merck

Fellowship.[51] Writing home to his proud mama and papa, he sheepishly admitted, "it thus appears that all of my worrying was quite unnecessary." With his finances and immediate future secured, he focused on more mundane issues such as his passport application, securing the proper clothing, and travel arrangements.[52]

EARLY ON THE MORNING OF September 11, 1950, Watson landed in Denmark, seasick from a rough voyage on the MS *Stockholm*, the smallest vessel of the Swedish–America Line and one especially prone to difficult voyages. (Six years later, in 1956, the *Stockholm* collided with the ill-fated Italian liner SS *Andrea Doria*.) Throughout the crossing, Watson swallowed Dramamine pills to counter the boat's incessant pitching and his constant need to vomit.[53] On his first day in Denmark, he wrote to his parents that Copenhagen was "wonderful," presaging Frank Loesser's hit song "Wonderful Copenhagen" by a year. He concluded the letter with the observation: "Somewhat to my surprise, Danish girls are the most attractive ones I have ever seen. The average face is not unpleasant to look at—quite a contrast to most of the States."[54]

Two days later, September 13, having rebounded from his sour stomach, Watson reported for work at the Institute for Cytophysiology, directed by Herman Kalckar. Kalckar, who was Jewish, had left Europe just before the Nazis invaded Denmark and spent most of the war years in the United States at Caltech, Washington University, and the Public Health Institute of New York City. He returned to Denmark after the war to rejoin the scientific powerhouse that was the University of Copenhagen. The king of the institution was Niels Bohr, whose "investigation of the structure of atoms and the radiation emanating from them" had won him the Nobel Prize in Physics in 1922.[55] The Kalckar–Bohr connection was especially close because Kalckar's younger brother, Fritz—who died suddenly in 1938, at the age of twenty-seven—had studied under Bohr. [56]

Paul Berg, a former Kalckar fellow and the 1980 Nobel laureate in Chemistry, described Kalckar as "a dreamer, often seeking novel explanations for paradoxical observations." Scientifically, he "was among the earliest to formulate the concept of high energy bonds as the form in which free energy was captured and stored during oxidative metabolism." For those who are hazy on such key biochemical principles, it might be helpful to recall the cellular powerhouse telegraphed in the form of a molecule known as ATP (adenosine triphosphate).[57] Kalckar was brilliant, "buoyant and fun-loving." He toasted every little discovery made in his lab, whether by himself or his fellows, with aquavit or Cherry Heering liqueur. His halting syntax, in both English and his native Danish, began at a baseline of "difficult to understand" and almost always descended towards the indecipherable.[58] Many of his colleagues believed he should have won the Nobel Prize but did not because his "personality and curiosity about a wide range of topics prevented him from focusing on one or two questions."[59]

Max Delbrück introduced Kalckar to the study of the genetics of phages during his year at Caltech in 1938.[60] Twelve years later, Kalckar hatched a plan to start a phage group of his own and recruited Watson and another Delbrück protégé named Gunther Stent to work at his institute. By the time both young scientists arrived in Copenhagen, however, Kalckar had changed his mind and asked Watson instead to focus on "the metabolism of nucleotides."[61] Watson, who neither possessed nor wanted to acquire the dexterity to conduct such delicate biochemistry research, immediately grasped that this project was a dead end. As Francis Crick later wrote, musing over an opening line for a memoir entitled *The Loose Screw* which he hoped to compose as a means of correcting the embellished record depicted in Watson's *The Double Helix*: "Jim was always clumsy with his hands. One had only to see him peel an orange."[62]

A week into their fellowships, on September 19, Kalckar dispatched Stent and Watson to the State Serum Institute, where they began a collaboration with Ole Maaløe, another former student from Delbrück's phage laboratory at Caltech.[63] Watson and Maaløe conducted a series of experiments in which they introduced a radioactive tracer to the bacteriophage's DNA, followed the virus through successive generations, and measured the radioactive DNA passed on to the progeny.[64]

Watson came to dislike Copenhagen and wrote home about his boredom and unhappiness there. In one letter, he described his purchase
of a used bicycle for 350 kroners, or fifty dollars, which allowed him
the singular joy of biking the 1.5 miles between the two institutes.[65]
This, however, was one of the few local customs he adopted. His only
social contacts were with those who spoke excellent English. In 2018,
he recalled of his Copenhagen period, "I never tried to learn Danish. I
wasn't really interested in Scandinavian culture. And when I was there,
I was only interested in DNA."[66]

On January 14, 1951, Watson wrote to his parents about the "miserable" climate marred by "rain and darkness." He longed for better weather
so that he might resume his daily bicycle rides and walks: "There is little
to do except work and read. These last few days I've been reading some
of [John] Steinbeck's short stories—'The Red Pony', 'The Long Valley.' I
like them considerably."[67] He found some additional comfort in going to
the movies. One night, he and Ole Maaløe saw the 1950 classic film noir
Sunset Boulevard. He was impressed by both Gloria Swanson's portrayal

Jim and Betty Watson in Copenhagen, 1951.

of the bizarre silent movie star Norma Desmond and Billy Wilder's crisp direction. He was especially drawn to the film because, when watching it, "I could easily imagine myself back in California."[68]

Fortunately, Watson's work with Ole Maaløe provided "enough data for a respectable publication and, using ordinary standards, [he] knew he could stop work for the rest of the year without being judged unproductive."[69] Convinced that biochemistry was not for him, he wasted many hours complaining to the other fellows about Kalckar's meandering research program, insisting to his peers in the lab, "we [will] never understand how genes replicate until we [know] the structure of DNA."[70]

In his letters home, however, Watson described Kalckar's generous mentorship. In early November 1950, Kalckar took Watson as his plus-one to a prestigious scientific gathering at the Royal Society of Denmark, the Danish version of the U.S. National Academy of Sciences. The Royal Society occupied a "very pretentious" building owned by the Carlsberg Brewery Foundation, and the membership consisted of "a very dignified group of men, most of them over 55 . . . [once] inside one gets the impression of a weekly men's club." The president of the society was Niels Bohr and "visitors to the meetings are very rare—only the speaker of the evening can bring a guest and only one. Kalckar spoke and took me along. I never felt so young in my life. Despite this feeling, I had a pleasant evening."[71]

At the meeting, Watson learned that Danish science was primarily funded by the wealthy Carlsberg Brewery Foundation. But, he crowed as only a young man can, "the controlling members of the Foundation are elected by the Royal Society and so in fact the largest industry in Copenhagen is owned by the scientists."[72] The Royal Society was not the only recipient of the Carlsberg Brewery Foundation's generosity. Niels Bohr and his family lived in the Carlsberg mansion, an Italian High Renaissance–inspired palazzo situated on the brewery's premises which Watson described as a combined "miniature palace" and museum filled with beautiful art, furniture, and plants. It had been built by the brewery's owner, J. C. Jacobsen, who, upon his death in 1887, willed it to be, as Watson recounted to his parents, "occupied by the most distinguished person in Denmark and will be occupied by Bohr until he dies. Already he has lived there 20 years."[73]

Shortly after the Royal Society meeting, Watson was "informed that Niels Bohr was coming" to a lecture he was to present at the university the following week. Imagine his pride and that of his parents when reading about it. None of the Watsons knew it at the time, but that afternoon marked an important historical occasion: when one of the men who theorized the structure of the atom came to listen to one of the men who would theorize the structure of DNA. After the presentation, Watson modestly wrote to his folks, "I worked quite hard preparing what I should say. I suspect I did not do too badly and Bohr seemed quite interested and joined in a somewhat lively discussion." In the letter, he gave equal time to raving about a film he saw that week: *The Ghost Goes West*, a 1935 British comedy directed by René Clair and starring Robert Donat. He deemed it "a most superb film [that] creates a most pleasant spirit."[74]

During the first few weeks of December, Watson was back to grousing again to his parents. This time it was about the overcommercialism of Christmas in Copenhagen, replete with tinsel-adorned trees and huge window displays at all the stores.[75] Everything changed on December 21

Copenhagen, the city of bicycles.

and Watson wrote home with understated excitement, "Kalckar is going to the Zoological Station in Italy (Naples) for April, May and June. I will probably go with him. It should be quite exciting. We will buy a car to transport us there."[76]

Far more than an opportunity to travel, this brief letter augured one of the most important journeys of Watson's intellectual life.

PART III

TICK-TOCK, 1951

God give me unclouded eyes and freedom from haste. God give me a quiet and relentless anger against all pretense and pretentious work and all work left slack and unfinished. God give me a restlessness whereby I may neither sleep nor accept praise till my observed results equal my calculated results or in pious glee I discover and assault my error. God give me strength not to trust in God.

—SINCLAIR LEWIS, *ARROWSMITH*[1]

[9]

Vide Napule e po' muore[1]

We have two highly gifted and very fine young biologists, Dr. James Watson (Bloomington & California Institute of Technology) and Dr. Barbara Wright (Hopkins Marine Station, Pacific Grove, California) who would like to join us. They are here on American fellowships (National Research Council). Do you think that this would be possible?

—HERMAN KALCKAR TO REINHARD DOHRN,
DIRECTOR OF THE NAPLES ZOOLOGICAL
STATION, JANUARY 13, 1951[2]

Should I say no!—? Certainly not! For a number of reasons—among which I only wish to point out that it would be a pity to handicap your teamwork. Besides, the United States have, for so many generations and so generously supported the Naples Station that I feel we have to repay this by liberally putting our working facilities at the disposal of American biologists, even if there are no American tables available here at the moment.

—REINHARD DOHRN TO HERMAN
KALCKAR, JANUARY 21, 1951[3]

These two letters—buried among hundreds of others in the acid-free boxes of a dusty Neapolitan archive—are the paper equivalents of an operatic overture, for it was in Naples that Jim Watson first heard Maurice Wilkins discuss the use of X-ray crystallography to determine the structure of DNA. Thirty years after that enchanted spring, Watson wrote his own letter to the director of the

Naples Zoological Station, as a means of burnishing the lamp of scientific mythology:

> Like many others, I went to Naples and its Zoological Station knowing that it had a treasured tradition and hoping that a little would rub off on me. Happily, this happened. There I met Wilkins and first realized that DNA might be soluble. So my life was changed, thanks to the use of the Station as a meeting place for the young.[4]

There was, however, a tawdry first act to this operetta, one in which Watson was only a peripheral player. Instead, the leads were the forty-two-year-old Herman Kalckar and a twenty-four-year-old marine biologist named Barbara Wright.

SLIM, ATTRACTIVE, AND ATHLETIC, Wright was born in Pasadena in 1927. Her parents—her father a science fiction writer and her mother a schoolteacher—divorced before her tenth birthday and she grew up in Pacific Grove, California. Like Watson, she loved tennis, exploring nature, and hiking in the mountains. Behind her heavy, black-rimmed glasses loomed deep-set, sienna-brown eyes which matched the color of her hair. Wright began her undergraduate work at nearby Stanford University with the intention of going to medical school. After receiving her bachelor's degree in biology with honors in 1947, she switched career lanes and stayed on in Palo Alto for a masters (1948) and a PhD in biochemistry and microbiology (1950).[5]

Watson and Wright had met a year earlier, in summer 1949, when they were both working in Delbrück's lab at Caltech. They vied for Delbrück's attention and disliked each other intensely. One weekend Watson, Gunther Stent, and a molecular biologist named Wolfhard Weidel decided to take a camping trip to Santa Catalina Island. Much to Watson's chagrin, Stent invited Wright to come along. Hiking down the steep Catalina Palisades, Stent and Weidel got lost. Wright and Watson managed to get themselves back to Avalon, the only town on the island. Frantic but unscathed, they reported their friends' disappearance to the local sheriff's office and spent the rest of the day in a police jeep riding "over the isolated regions of the island in search of them."[6] Watson wrote to his parents about the adventure on August 15, 1949: "Fortunately, they were able to extract themselves from the cliffs and returned to Avalon without our help. The sheriff was a most likable person and I had a most spectacular ride over the island."[7]

During the search, a gawky Watson, two years Wright's junior, kept trying to impress her with his academic accomplishments and sightings of white crowned sparrows and Clark's nutcrackers. She wasn't buying what he had to sell. Watson reported to his parents that he lost his reading glasses during the misadventure. If we read between the lines, it seems that Watson lost an ounce or two of his pride, too, which may have contributed to his cattiness when Barbara Wright's name popped up in conversation.[8]

※

WRIGHT BEGAN HER WORK in Kalckar's laboratory on December 1, 1950, ten weeks after Watson's arrival there. Her fresh-faced, American beauty soon drove the Danish biochemist to distraction. Whenever she entered the lab, Kalckar transformed from an ignored, frustrated husband into the life of the party, much to Watson's youthful disapproval. In mid-December, C. J. Lapp, the director of fellowship programs at the National Research Council, wrote to Watson asking if he had met the newly arrived "Miss Barbara Wright . . . all of the evidence in our office indicates that Dr. Wright is both an accomplished scientist and a charming person."[9] One can only speculate as to how Watson reacted to Lapp's letter, but he could not contain his jealousy to Max Delbrück: "In fact, Herman seems more interested in Barbara's work than his own—he considers it very good and that Barbara is also very good." Watson was confident that Kalckar would eventually see through her pretty looks, "since Ole and I have yet to see anything concrete in her Pacific Grove thesis." The young man found Kalckar and Wright to be polar opposites. "While he gives an initial impression of vagueness," he wrote to Delbrück, "he can in fact be very exact. Barbara on the contrary, strikes people as being very methodical, but on closer inspection is probably rather inexact."[10]

Several weeks of hidden lust elapsed before Kalckar's wife, a cold and prudish musician named Vibeke "Vips" Meyer, suspected anything was amiss. Her husband's roving eye should not have been that big a surprise. The couple had not shared a marital bed for years, and Kalckar stayed in the marriage because Meyer's family was well connected in the interlocking cultural and political circles of Danish society. By Christmas, Kalckar could no longer keep the love affair a secret. His conversations, which had once been "incomprehensible," became all too clear when he told his staff that the "marriage was over, and he hoped to obtain a divorce."[11]

During the second week of January, Kalckar and Wright fled "for 10 days to Norway." When they returned, Kalckar moved into Wright's apartment. On March 22, Watson finally informed Max Delbrück of the drama in Copenhagen: "I feel I can break the silence which we felt bound to . . . Herman, to my great amazement, told me he was in love with Barbara and did not know what would happen to Vips, Barbara, or himself." He went on to explain that, in the weeks before this confession, Herman's "general appearance was quite frightening due to lack of sleep and food" and, even though Kalckar diagnosed himself as having active tuberculosis, Watson "suspected otherwise. In this very confused state, he told virtually all of his friends of his feelings. The unknown factor to us was Barbara's reaction. She seemed very unhappy but would not talk with anyone."[12]

Deeply disappointed by Kalckar's behavior, Watson found "it difficult to describe the very morbid feeling which pervaded Herman's lab during this interval. The disintegration of his normal reactions resulted in a complete loss of morale in the lab and virtually no work was accomplished for over 2 months by anyone." Watson and Stent escaped this noxious atmosphere by spending their days at the State Serum Institute.

But it was too late to escape the spreading stain. Kalckar's torrid affair, Watson told Delbrück, was the talk of the entire Copenhagen scientific community:

> It is difficult now to predict how this story will end. At times, it has seemed like a very bad Hollywood tragedy. This pessimism was probably unjustified since Herman is slowly regaining his former charm and sense of balance. He is leaving with B.W. in two weeks for 3 months at Naples (Zoological Station) and we hope he will return in his former state.[13]

Before she even packed her bags for the trip south, however, Wright learned she was pregnant. Vips Kalckar now had few choices other than to accede to a divorce.

The irony, of course, is that when Kalckar first asked Watson, in December, to join him and Wright in Naples the following May, Watson had no idea that he was being invited to act as a beard. At the time, the affair was not yet public. The middle-aged biochemist concealed the fact that he had already booked a romantic villa for two overlooking the Bay of Naples, while Watson was to fend for himself in one of the old boarding houses near the Zoological Station. With this ruse of inviting both his two newest postdoctoral fellows to help him conduct research, Kalckar could declare that his Naples expedition was strictly legitimate. Desiccated gossip aside, one cannot help but smile at the paradox that the unraveling of the double helix of DNA began with the coupling of Kalckar and Wright.[14]

THE STAZIONE ZOOLOGICA DI NAPOLI was founded in 1872 by Anton Dohrn, a German naturalist and acolyte of Charles Darwin and Ernst Haeckel. His generation of zoologists was devoted to advancing Darwin's evolutionary theories into a body of scientific fact.[15] Like many marine biologists, he was attracted to the rich aquatic life and balmy

The Naples Zoological Station.

climate of the Bay of Naples. Applying a strong dose of Teutonic charm, Dohrn convinced the city council to grant him a prime location in the center of the Villa Comunale, once a royal park, for the construction of his proposed zoological station.

Dohrn designed the Stazione with a superb aquarium on the ground floor to attract the public and create a steady source of revenue to keep the place running. On the floors above were a series of laboratories run on a "table system" which Dohrn is credited with creating. Specifically, laboratory benches, or tables, were rented out to government research agencies, universities, and scientific organizations. The annual table fee allowed each contributing institution to send one scientist per year; this sum could also be divided monthly, seasonally, or semi-annually. Every evening, the resident scientists filled out orders for the various sea creatures they wished to study and, early the following morning, the Stazione's fleet of boats and fishermen went out to catch them and bring them back the laboratories.[16] The University of Copenhagen Institute for Cytophysiology was one of many institutions that rented a table.[17]

Anton Dohrn's son, Reinhard, took over the reins of administration in 1909, and had the extra chores of cleaning up the damage incurred by two world wars. Beginning in 1947, thanks to a $30,000 grant from UNESCO, the Stazione hosted annual symposia on topics related to genetics and embryology for some of the best biologists in Europe.[18]

DESPITE ITS REPUTATION FOR SUNSHINE, "for [his] first six weeks in Naples [Watson] was constantly cold." He exhibited a complete lack of interest in marine biology and could barely endure the drafty Stazione, let alone his poorly heated, "decaying room atop a six-story nineteenth century house."[19] As he rambled along the narrow, twisting, cobblestone streets, he was repelled by the squalor of post–Second World War Naples. On April 17, 1951, he wrote his parents:

> Naples is very different from Milano. Beautifully situated on the water and dominated by Vesuvius, it is an extremely ugly city—both intrinsically and due to war damage. The entire city can be described as a slum and the people are wretchedly poor, living in slums which make the Negro section of Chicago look almost pleasant in comparison. The city is large (over 1,000,000 people) and very dirty.[20]

Two weeks later, on April 30, Watson wrote to his sister that he was acclimating to Naples and begrudgingly admitted "the people though looking very similar and dirty have a civilization of their own which is definitely not amoral." On the weekends, Jim took trips to Capri, Sorrento, and Pompeii so that when his sister arrived in the middle of May he could escort her about the sights with confidence and skill. While Wright and Kalckar were ostensibly working on purine metabolism in sea urchin eggs, Watson was listed in the Stazione's records as conducting "bibliographic work."[21] He thus informed his sister: "I have been spending most of my time reading and writing. I have been away from my Ph.D. work long enough so that I can now write it up without complete boredom."[22] He had open access to the Stazione's library, which was well stocked with over 40,000 volumes and every leading biological journal then published in English, Italian, and German. Many of the journals in the collection went back to their first issues, including all the "journal articles from the early days of genetics."[23] High above the stacks, set within a series of friezes and pilasters designed by the sculptor Adolf von Hildebrand, were four colorful frescoes by the Post-Impressionist painter Hans von Marées depicting, in the artist's words, "the charm of life on the sea and on the shore."[24]

At this point in his life, Watson was worried about his future prospects. Subject to too many daydreams about "discovering the secret of life," he was unable to conjure up "the faintest trace of a respectable idea."[25] He did find the time to rewrite the manuscript he and Ole Maaløe had produced during the winter of 1950–51, each iteration of which Watson sent to Pasadena so Max Delbrück could edit and, eventually, sponsor it for publication in the *Proceedings of the National Academy of Sciences*. In this study, Watson and Maaløe radio-labeled the phosphorus in parental phage virus particles, which is present in the virus's DNA,

The library of the Naples Zoological Station.

and extracted it in the progeny. The two men hoped they had determined a new method to replicate the famous Avery experiment. But because their yield of radioactive phosphorus was only about 30 percent, Delbrück changed the term "genetic material" to "virus particle" in the draft manuscript—indicating that in 1951, the world authority on viral genetics was not "ready to restrict the genes of viruses to viral DNA."[26]

The most pressing issues on Watson's mind, however, were neither his supervisor's ongoing love affair nor the writing of scientific papers. On March 6, 1951, he received a letter from Local Draft Board No. 75 ordering him to report to Chicago within three weeks for a pre-induction physical examination. To stall the nation's armed services, he recruited Kalckar, Luria, and Delbrück to endorse his application for a deferment from the U.S. Army (which he ultimately received)—no inconsequential issue when the United States was participating in what was then called the Korean conflict. He also needed to reapply for another year of the Merck Fellowship in order to continue his research abroad. Between these tasks, he was barely staying afloat in a sea of worry.[27]

The highlight of Watson's sojourn at the Stazione was the UNESCO-sponsored symposium "The Submicroscopic Structure of the Protoplasm," held on May 22–25, 1951. Academic conferences are notorious for inducing boredom, if not rigor mortis, among those attending, as speakers drone on interminably, reading word for word from prepared manuscripts and the text of the slides projected on the screen behind them. All the while, the audience pretends to listen, heads nodding up and down like the bows of an orchestra's string section in a symphonic cadence. Only rarely does an inspiring lecturer come along with an entertaining and edifying speech. Most scientists prefer the published page, not only because they enjoy seeing their name in print but, more importantly, because it is the only way to ensure priority for their discoveries. In the modern world of publish or perish, few scientists become great ones simply by giving lectures.

The keynote speaker of the Naples symposium was the man who inspired Pauling to crack the structure of proteins: William Astbury of the University of Leeds.[28] The proteins on which Astbury focused—wool, cotton, keratin, and animal hair—were composed of "long, chain-like molecules" that were easily stretched and amenable to X-ray crystallography. Although DNA was hardly a natural fiber used by the textile industry, it, too, was long and stretchy enough to be amenable to Astbury's X-ray diffraction techniques. For more than a decade, Astbury had been sniffing about DNA's structure without much success.[29]

In 1938, Astbury and his doctoral student Florence Bell published the first X-ray photographs of DNA fibers. While the images were somewhat blurry, they described DNA's nucleotide structure as looking like a "pile of pennies."[30] Astbury's updated 1947 paper on "X-ray studies of nucleic acids" correctly estimated the distance between the nucleotides to be roughly 3.4 nanometers with a "large structural repeat of some sort at about every 27 Ångstroms." Slightly changing his descriptive metaphor, Astbury "inferred that the nucleotides lie immediately on top of one another like a great pile of plates *and are not disposed spiralwise round the long axis of the molecule*" (italics added).[31]

Although Watson was eager to attend the conference, he was disappointed by the jolly, roly-poly, bald and bug-eyed professor from Leeds. He found Astbury to be a dinosaur who preferred drinking Scotch whisky, scratching out Mozart tunes on his violin, and telling dirty jokes to discussing science.[32] As a "protein man," Astbury was unwilling to eliminate proteins from the secret-of-life equation. In his lecture, titled "Some Recent Adventures Among Proteins," Astbury hedged his bets by elaborating a "nucleoprotein theory," whereby proteins dominated in viral replication but it was a "fair conclusion that nucleic acid is essential to the process, and in fact to all processes of biological duplication."[33] Watson admits to snoozing through most of Astbury's lecture.[34] Sixty-seven years later, in 2018 he was still dismissing Astbury's words as "not very inspiring."[35]

The invited speaker Watson most wanted to meet was John Randall. The King's College biophysicist was to discuss his team's MRC-sponsored work on the structure of nucleic acids. This MRC grant, not incidentally, was one that Astbury had narrowly lost out on securing.[36] Still, "the odds . . . were against any real revelation," as Watson later

recalled, because "much of the talk about the three-dimensional structure of proteins and nucleic acids was hot air." Although this work had
been pursued for nearly two decades, "most if not all of the facts were
soft. Ideas put forward with conviction were likely to be the wild products of wild crystallographers who delighted in being in a field where
their ideas could not be easily disproved." Worse, few biochemists,
including Herman Kalckar, understood the jargon-filled complexities
offered up by the X-ray crystallographers, and fewer still were willing to
believe their assertions. For Watson, "it made no sense to learn the complicated mathematical methods in order to follow baloney. As a result,
none of my teachers had ever considered the possibility that I might do
post-doctoral research with an X-ray crystallographer."[37]

Much to Watson's disappointment, Randall was a no-show, having
canceled at the last minute.[38] He threw his assistant director, Maurice
Wilkins, a bone in the form of a free trip to Naples for the price of
presenting a lecture.[39] If one had placed a bet that morning on which
speaker would spark the audience's imagination with something truly
remarkable, Maurice Wilkins would not have been tipped to be that
man. Yet this is exactly what happened.

Before traveling to Naples, Wilkins had developed a new method of
preparing the Signer calf thymus DNA for X-ray crystallography. Initially, he had simply placed a bit of the substance, which looked like
the discharge from a rheumy nose, onto a glass microscope slide. Using
another glass slide as a spatula, he spread the stuff out into a thin film.
As he recounted in his 1962 Nobel lecture, with each successive dip of
a glass rod into the jar containing the snot-like elixir, he could see that
a "thin and almost invisible fiber of DNA was drawn out like the fila-
ments of a cobweb. The perfection and uniformity of the fiber suggested
that the molecules in them were regularly arranged."[40] The innovation—
spinning a thin filament of the substance rather than merely smearing
it onto a slide—introduced a critical next step in the experimental process. After Wilkins stretched the fibers, his graduate student Raymond
Gosling "sat like a little spider tying [them] round a bent wire [at first a
paper clip and, later, a more elegant tungsten wire] and pushing them

together and gluing the two ends so that [he] had a multi-stranded spec-imen."[41] Wilkins and Gosling carefully positioned this setup in front of an old Raymax X-ray camera they'd found in the chemistry department basement and filled its chamber with hydrogen to reduce background scatter. There were more adjustments to be made, but the result was a photograph with a crisp column of horizontal bands that was far sharper than Astbury's 1938 "pile of pennies" photos.

After developing one of these images in the department's smelly dark-room, Gosling joyfully recalled, "I went back down the tunnels over to the Physics Department, where Wilkins used to spend his life, so he was still there. I can still remember vividly the excitement of showing this thing to Wilkins and drinking his sherry . . . by the gulpful."[42] This strung-out preparation—which only improved as Wilkins gained more hands-on experience—allowed for a more easily discernible structure under X-ray diffraction and may have been his greatest contribution to the DNA steeplechase.

On the morning of May 22, 1951, a bored Watson sat in the back row reading a newspaper as Wilkins spoke in a diffident manner about "ultraviolet dichroism and molecular structure in living cells." Eventu-ally he moved on to discussing nucleic acids, and what he had to say did "not disappoint" Watson. When he projected his strikingly detailed pic-ture onto the screen, Watson looked up—and both the newspaper and his jaw dropped. Although "Maurice's dry English form" failed to reg-ister the enthusiasm of his remarkable observation, he clearly "stated to the audience that the picture showed much more than previous pictures and could, in fact, be considered as arising from a crystalline substance. And when the structure of DNA was known, we might be in a better position to understand how genes work."[43]

Watson was not the only member of the audience captivated by Wilkins's spectacular data.[44] Dohrn immediately jotted down notes for an effusive letter he later sent to Randall: "Thank you for having sent down your collaborator Wilkins; his paper met with greatest interest and as he speaks rather slowly, also the non-English part of the audi-ence could follow, it was a great success."[45] Astbury, too, praised Wilkins, announcing to all who would listen that the "pattern was much better than anything he had got."[46]

At the cocktail party following the day's lectures, Wilkins struggled to make small talk and match Astbury drink for drink. Watson watched from afar, obsessed by the notion that if "genes could crystallize . . . they must have a regular structure that could be solved in a straightforward fashion." Before finishing his first gin and tonic, Watson knew he could not waste another instant in Kalckar's adulterous laboratory. Copenhagen was the scientific equivalent of wandering aimlessly in the desert. The way to the Promised Land was through X-ray crystallography and by convincing Wilkins to take him on at the King's College Biophysics Unit. But before he could approach the mild-mannered physicist, "Wilkins had vanished."[47]

On the final day of the symposium, Saturday, May 26, 1951, the Stazione staff organized a field trip for the conferees. Their destination was the ancient temples of Paestum, a once important Greek city on the coast of the Tyrrhenian Sea, in the former Magna Graecia—better known today as Campania. These majestic ruins are not far from bucolic Salerno, an agricultural wonderland of livestock and *caseifici*, or cheese factories, producing tons of *mozzarella di bufala* each year. Visitors to Naples almost always make a beeline for Pompeii. Each year, more than 2.5 million tourists marvel at the remains of a once industrious city at the foot of Mount Vesuvius—which in 79 AD erupted all over Pompeii and its people, thus creating the definitive description of a bad day. For those seeking the beauty of the sea, there are boat excursions to the islands of Capri and Ischia. Few travelers, in comparison, make the 95-kilometer journey to Paestum's three magnificent Doric-columned temples.[48]

As the group was boarding the tour bus, Watson attempted to start a conversation with Wilkins. But before he could "pump Maurice," the driver abruptly ordered everyone to take their seats. Wilkins scampered away from the odd American to sit close to the man he most wanted to impress, Professor Astbury. As the bus made its way along the narrow, twisting coastal road, the international collection of biologists, biochemists, physicists, and geneticists passed the time gossiping, laughing, and

Paestum: the Second Temple of Hera.

talking shop. Astbury was one of the loudest participants in the cacophony, sharing ribald jokes and nips of Scotch whisky from his battered, sterling silver flask.

A few rows behind, Watson sat silently next to a pretty young woman, properly dressed in a starched, pink dress. Her white-gloved hands clutched a white leather handbag and her long, blond, flipped hair was topped by a flimsy, pink pillbox hat. This was Watson's younger sister, Elizabeth, the woman he most adored and respected. She had arrived in Italy a few days earlier to join him for a tour of Europe and, thereafter, possible matriculation into either Oxford or Cambridge.[49] All the way to Paestum, in which he had zero interest, Watson schemed on how best to approach Wilkins and ask if he could join his laboratory.

Soon after they arrived, the group scattered across the vast archeological site to enjoy what one of the participants extolled as a "marvelous excursion."[50] Watson parked his posterior on one of the square, squat stones at the base of the second (and best preserved) Temple of Hera. Perhaps inspired by the temple's ancient gods, a lightning bolt of an idea struck him. Wilkins appeared to be attracted to Elizabeth Watson and "soon they were eating lunch together." This pairing pleased Watson no end because, for many years now, he "had sullenly watched Elizabeth being pursued by a series of dull nitwits." His joy was inspired by more than the wish to see her wed to someone other than "a mental defective." He imagined Wilkins falling in love with his sister and, in good brother-in-law fashion, offering him a position to collaborate on the King's College X-ray work on DNA.[51]

In Watson's telling, Wilkins excused himself from the brief encounter, an act Watson misinterpreted as the result of "good manners and assumed that I wished to converse with Elizabeth." Unfortunately, once they returned to Naples, no further conversation about DNA was to be had. Wilkins nodded Watson's way and took his leave. Watson's tentative experiment in pimping out his sister failed miserably: "Neither the beauty of my sister nor my intense interest in the DNA structure had snared him. Out futures did not seem to be in London. Thus, I set off to Copenhagen and the prospect of more biochemistry to avoid."[52]

Memoirs are slippery, sly sources for documenting historic events. Hence, Wilkins's 2003 recollections of the Paestum trip do not quite correspond with Watson's more famous telling of this tale. During his first meeting with Watson, Wilkins understood very little of his flat, Midwestern-accented outpouring on "genes and viruses . . . I did not know much about bacteriophages and could not make much sense of what he was telling me." Although he remembered Watson as one of the more interesting attendees at the conference, Wilkins denied a flirtation with Elizabeth: "His sister was with him but I do not remember seeing her—in any case, I was somewhat intoxicated by all the beauty [of Paestum] around me!"[53] As soon as Wilkins returned to London, however, he told Raymond Gosling that there was no possibility of taking the American on board. "Wilkins was afraid of him," Gosling recalled later. "He's quite scary, old Jim, on full flight."[54] In another version, Wilkins called Watson a "gangly young American" and instructed Gosling that if Watson ever showed up at King's, he was to tell him that Wilkins "had left the country."[55]

Nonetheless, Watson's new research direction was clear: he must work with the biophysicists and X-ray crystallographers. In Naples, he quickly divined that applying their methodology to the structure of DNA was not only his future, it was the future of molecular biology. Roadblocked against sliding into the King's College unit, there were only two other scientific plays: the first entailed completing his fellowship under Kalckar and then traveling west to Caltech in the hope that Linus Pauling would teach him how to become an able X-ray crystallographer. Watson rejected this path as quickly as it popped into his fertile mind because "Linus was too great a man to waste his time teaching a mathematically deficient biologist."[56] The second option was even more risky, because it involved breaking the contractual terms of his fellowship—which required him to stay in Scandinavia—and finagling his way into the Cavendish Laboratory Biophysics Unit at Cambridge. After constructing a calculus of options, Jim Watson chose Cambridge.

From Ann Arbor to Cambridge

*I promised you I'd write the Committee in August and failed
to do so. Therefore, I am to blame. You goddam bastard, you
wrote the silliest letter to the Committee about it.*

—SALVADOR LURIA TO JAMES D.
WATSON, OCTOBER 20, 1951[1]

In July 1951, the University of Michigan at Ann Arbor hosted an international postgraduate course on biophysics. Every summer during the decade leading up to the Second World War, the Michigan physics department had presented summer programs in theoretical physics taught by such luminaries as Niels Bohr, Enrico Fermi, and J. Robert Oppenheimer. Invitations to these summer symposia were coveted by physicists on both sides of the Atlantic. The origins of a long catalogue of important discoveries and publications can be tracked to these gatherings where "the distinctive aroma of genius" was in the air.[2]

The 1951 curriculum on biophysics was organized by Gordon Sutherland, a professor in Ann Arbor from 1949 to 1956, by way of Cambridge.[3] For this course, he gathered together eight superb biophysicists to serve as instructors, including Salvador Luria of Indiana, Max Delbrück of Caltech, and John Kendrew of the Cavendish Laboratory. His goal was to bring "physicists and biologists together so that the former would appreciate the biological problems to which the physical methods should be applied and the latter would hear of some of the newer tools and techniques in physics which can be used in biological research."[4]

Like many other American campuses, the University of Michigan in Ann Arbor experienced a huge post–Second World War growth spurt, evidenced by the erection of new buildings, laboratories, and classrooms.

American academia bathed in a river of money, fed by a steady stream of federal government research grants, defense contracts, and tuition dollars for a parade of new students benefiting from the Serviceman's Readjustment Act of 1944, known as the G.I. Bill. The University of Michigan was, as Sinclair Lewis once described it, a place where "buildings are measured by the mile . . . It is a Ford Motor Company, and if its products rattle a little, they are beautifully standardized, with perfectly interchangeable parts."[5]

The weather that summer was sunny and hot enough to make one forget about the bitter winters lurking around the corner of every calendar. The relatively few students still in Ann Arbor ambled across the

The University of Michigan, c. 1950.

central campus's forty acres of crisscrossing diagonal footpaths (always referred to as "the Diag") and into the red-brick and limestone buildings, fronted by the occasional grand set of pillars. In between classes, they sat on the verdant grass in the shade of the Diag's burr oak and American elm trees, which were arrayed in parallel rows along the footpaths like a wooden version of the Michigan marching band. The professors there were either stuck in town teaching undergraduates who had failed their classes earlier in the year or were conducting special courses for professional colleagues, such as the offering on biophysics. The seasonal quiet was pierced every fifteen minutes, from 9:15 a.m. to 9 p.m., by the echo of Westminster quarters deep within the Burton Bell Tower, a 212-foot, limestone-clad *art moderne* sentinel looming over the Diag. Most afternoons, the university's carillon, the fourth heaviest in the world, plunked out tunes on its fifty-three bells.[6]

JIM WATSON AND HIS SISTER were far away from southeastern Michigan that summer, making a circuitous return from Naples to Copenhagen by way of northern Italy, Paris, and Switzerland. Each night, they read for pleasure. Watson turned the pages of Harvard philosopher George Santayana's bestselling 1936 "memoir in the form of a novel," *The Last Puritan*, which tells the story of a scion of an old Boston family whose puritanism and genteel nature is at complete odds with twentieth-century American culture. Identifying with its protagonist, Watson wrote to his parents that this now all-but-forgotten book "was quite superb especially the earlier sections" describing the main character's ancestry and boyhood.[7]

For work, he took a side trip to Geneva and spent a few days with Jean Weigel, a Swiss phage biologist he had met during Max Delbrück's Cold Spring Harbor Phage Group summer courses in 1949 and 1950. Weigel, who had recently returned to Switzerland after spending the winter term at Caltech, told him that Pauling had just solved the structure of proteins. If he was correct, Pauling would become the first scientist to describe the configuration of a "biologically important macromolecule."

As he listened to Weigel recount Pauling's latest triumph, Watson imagined the grand announcement as if he were there when it happened: "A curtain kept his model hidden until near the end of his lecture, when he proudly unveiled his latest creation. Then, with his eyes twinkling, Linus explained the specific characteristics that made his model—the α-helix—uniquely beautiful."[8] Weigel, who had little experience in X-ray crystallography, was unable to answer the spray of questions Watson aimed his way. He told Watson that several colleagues "thought the α-helix looked very pretty" but they were all waiting to read Pauling's paper in the *Proceedings of the National Academy of Sciences* to be certain.

In our age of instant global communication, it is difficult to comprehend how slowly information traveled a half a century or more ago. During the Atomic Age of the early 1950s, the latest issues of scientific journals like *PNAS* still underwent a laborious process of typesetting, proofing, printing, and binding before being sent via U.S. Mail trucks, trains, and planes to libraries and readers across the United States. Bundles of issues were transported across the ocean by steamship and, six weeks later, delivered to European subscribers. Thus, when Watson returned to the University of Copenhagen in July 1951, the librarian there had only recently placed the April issue of *PNAS* on the shelves. Watson practically pulled it from her hands, devoured its contents, and then read them again; a few weeks later, he did the same with the May issue of *PNAS*, which contained "seven more Pauling articles." As Watson later recalled, "most of the language was above me, and so I could only get a general impression of his argument. I had no way of judging whether it made sense. The only thing I was sure of was that it was written with style."[9] Even at this inadequate level of comprehension, Pauling's latest discovery sent Watson into a tailspin of worry. What if Pauling applied his "rhetorical tricks" to DNA before he himself had a chance to solve the puzzle?[10]

By late July, most of the Kalckar lab members had returned from their summer vacations, which helped assuage Watson's loneliness. As he told them of his decision to leave Copenhagen for Cambridge, he offered a menu of excuses, including the effects of Copenhagen's damp, cold weather on his mental health, his disdain for biochemistry, and his Naples conversion to the wonders of X-ray crystallography. Writing to his parents on July 12, he gave a self-edited version of his desire to move to Cambridge in the fall: "I feel that I have largely exhausted the possibilities of Copenhagen science. Cambridge is probably the best university in Europe from my viewpoint so I will probably go there in late September or early October."[11] Two days later, on July 14, he told his sister another reason behind all these very good explanations: his deeply-held disgust with Herman Kalckar. Young men can be infuriatingly judgmental about the scruples and foibles of love, and Watson was definitely of this mold. He was unable to tolerate Kalckar's adulterous affair, let alone work side by side with a man who sullied his equivalent of a church—the laboratory. The mere sight of Kalckar and the visibly pregnant Barbara Wright holding hands was, for Watson, "so essentially depressing."[12]

WATSON'S FUNDING FROM THE Merck–National Research Council fellowship was due to run out in September 1951. Under the terms of the fellowship agreement, he could apply for a second year of support. Arrangements had already been made for him to work at the Karolinska Institutet in Stockholm with the cell biologist and geneticist Torbjörn Caspersson, who studied the biochemistry of nucleic acids and protein synthesis.[13] After his Naples epiphany, however, Watson viewed the transfer to Sweden as a waste of time. If he was to change course in the middle of his fellowship, however, he needed to write a proposal quickly and get it to the fellowship board for approval.[14] Here, the muse of serendipity— so crucial to this narrative—once again intervened.

Far away and unknown to Watson, at the University of Michigan summer course on biophysics, John Kendrew and Salvador Luria transformed Watson's hope of moving to Cambridge into a plan of action. On a humid late July evening, the two scientists adjourned from the classroom for an informal round of social drinking. A quarter century afterward, Kendrew recalled how the meeting changed the course of

Watson's life without him even being present: "over a glass of beer, [I] said to Luria, 'We're expanding, and looking for bright students, do you know of anybody, and he said, 'Well, there's a fellow called Watson, he's at the moment in Copenhagen and unhappy because his supervisor's changing wives.' "[15]

John Kendrew, c. 1962.

In Watson's day, moving the location of one's research fellowship, let alone changing the nature of the work proposal, was highly irregular, if not strictly prohibited. With few exceptions, requests of this nature were denied, since most fellowship boards considered such moves to be a sign of immaturity, a deficit of seriousness about one's science, and poor form on the part of the applicant. In an era of strict, top-down management of students by faculty, Watson recalled in 2018, "it just wasn't done."[16]

In his letters home during August, Watson described his intention to move his work to Cambridge as a fait accompli. On August 21, he told his parents that he was winding up his experimental work in Copenhagen: "I have definitely decided to go to Cambridge for the coming academic year, as I know I have been assured of space in a laboratory. I will probably leave Copenhagen for good around the middle of October."[17] Only a week later, on August 27, he wrote that he was eager to attend the Second International Poliomyelitis Conference, to be held in Copenhagen in early September, before making his move to Britain.[18] In the same letter, he told them that, on the advice of Max Delbrück, he was applying for a National Foundation for Infantile Paralysis fellowship to fund his projected biophysics work and asked his parents to obtain his transcripts from the University of Chicago and Indiana University and send them along to the NFIP offices in New York City.[19]

At the opening session of the polio conference, directly after introductory speeches by Niels Bohr, the honorary president of the conference, and Basil O'Connor, Franklin D. Roosevelt's former law partner and the president of the NFIP, Max Delbrück delivered a plenary lecture on virus multiplication and variation. In between sessions, Watson interacted with many of the world's top virologists, including the powerful Thomas Rivers of the Rockefeller Institute; Andre Lwoff of the Institut Pasteur in Paris, who would win the 1965 Nobel Prize in Physiology or Medicine for his work on the genetic control of enzyme and virus synthesis; John Enders of Harvard Medical School, who, with his students Frederick Robbins and Thomas Weller, won the 1954 Nobel Prize in Physiology or Medicine for developing methods of culturing poliomyelitis viruses in various tissues; and Thomas Francis, Jr., of the University of Michigan, who would soon run the largest ever vaccine field trial to test the Salk vaccine and, in 1955, announce it to be "safe, effective and potent."[20] "From the moment the delegates arrived," Watson recalled, "a profusion of free champagne, partly funded by American dollars, was available to loosen international barriers. Each night for a week there were receptions, dinners and midnight trips to waterfront bars. It was my first experience with the high life, associated in my mind with decaying European aristocracy."[21]

Watson was hardly the only one making connections at the meeting. The same conference marked the beginning of a productive working relationship between Basil O'Connor and Jonas Salk. On the steamship back to the States, the two men became quite friendly, discussing methods of vaccination at the captain's dining table and while lounging on the first-class deck.[22] Huge NFIP grants flowed Salk's way soon after they disembarked in New York.

AFTER THE CONFERENCE, Watson "went off to England in excellent spirits" to meet with Max Perutz and, on September 15, wrote a long and reassuring letter to his parents:

> My decision to move to Cambridge was due to the presence of some excellent physicists who work on methods of determining the structure of very complex molecules. In the future this work

will have a strong bearing on our ideas in virus research and so I thought it might be worthwhile to learn their techniques while in Europe . . . I will work in the Cavendish Laboratory—a quite famous lab where many of the important discoveries of physics were made. My work will be half biology and half physics but in practice will be largely physics and should involve considerable mathematics. I will thus, in fact, be starting school again and at present I feel in a similar way to when I went to Bloomington 4 years ago . . . it is in a way very pleasant to feel that a vast amount of knowledge can be quickly obtained by reading and studying . . . and so I will be quite happy to start school again.[23]

Watson left it until early October to inform C. J. Lapp of the National Research Council about his migration. Insisting that biochemical methods alone were not adequate to determine the role nucleic acids play in genetics, Watson wrote to Lapp, "I feel that my future role as a biologist would be greatly expanded if I could have the possibility of studying in Dr. Perutz's laboratory during the coming academic year."[24] He left out one critical fact: he was writing this letter while sitting only a few doors down from Perutz's office in the Cavendish Laboratory.

By long-distance telephone, Watson aligned the support of Herman Kalckar, who wrote a laudatory letter to the NRC on October 5, 1951— the day Watson arrived in Cambridge to begin his work there—stating that he had encouraged Watson to move to Cambridge and that Watson's request to study with Max Perutz "merits full support."[25] Eleven days later, on October 16, Watson wrote to his sister about the pushback from the new chairman of the Merck Fellowship board at the NRC, his old University of Chicago professor Paul Weiss, who still resented Watson's inattention during his lectures a few years earlier and now had the perfect opportunity to pay back his former student: "They cannot see why I want to leave Copenhagen and so do not approve of my Cambridge idea. I will leave the matter to Lu [Luria]. As he wanted me to work for Perutz, I know he will fight for me. I do not intend to worry about the matter."[26]

In reality, there was a great deal to worry about. Neither Weiss nor Lapp looked kindly upon their cheeky fellow's flighty actions, and the fellowship board had every right to deny his request. To calm the waters, Salvador Luria wrote a letter on October 20 (as he had mapped out in separate letter he wrote to Watson on the same day) taking full blame for the transfer, claiming to have been the one who arranged it and apologizing for dropping the ball by forgetting to inform the bureaucrats in Washington. Luria described Watson as a mere "boy" on whom he and Delbrück "placed the greatest hope for developing the work on reproduction of viruses and biological macromolecules along new and unexplored lines." The letter soon devolved into a morass of "white lies" about how Watson's work at Cambridge would advance his previous year's virology research. Luria told an outright fib in saying that Watson would work primarily under Dr. Roy Markham, a specialist in viral nucleoproteins and turnip yellow mosaic virus, at the Molteno Institute for Research in Parasitology, rather than at the Cavendish.[27] Hence, the plan was "far from being a drift in the wilderness" and was, instead, "a considerable search for the type of preparation that may improve his usefulness to biology."[28] Widening his net of deception, Luria contacted Markham and persuaded him to take part in the subterfuge, which Markham described as a "perfect example of the inability of Americans to know how to behave. Nonetheless, I promised to go along with this nonsense."[29] As Watson recalled, "armed with the assurance that Markham would not squeal, I humbly wrote a long letter to Washington, outlining how I might profit from being in the joint presence of Perutz and Markham."[30]

The subterfuge appeared to serve its intended function. Weiss wrote to Watson on October 22 that he now understood that Watson's plan to learn about molecular biology at the Cavendish was "merely coincidental to other work on virus nucleoproteins to be carried out at the Molteno Institute, more closely related to the line you have been following thus far." In the same letter, Weiss asked for more details about the proposal, including Watson's intended date of departure from Copenhagen.[31] But Jim Watson had in fact already left Denmark, and Weiss's letter had to be forwarded to him in England. Watson took false comfort in how well

Luria's plot had worked until November 13, when he received a further letter of reversal from Weiss that was neither kind nor permissive; in Watson's words, Weiss "did not play ball."[32]

There was one ray of celebration in the midst of this scuffle. On October 29, only eight weeks after his heady face time with the polio movers and shakers, the National Foundation for Infantile Paralysis informed Watson that he had won a fellowship to begin the following academic year, 1952–53.[33] Thus, by 1955, the NFIP could rightfully claim to have funded both the first successful polio vaccine and the key work that went into identifying the double helix structure of DNA.

Good news aside, there was still the NRC stalemate to resolve. On Luria's advice, and with Max Perutz's blessing, Watson wrote a long letter to Weiss on November 13, making a rare display of humility. He apologized for the circumstances surrounding his move to Cambridge but insisted that his motives were scientifically sound and pure. He told Weiss he was motivated solely by the potent combination of working at both the Molteno Institute and the Cavendish Laboratory, where he could collaborate on determining the structure, rather than the metabolism, of viral nucleic acids, which "might lead us more directly to the mechanism of replication."[34]

A week later, on November 21, Watson received an even harsher letter from C. J. Lapp, who by now had figured out the irregularities surrounding Watson's move to Cambridge. Watson, unwilling to accept responsibility for the move, replied that Lapp's letter came "as a great shock to me" and, according to plan, blamed Luria: "I did not come here on my own initiative but on the advice of Dr. Luria and applied immediately after a place in a Cambridge laboratory had been offered to me. I feel, however, that perhaps I should have given you a fuller account of the events which led me to this step." He added that Dr. Kalckar's "domestic difficulties" hampered his scientific work in Copenhagen because "I did not find the encouragement and advice I had expected."[35]

Although Watson had received a $3,000 grant from the National Foundation for Infantile Paralysis, these funds were meant to support work with Delbrück at the California Institute of Technology the following year, not his time in Cambridge. The NFIP award was a terrific safety net but, as he told his parents on November 28, he still wanted to pursue work at the Cavendish: "I will accept [the NFIP grant] with the possible personal reservation that I may want it postponed for 6 months to a year. My personal plans are necessarily vague."[36] The same day, he wrote a pessimistic letter to his sister about the chances of the Merck Fellowship board coming through to support his work in Cambridge that year.[37] The following week, on December 9, Jim informed Max Delbrück of his bleak situation: "I am still in very hot water concerning my move to Cambridge. The Merck Fellowship Board (Paul Weiss) is bloody mad and so any papers I publish may carry the identification 'one time fellow at the National Research Council.'" At least, Watson continued, he had yet to be "officially fired," and there existed a possibility of getting the equivalent of one-third of his NRC stipend from the Cavendish. He concluded confidently, "However, in no way do I regret my hasty move to Cambridge. Herman's lab was just plain depressing."[38]

On January 8, 1952, upon returning from his winter break at the posh Scottish estate of a colleague, Watson wrote an update to his parents as he "thought you might be more concerned than I am . . . I admittedly have no respect for the authority of Paul Weiss who is a most unpleasant man. I have definitely benefited from coming to Cambridge greatly and so basically do not regret my premature departure from Copenhagen. I was feeling intellectually quite stagnant in Copenhagen."[39]

When Weiss's office informed Watson that any work he wished to conduct in Cambridge would have to be treated as a new application, he duly completed a new grant request and sent it in by the requested date of January 11, offering several more transatlantic apologies to the petty powers that were. One week later, on January 19, he confessed to his parents that the "fellowship complication" was "seriously affect[ing] my enjoyment of Cambridge." He tried to comfort them by noting he still had seven hundred dollars "to live upon" and did not need any money from his "understanding parents."[40]

Finally, on March 12, Watson was formally rapped on the knuckles for going "to Cambridge to work on molecular structure analysis without knowledge or consent of the Board."[41] The Merck-NRC Fellowship board relented a bit by awarding him eight months of financial support at Cambridge, instead of the original twelve-month fellowship at Copenhagen. Calculating that he had enough money to keep himself sheltered and fed—from the new grant, added to what he had saved during his year in Copenhagen—Watson politely accepted the NRC's new terms. To Luria, however, he described Weiss as a "bloody bastard." Professor Luria corrected his former pupil with an even sharper assessment: "As for Paul Weiss, I incline to agree with your definition, although being less British than you are, I would call him a 'damn-son-of-a-bitch' rather than a 'bloody bastard.' "[42]

In Watson's overly triumphant *The Double Helix*, he depicts the bureaucrats at the National Research Council as incompetent fools for not fully investing in his DNA work and, hence, losing a magnificent line of credit on the famous paper Watson and Crick would publish a little more than a year later. Seen with the 20/20 vision of hindsight, there is a great deal of truth to this. Watson understood where the future of genetics lay and he wanted to be an integral part of it, regardless of the rules enforced by the stodgy guardians of fellowship dollars. In Watson's mind, these unimaginative administrators took unbridled joy in stifling his creative genius. They did not yet understand that Watson was *Watson*, even if Watson apparently did. His unwavering self-confidence and single-minded ambition represented both his best and worst qualities. Flouting the terms of an agreement he'd made, Watson boldly situated himself wherever the scientific action was to be found with "the very few people [who] are willing to attack both the necessary physics and biology" and solve the complex molecular structures of proteins and DNA.[43] It is the rare twenty-three-year-old postdoctoral fellow who rolls the academic dice in this fashion and wins. In real time, of course, the NRC administrators had no idea how events would unfold. They considered Watson's actions to be little more than a shabby short-circuiting of contractual obligations by an immature young man. Believing they needed to make an example out of him, they punished him, proving that forward-looking administrators were as rare then as they are today.

An American in Cambridge

From my first day in the lab, I knew I would not leave
Cambridge for a long time. Departing would be idiocy, for I
had immediately discovered the fun of talking to Francis Crick.

—JAMES D. WATSON[1]

ambridge was the most beautiful place Jim Watson had ever seen. He was enthralled by its brick and limestone Gothic college buildings, with their great halls, chapels, spires, and green lawns. Neither the granite halls of the University of Chicago and Indiana University, nor the palm trees of Caltech and woodsy waterfront of Cold Spring Harbor, had prepared him for the perfection that was now his to enjoy simply by showing up for academic duty. It was in Cambridge, after all, where Watson found his beautiful explanation of all living things.

Having successfully escaped cold, gray Copenhagen, Watson had no intention of working on the biochemistry of viruses with Roy Markham at the Molteno Institute. John Kendrew had already sent word to Max Perutz from Ann Arbor about the impending arrival of Salvador Luria's protégé, describing Watson as a bright young man who could contribute an extra pair of hands to their Medical Research Council Biophysics Unit. The biological structure Perutz and Kendrew sought to map was hemoglobin, the protein in red blood cells that carries oxygen from the lungs to the most distant organs and tissues of the body. The two biophysicists were also studying myoglobin, a simpler but similar iron- and oxygen-binding molecule in the muscles of most vertebrates and almost every mammal.[2]

Perutz's short stature, bald dome, thick glasses, and Austrian-accented

English made him seem much older than his thirty-seven years. His gentle demeanor and kindness masked an acute hypochondria and several odd phobias, such as avoiding candlelit restaurants, unripe bananas, and mineral water as threats to his well-being. The scion of a wealthy Jewish family that had earned its fortune by introducing mechanical looms and spinning machines into the Viennese textile industry, Perutz

Max Perutz, c. 1962.

had enjoyed a privileged youth. He matriculated into the University of Vienna in 1932, where, instead of studying law as his parents desired, he "wasted five semesters in an exacting course of inorganic analysis" before becoming energized by organic chemistry and biochemistry.[3]

The fact that his parents had had him baptized into the Catholic Church would hardly have protected Perutz from Hitler's deadly anti-Semitic policies. Fortunately, he was intellectually fascinated by the discovery of vitamins by Sir Frederick Gowland Hopkins of the University of Cambridge, for which Hopkins won the 1929 Nobel Prize in Physiology or Medicine. Thus, in 1936, Perutz left Vienna for the doctoral program at Cambridge. Instead of working with Hopkins, however, he found a place studying under the charismatic J. D. Bernal and the influential Cavendish professor Sir William Lawrence Bragg. When selecting a dissertation topic, Perutz asked Bernal how he might help determine the building blocks of living cells; Bernal answered oracularly, "the secret of life lies in the structure of protein, and X-ray crystallography is the only way to solve it."[4] After the Anschluss (Hitler's invasion of Austria in March 1938), Perutz's parents fled to Switzerland. In 1939, Bragg arranged for Perutz to win a Rockefeller Foundation fellowship, which launched his academic career and allowed him to move his parents to England. Little wonder, then, that in 1981 Perutz reflectively wrote, "It was Cambridge that made me, not Vienna."[5]

OUT OF THE BLUE on a September afternoon, as Max Perutz later recalled, "a strange young man with a crew-cut and bulging eyes popped through my door and asked, without saying as much as hello, 'Can I come and work here?' "[6] Watson recalled being far more apprehensive about studying in the world's most storied physics laboratory. Perutz agreed to take him on and lent the young man a physics textbook, reassuring him that "no high-powered mathematics would be required" for the research he had in mind. He then gave Watson a modest explanation of his recent work confirming Pauling's α-helix and, inadvertently, stunned Watson by adding that it had taken him only twenty-four hours of labor to do so. Watson recalled, "I did not follow Max at all. I was even ignorant of Bragg's Law, the most basic of all crystallographic ideas." After filling his new recruit's head with incomprehensible formulae and terminology, Perutz took Watson on a walk "through King's, along the backs, and through to the Great Court of Trinity," a trek Watson would make countless times over the next year and a half. For the rest of his life, Watson regaled listeners with the same line: "I had never seen such beautiful buildings in all my life, and any hesitation I might have had about leaving my safe life as a biologist vanished."[7]

Afterward, Perutz and Watson inspected some of the "nominally depress[ing] . . . damp houses known to contain student rooms." To Watson, many of these dwellings were redolent of "the novels of Dickens," but he felt "very lucky" to find a semi-suitable room in a two-story house on Jesus Green, only a ten-minute walk to the Cavendish.[8] The following morning, he was introduced to Bragg, whom he initially dismissed as a "Colonel Blimp," an academic fossil "who spent most of his days sitting in London clubs like the Athenaeum . . . in effective retirement and would never care about genes."[9] (He was referring to the pompous, plump, jingoistic British comic strip character drawn by cartoonist David Low who was, in 1943, the source for a popular film, *The Life and Death of Colonel Blimp*.) Only after he had looked up Bragg's curriculum vitae did he realize that Bragg "is very good . . . in fact a Nobel Prize in physics."[10]

Watson met John Kendrew, Perutz's right-hand man and collaborator, after Kendrew's return from the United States. Kendrew, the son of an Oxford professor of climatology, was a first-class honors graduate of Trinity College, Cambridge, and a wing commander and Second World War hero who had helped develop radar at the Royal Air Force's Military Research Establishment. After the war's end, Kendrew resumed his studies at Cambridge and in 1949 earned his doctorate under Bragg. His thesis was on the differences between fetal and adult hemoglobin in sheep; his discoveries had a profound impact on the practice of human neonatal and pediatric medicine.[11]

Having been accepted into the Cavendish fold, Watson needed to take a short trip back to Copenhagen to collect his few possessions and inform Herman Kalckar about his "good luck in being able to become a crystallographer."[12] A few hours after bidding farewell to his former chief, Watson was back on another train, this time headed south. Bored by the monotonous landscape, he dozed off and dreamed in his second-class compartment. In 2018, when asked about this trip, he could not recall his precise thoughts but said that "there was no way I could have imagined how momentous the next eighteen months would be." At the same time, he did understand that the clock was ticking rapidly and that he had a relatively short time before his Cambridge coach would turn into a pumpkin and he would have to return to America, with or without a major discovery under his belt.[13]

Fortunately, Watson had enough of a bankroll to live on for the year, even after generously "cover[ing his] sister's recent purchase of two fashionable Paris suits," and assuming that no more money was forthcoming from the National Research Council.[14] For the first few months, he pinched his shillings to make ends meet. His rent at the Jesus Green lodging house included an adequate breakfast each morning. Still, the landlady had strict rules for her tenants that rankled Watson. She was, as he wrote to his parents on October 16, "eccentric and does not believe in any noise."[15] Despite her express orders, Watson, routinely violated her request that he remove his shoes if he came home after 9 p.m., "the hour at which her husband went to sleep." He often forgot "the injunction not to flush the toilet at similar hours and, even worse, [he] frequently went out after

10 p.m.," when all of Cambridge was closed and his "motives were suspect." After less than a week, he realized that he would not live there for
long, and within a month, the landlady "threw [him] out" for good.[16]

John Kendrew and his wife, Elizabeth—an ill-suited couple who pursued what was then politely referred to as independent lifestyles—rescued
him.[17] They gave him a room at the top of their tiny terraced house on
Tennis Court Road, directly across the street from the Downing Street
complex of science buildings, which included the Sedgwick Zoology
Laboratory, the Earth Sciences Museum, and the Molteno Institute. The
Kendrews' home was "unbelievably damp and heated only by an aged
electric heater" but they charged him "almost no rent." He later recalled
that it looked like the perfect setting for contracting tuberculosis. Yet
given his tight budget, he accepted their kind offer and moved into Tennis Court Road until his finances improved.[18]

After a few days of working in Cambridge, Watson knew "as many
people as [he] knew during his entire stay in Copenhagen." He joked to
his parents about his new-found popularity, "it is very helpful to know
the native language!" Confessing his insecurities about working in a field
where he knew "much less than anyone else," he described his state of
mind as "again the complex of a student." Fortunately, there were no
examinations, so the pressure he felt was all internal and relieved by
many trips to the library "to read until I become bored." He did find
some relaxation playing squash and tennis. He also befriended a nuclear
physicist named Denys Haigh Wilkinson, who studied the mechanisms
of bird migration and navigation, and spent many of his first weekends
in Cambridge with Wilkinson exploring the countryside and local sew-
age disposal sites looking for shore birds, a list of which he dutifully
reported to his father: "snipe, Kentish plover, golden plover and countless lapwings."[19]

☙

FOR NEARLY THREE WEEKS, Watson worked directly under John Kendrew. His chief task was to run between the Cavendish and the local
slaughterhouse to fetch heavy buckets filled with horse hearts, precari-

ously placed on shaved ice, for the extraction of myoglobin.[20] Even as a young man, it was clear that Watson was no experimentalist. He was too clumsy and impatient to carry out the dexterous maneuvers most scientific experiments require. This was especially true for biological material, for which, as the surgeon Sherwin Nuland once noted, "the gentle touch is crucial. Sensitive tissues do not respond well when handled roughly . . . living biological structures tolerate very little abuse and are quick to express their displeasure when treated with less than the consideration that Mother Nature has made them accustomed to."[21] Watson did not, and never would, acquire "the gentle touch" needed to handle "living biological structures." He frequently damaged the equine cardiac muscle so severely that Kendrew was unable to properly crystallize and visualize the specimen's molecular structure. Such ineptness proved to be another stroke of good luck for Watson. Had he been able to delicately grip the horse hearts, Kendrew might have permanently assigned him to that pursuit. Instead, Kendrew saw the futility of allowing him near the specimens, concluded that "long-slog stuff was just not his thing," and set him free to while away his days with a Cavendish pariah named Francis H. C. Crick.[22]

WHAT MAKES FOR A FRUITFUL COLLABORATION is as mysterious as what makes for a successful marriage. Watson immediately "discovered the fun" of interacting with Francis Crick. "Finding someone in Max's lab who knew that DNA was more important than proteins was real luck," he recalled, but "as long as no one nearby thought DNA was at the heart of everything, the potential personal difficulties with the King's lab kept [Crick] from moving into action with DNA."[23]

In 1988, Crick recalled first hearing of Jim Watson's arrival in Cambridge from his wife, Odile, who greeted him at the door one evening with the words, " 'Max was here with a young American he wanted you to meet and—you know what—he had no hair!" Watson's crew cut was, he wrote, "then a novelty in Cambridge. As time went on, Jim's hair got longer and longer as he took on the local coloration, though he never got so far as to sport the long hair men wore in the sixties."[24] The following day, when they actually met, he and Watson "hit it off immediately, partly because our interests were astonishingly similar and partly, I suspect, because a certain arrogance, a ruthlessness, and an impatience with sloppy thinking came naturally to both of us."[25]

The two scientists eschewed all "good manners," which Crick defined as the "poison of all good collaboration in science"; instead, they chose complete candor and, if necessary, rude responses to any idea or solution that either adjudged to be twaddle.[26] Crick's insider perspective meshed well with the young American's outsider view. Because Crick was so warm, brilliant, silly, and cynical toward authority of all shapes and sizes, he represented for Watson an infinite source of joy. Watson told Max Delbrück that Crick was "no doubt the brightest person I have ever worked with and the nearest approach to Pauling I have ever seen—in fact he looks considerably like Pauling. He never stops talking or thinking and since I spend much of my spare time in his house (he has a very charming French wife who is an excellent cook) I find myself in a state of suspended stimulation."[27]

Perutz and Kendrew carried out much of their work in a tiny office on the ground floor of the Austin Wing, with a desk squashed into the corner of their anteroom for Crick. Shortly after Watson joined the group in the fall of 1951, a room on the floor above, number 103, became available. Bragg suggested that Perutz exile Crick to the cubicle, as the biochemist Erwin Chargaff later posited, in a "futile attempt to escape [Crick's] armor-piercing voice and laughter."[28] Crick recalled the matter more graciously: "One day, Max and John, rubbing their hands together, announced that they were going to give it to Jim and me 'so that you can talk to each other without disturbing the rest of us.'"[29] The room was adjacent to the stairwell leading down to the building's exit. Like many of the other rooms in the Austin Wing, number 103 was a stark 20-by-18-foot rectangle with a 13-foot-high ceiling. It had whitewashed brick walls "over which [ran] a few broad wooden laths to one of which the first diagrams of DNA was pinned. Two large metal-framed windows look[ed] east into a clutter of other buildings."[30]

In their newly-assigned lair, Crick offered Watson a series of tutorials on X-ray crystallography. He was an excellent teacher. On November 4, 1951, Watson wrote to his parents that the subject was "really not as difficult as it looks and I alternate between reading and doing some fairly routine biochemical work." He added that Crick acted as a brake when his brainstorming took a wrong turn, and that his new friend encouraged him to "become more well-rounded" in his reading habits. He closed his letter with love, saying, "I'm in a lab where many exciting events are now happening and which I feel should have an important consequence in the biologist's way of thinking."[31]

The relationship was hardly one-sided, with Watson encouraging Crick how to master the biology of living molecules. As Anne Sayre noted, Watson also had a knack "for keeping Crick's mind on the problem at hand." Crick was a scientific volcano, constantly erupting with stunning ideas and concepts, but at this point in his career he had not yet developed "the kind of doggedness that led him to settle down and pursue any one of them to a neat conclusion. . . . Jim nagged Francis, and it helped."[32] But, as Crick pointed out years later to the author Isaac Asimov, the collaboration was far more nuanced than it appeared at first glance: "There's a myth that exists that Jim was the biologist and I was the crystallographer and that just won't stand up to critical examination. We both did it together and switched roles and criticized each other, which gave us a great advantage over the other people trying to solve it."[33] The most important aspect of their work together, Watson recalled, was that "with me around the lab always wanting to talk about genes, Francis no longer kept his thoughts about DNA in a back recess of his brain." As a result, almost all of their discussions focused on the structure of genes and DNA.[34] But at this point it was still just talk. Since his arrival in Cambridge in the fall of 1951, Jim Watson had been gnashing his teeth in frustration over the division of labor between King's College, London, and the Cavendish and the wide acceptance of DNA as Maurice Wilkins's "personal property."[35]

One afternoon, while walking around the Great Court of Trinity College, Crick suddenly figured out what they needed to do in order

Watson and Crick walking along the backs of King's and Clare colleges, 1952.

to both solve the puzzle and avoid poaching on Wilkins's laboratory work. It was really quite straightforward, he impressed upon Watson: they would "imitate Linus Pauling and beat him at his own game." They would build a stochastic model of DNA, using deduction and the process of thoughtful elimination. Crick taught Watson "that Pauling's accomplishment was a product of common sense, not the result of complicated mathematical reasoning." They only needed to rely upon the "simple laws" of quantum and structural chemistry. Staring at X-ray diffraction patterns was all well and good, both Watson and Crick agreed, but "the essential trick, instead, was to ask which atoms like to sit next to each other."[36]

Most physicists relied upon pencil and paper or a blackboard to produce their theoretical musings. But Crick and Watson's "main working tools were a set of molecular models superficially resembling the toys of preschool children." Their task was to "play" with these molecular models and "with luck, the structure would be a helix. Any other type of configuration would be much more complicated. Worrying about complications before ruling out the possibility that the answer was simple

would have been damned foolishness."[37] There was just one major problem with this approach: they needed, at least, some X-ray data to confirm
their theoretical notions of "which atoms like to sit next to each other."
At this point in time, however, neither of them had access to the exciting
work unfolding at King's College, London.

IN THE FIRST WEEK OF NOVEMBER, Crick invited Maurice Wilkins to
Cambridge for a weekend of dining, gossip, and relaxation. The real
purpose was to seek his blessing for their building out a model of DNA.
The star attraction was a Sunday joint of roast beef, expertly prepared by
Odile with the perfect amount of garlic, thyme, salt, and pepper, served
with plates of boiled potatoes simmered in butter, mint, and chives, a
fine Yorkshire pudding, and, because it was England, a heaping bowl of
overcooked, mushy peas. Before Crick could carve the first slice of meat,
Wilkins pessimistically informed them that Pauling's "model building
game" would never solve the structural puzzle of DNA. Stuck on the
misconception that DNA was a coiled helix of three polynucleotide
chains, Wilkins insisted that much more X-ray crystallographic analysis
was needed before a useful model could be built.[38]

Unfortunately, gathering this data depended entirely upon the cooperation of an "easily upset and hostile" Rosalind Franklin. Wilkins pejo-
ratively referred to her as "Rosy," a nickname she despised and which
Watson and Crick eagerly adopted. Throughout the dinner, Wilkins
bemoaned that his relationship with Franklin was growing worse daily.
He complained that she had made him hand over to her "all the good
crystalline DNA"—the treasured Signer samples he had obtained from
Bern—and that the poorer DNA samples he was forced to use simply
"did not crystallize." Worse, she demanded that she be the only one
allowed to use the King's X-ray machine on DNA—a "bad bargain."
John Randall had too readily acceded in order to calm the situation and
keep her out of his office.[39]

There was one bright prospect offered by Wilkins: Watson and Crick
might be able to grab a glance at Franklin's X-ray diffraction pictures a

few weeks hence, on November 21, when she was scheduled to give a seminar at King's on the progress of her research. Crick quickly realized that attending the seminar was essential, because "Jim and I never did any experimental work on DNA, though we talked endlessly on the problem."[40] As Watson recalled, "the crux of the matter was whether Rosy's new X-ray pictures would lend any support for a helical DNA structure."[41]

Crick announced that he was already committed to be in Oxford the afternoon following Frankin's seminar, for an important meeting with Dorothy Crowfoot Hodgkin, the crystallographer whose solving of the molecular structures of vitamin B12, penicillin, and insulin would win her the 1964 Nobel Prize in Chemistry.[42] Though this did not preclude him attending Franklin's presentation, Crick told Wilkins he preferred to focus on preparing for the meeting with Hodgkin, which was to discuss his new paper on helical theory. As a result, and with Wilkins's assent, Watson would attend it alone, then accompany Crick to Oxford so that he could brief Crick about Franklin's work during the hour-long train journey from London.

In the days leading up to Franklin's seminar, Watson pored over his crystallography textbooks and notebooks, redoubling his efforts to understand the complex physics she would, undoubtedly, reference. As he later recalled, with characteristic determination and competitiveness, "I did not want Rosy to speak over my head."[43]

The King's War

What terrible things were done to Rosalind Franklin? She was provided with the best DNA, exclusively. She was provided with Gosling as a research student. She was provided with the Ehrenberg fine-focus X-ray tube exclusively. When she wanted a special camera built in the workshops, that was done by a very good man. She was provided everything—except the right to have lunch in the lunchroom! If there had been any impediment to her work, that would have been something to regret. If you wanted to put the matter the other way around—she refused to join the effort of the group; she took the best facilities, then hogged the problem.

—MAURICE WILKINS[1]

The first two years at King's College were troubled by petty rivalries and jealousies that [Rosalind] felt acutely. Her mind was clear, incisive and quick thinking, and her methods and conclusion often unconventional and original. Like most pioneers of thought she met opposition and when, as often happened, she could not persuade her colleagues to follow at her pace she was apt to become impatient and despondent.

—MURIEL FRANKLIN[2]

Many versions of the fraught Franklin–Wilkins relationship place the blame on Franklin. She was too aggressive. She was too proprietary with her research. She was too independent. She was too stubborn. She was too antagonizing. She was too feminine, or not feminine enough. She was too prickly and difficult. She was too unwilling to work with others. She was too focused on obtaining hard data to substantiate theory, rather than the other way around. She was too upper-class and condescending. Even more offensive in an Anglican nation with only 400,000 Jews (a mere 0.8% of the population), she was too Jewish.[3] On and on, the "She was *too* . . ." arguments have creeped along, depending on the wounded perspective of the teller of the tale.

Since Rosalind Franklin's untimely death, several unsubstantiated rumors of a misdirected romance with Wilkins have circulated. In 1975, John Kendrew offered a stream of left-handed compliments about her physical appearance ("I would describe her as attractive rather than unattractive. She didn't dress all that badly. Jim is quite wrong about that"), her "tough" intellectual rigor ("if she thought somebody was talking nonsense she would say it, in even more direct terms than Francis does"), and the fact that he never found her difficult ("but then, of course, I wasn't working with her, so that doesn't prove anything; but . . . I always found her very easy to get on with, and a very pleasant person"). What was especially out of character for this tight-lipped, cautious man was his "private theory . . . I always supposed that Rosalind made a pass at Maurice, and Maurice didn't respond . . . and that was the origin of the trouble." Admitting this was purely speculation on his part and that things could have happened the "other way around, or not happened at all," Kendrew maintained that "the difficulty they had was something deeper in human terms than the problems of working with one another; that there was some emotional something which had happened, or had failed to happen."[4]

Francis Crick, too, explained the problem as one of unrequited love, but the other way around. Wilkins, he observed, spoke about her "constantly" and was obsessed with her. "We all think Maurice was in love with her . . . And Rosalind really hated him . . . either because he was stupid, which was a thing which always annoyed her, or else something happened between them . . . [There was] a big love-hate thing, very strong."[5]

Geoffrey Brown and Raymond Gosling also found it difficult to

ignore her allure. To Brown, Franklin was "beautiful, like a goddess."[6]
Gosling, too, was mesmerized by her looks: "She had quite a good figure but was definitely rather more thin than rounded shall we say . . .
[she was] often really very beautiful, especially when she was excited or
angry." Gosling was charmed by the quirks of her nature as well. Under-
neath her "professional shell," he insisted, she was "a delightful, relaxed
human being . . . [but] she was no run-of-the-mill ordinary bird . . . [she
was] slightly eccentric in the sense that she wasn't average. She didn't
behave in the way that average people behave . . . she was a very intense
person, to the point of being eccentrically so. And sometime[s] I must
confess, I think that she was a very attractive person who would have
quite liked to have been less eccentric and less determinedly academic
and have done [the] sorts of things like Watson and Crick. She didn't
have a great deal of small talk. She was a pretty determined lady."[7] Gosling also followed her "very full social life. I mean, I know for a fact
that at one stage, I think, she was going out with the first violin of the
London Philharmonic. Now, that is a cut above the beer-drinking chaps
like us who were sitting in Finch's [a local pub]."[8] More to the point,
Gosling, like Crick, "always supposed that Wilkins was very attracted by
Rosalind and had sometimes suspected that Rosalind was attracted by
Wilkins and that their mutual animosity had something to do with this
supposed mutual attraction."[9]

In the decades after Franklin's death, Maurice Wilkins did little
to quell the rumors of his romantic feelings for her. In 1970, Wilkins
recalled his first impression of Franklin as being bright and intense and
"of course, she was rather good-looking, you know."[10] Six years later,
in 1976, while previewing and heavily marking up the manuscript of
Horace Freeland Judson's book *The Eighth Day of Creation*, Wilkins
objected to Judson's description of Franklin's nose as "fleshy." In broad,
black fountain pen strokes, he admonished Judson, "Rosalind Franklin's
nose was not fleshy! She was a handsome girl."[11]

In recent years, however, not every historian has subscribed to these
theories of attraction. Franklin's biographer Brenda Maddox believed
that Franklin was unaware of or uninterested in Wilkins's attentions,

preferring safer, nonsexual relationships with married men, such as her
Parisian colleague Jacques Mering, or men much younger than her, like
Gosling and Brown. Wilkins never had a chance, Maddox thought,
because Franklin only respected men who were determined and brilliant.[12] More pointedly, Jenifer Glynn, Rosalind's younger sister, characterized these gossipy notions of *amour fou* between Franklin and Wilkins
as the "stupidest explanations I have ever heard."[13]

WHEN ROSALIND FRANKLIN FIRST REPORTED to the King's Biophysics Unit on January 8, 1951, she met only with Randall, Gosling, Alec
Stokes—a theoretical physicist who worked with Wilkins—and Louise
Heller, a graduate medical physicist from Syracuse University who was
volunteering at the lab. Wilkins was still away on holiday, hiking the
hills of Wales. From his eyewitness point of view, Raymond Gosling
explained: "Had Maurice been in the lab, he would have been at that
meeting. All sorts of things might have gone differently." As it did occur,
Franklin was naturally nervous and made sure to ask her new boss a few
appropriate questions regarding the research she was expected to do.
Randall basically told her, "here were the X-ray photographs, there are
lots of spots on them, get some more, and solve the structure of DNA
from the X-ray-diffraction pattern. And that's what Rosalind from then
on tried to do."[14]

Franklin's first task was to purchase the radiographic apparatus
needed to obtain the best possible X-ray photographs. From her time
at the British Coal Utilization Research Association and in Paris, she
knew which manufacturers to approach and who to avoid, what were
the best prices, how to adjust the specifications on the devices available,
and, because DNA was so difficult to properly photograph, what other
adjustments might need to be made in order to meet her research needs.

Several months earlier, Wilkins realized that the equipment they had
on hand at King's was "unsuitable for the particular work [they] needed
to do." The X-ray machine he initially used was on loan from the Admiralty, and they wanted it back. Wilkins was more than eager to be rid

of the antiquated apparatus, which was too clumsy and big for the delicate DNA fibers he was spinning. He considered purchasing a "rotating anode X-ray generator to give a very powerful beam," but, soon after visiting the Birkbeck College crystallography laboratory, he became interested in a "new fine-focus X-ray tube" designed by Werner Ehrenberg and Walter Spear which concentrated X-rays on very small specimens. With this device, Wilkins found he could better control the ambient humidity and photograph solitary DNA fibers only one-tenth of a millimeter in width. Instead of selling a manufactured tube to Wilkins, Ehrenberg generously donated his prototype.

Franklin liked the Ehrenberg tube, too, and proceeded to draw a design for a tiny, tilting camera, along with a vacuum pump to keep air out of the photographic field, which the King's physics workshop promptly fabricated. Gosling and Franklin then figured out how to attach a brass collimator for the introduction of a pinpoint beam of X-rays into the camera, with the specimen precariously mounted at its center. "The only unusual feature," Gosling wrote, "was the condom which I had carefully fitted to the brass collimator to reduce the loss of hydrogen" that was passed through the camera to keep the humidity level consistent so that the DNA specimen would not dry out. There is no record of what Franklin thought of this rubbery feature, which dated to several months before her arrival, when Gosling and Wilkins were fiddling with their apparatus. One afternoon, while struggling to stop the humidity loss problem, Wilkins had pulled out from his pocket a packet of Durex condoms—a popular British brand, whose trade name, coined in 1929, was an amalgam of *du*rability, *r*eliability, and *ex*cellence—and suggested to Gosling, "Try this."[15] The condoms minimized an uneven spread of air across the photographic field, thus preventing a foggy appearance on the film over the long periods needed to get "a reasonable diffraction pattern from such a weakly-scattering specimen."[16]

Wilkins later claimed to have had a positive relationship with Franklin during their early days of working together. Upon returning from his holiday, his first order of business was to meet with her so "that Rosalind be put in the picture as soon as possible." In his memoir, he did not recall declaring that she was his assistant, which, according to others, so rankled her and surprised those working at King's. Instead, he described a far more independent arrangement. He assigned her space in the basement of the building, "somewhat separate from our new main lab," so that she and Gosling "would have peace and quiet there for making the laborious calculations, involving a system of cards, that were necessary, in those pre-computer days, for turning the patterns on X-ray diffraction photographs into the three-dimensional structures of molecules."[17]

Early in Franklin's stay at King's college, Wilkins visited her in her "poky little room," with her desk arranged so that her slender back was the first thing a visitor saw. When she turned around, Wilkins was surprised to find "that she was quietly handsome with steady, watchful, dark eyes." As the conversation inevitably segued to their research interests, he adjudged her to know "what she was talking about." When she stood up, he was surprised to discover that she was shorter in stature than he had imagined given her "authoritative" and confident demeanor.[18] He also recalled being baffled by her placement of a "small mirror on the wall so that it faced her as she sat at her desk. It was too small to allow her to see who was behind her at the office door, and at the time I wondered whether she was anxious about her appearance." Before long, he found her intense, brusque manner to be unsettling, such as when she waved him into her office without looking up and gestured to him to sit down while she completed a task—an action he misinterpreted as bad manners. Clumsy encounters aside, he later claimed that he was initially "confident that she would make a good colleague."[19]

During her first weeks at King's, Franklin worked on completing a few papers generated from her Paris laboratory research. On Saturdays, several of the unmarried staffers worked a half day in the laboratory and then ate lunch at the Strand Palace Hotel, which was just down the street from King's College, toward Trafalgar Square. Often there were several physicists at the table, and on a few occasions, according to Wilkins, "Rosalind and I were the only two." In his telling, they discussed "various topics other than science," including politics, the threat of nuclear war"—an issue Wilkins was very concerned about, given his service during the Second World War—and neutralism, a political phi-losophy suggesting that Britain could become a noncombatant in the Cold War, which held great appeal for Franklin and which Wilkins dismissed out of hand as "clearly silly." He recalled finding her pleasant to talk to even if "occasionally a small spikiness would show," such as when he remarked how much he was enjoying a dish of fruit and cream and she "replied coldly, 'But it was not real cream.'"[20] As their interactions progressed, he may have hoped for a deeper friendship with Franklin, but claimed not to have desired a romance with her because he was more attracted to "shy young women."[21] Wilkins's descriptions of these intimate lunches do, however, come spiced with a soupçon of questionability. Sylvia Jackson, another physicist at King's, later recalled how the King's physicists—Wilkins, Jean Hanson, Alec Stokes, Angela Brown, Willy Seeds, and herself—went to the nearby Strand Palace each Saturday, or occasionally "to a pub in Epping by car." Rosalind "did not basically take part in that sort of thing. Too intent . . . She was absolutely dedicated, a tremendously hard worker; she strode along rather quickly; she was enormously friendly if you gave her half a chance. But I found her *formidable*" (emphasis in original).[22]

FRANKLIN'S FRIEND AND BIOGRAPHER Anne Sayre has famously argued that a major slight Franklin endured at King's College was not being allowed to have lunch or tea in the senior common room, which was restricted to men only. She and the many other female workers at the Randall lab ate in a smaller, gender-integrated dining room down the hall, where most of the younger staff members gathered for their noonday meals and tea breaks.[23] Raymond Gosling reports that the senior men, including Wilkins and Randall, preferred the male-only common room because it had the advantages of being "bigger [and] there was faster service."[24] Franklin, who was sensitive to such rebuffs given her minority status in both gender and religion, objected to this arrangement. It represented one more barrier in a field so dominated by men that some have called it "a scientific priesthood."[25] For years, Wilkins could not or would not understand her side of the boulevard with respect to this point. When accused of anti-feminism (or anti-Semitism), he heatedly denied any bigotry and added that he would be "horrified to think otherwise."[26]

Honor Fell, Crick's old boss at the Strangeways Laboratory in Cambridge and a Senior Biological Advisor to the King's College MRC biophysics research program, made weekly visits to the London lab. She was oddly dismissive of the war between Franklin and Wilkins: "I *do* know a great deal about that unit. And I never saw a sign of sex discrimination. Of course, Franklin was a rather difficult character . . . They were *both* rather difficult people. But I don't believe she was discriminated against because she was a woman. I never saw any sign of it—and I'm sure I would have done . . . There was always this rumpus between them, a chronic row between them . . . I would say that if Dr. Franklin suffered any injustice, it was as a person and not because she was a woman. She and Wilkins were just totally incompatible temperamentally" (emphases in original).[27]

The work environment for women at King's was more nuanced than reported by either Sayre or Fell. Sylvia Jackson insisted that "the place was if anything more open for women than any other British research institution."[28] When Franklin began her work there in early 1951, nine of the thirty-one scientists working in the Biophysics Unit were women, an astounding number considering how few female physicists there were in England in the 1950s. Between 1970 and 1975, the journalist Horace Judson took great pains to interview or correspond with almost all of these women and concluded "those of Franklin's colleagues at King's who were women unanimously reject the view that her troubles there arose because she was shut out as a woman." He acknowledged that they spoke with him "decades after the events" and that they understood the systemic barriers they faced that men did not. "Nonetheless," he wrote, "they reject as unhistoric and anachronistic the use of Rosalind Franklin as an emblem for the condition of women in science."[29] Many women today, when reading such a glib dismissal of these hurdles, might shake their heads in disbelief. Although there was value in the fact that women could earn a degree or work in a laboratory at King's, Sayre (and many others) insisted that it remained very much a patriarchal world: "Rosalind was not a man. She was unused to purdah and often it offended her."[30]

A far more metastatic symptom of the misogyny at King's was the barrage of name-calling and practical jokes Franklin endured—goings-on that today would be condemned as harassment. The chief instigator of this was Willy Seeds, a sarcastic, overweight Dubliner who was developing new methods of reflecting microscopy and ultraviolet microspectography for the study of nucleic acids and nucleoproteins.[31] He came up with nicknames for lab members that bordered on the malicious and had an uncanny power to stick, especially when the individual protested.[32] It was Seeds who christened Franklin "Rosy," an appellation no one dared use to her face.[33] Years later, over lunch with Crick at the Eagle, Franklin met Dorothy Raacke, an American marine biologist. Raacke politely asked Franklin how she would like to be addressed. " 'I'm afraid it will have to be Rosalind,' was the reply, pronounced in two quick syllables; then with eyes flashing, 'Most definitely, *not* Rosy'."[34]

Her prickliness and hypersensitivity to such jokey ploys hardly helped

King's College Biophysics Unit interdepartmental cricket match, c. 1951. Left to right: John Randall (in top hat) and his wife, Doris, Ray Gosling (back to camera), an unidentified woman, Maurice Wilkins, and (in top hat) Willy Seeds.

matters. Late in his life, Seeds defended himself with the explanation that he coined nicknames for virtually every staff member in the lab in order to "demystify superiors" and work around the custom of calling men by their surnames and prefacing a woman's surname with "Mrs." or "Miss."[35] Wilkins also adopted the use of the name "Rosie" (as he usually spelled it), a tacit approval of Seeds's behavior given his status as assistant director of the unit. Yet, in 1970, he claimed that he never referred to Rosalind as Rosy or Rosie: "I'm not one myself for nicknames."[36]

The animosity between Seeds and Franklin went deeper than mere name-calling. The two fought bitterly over the tilting camera she designed for her research; she antagonized Seeds by shooting down his suggestions on how best to construct the device. On the spurious grounds that she was wasting the scarce resources of the lab's work-shop, Seeds decided to wage a low-level war with her. One evening, hav-ing discovered that each night she covered her delicate apparatus with a heavy black oilcloth, he snuck into her workspace and hung a sign say-ing "Rosy's Parlour," suggesting a "hint of the gypsy, the alien and the occult show[ing] the swart associations that Rosalind stirred at King's." Predictably and understandably, Franklin exploded and called them all "little schoolboys." Just as predictably, the "little schoolboys," including Wilkins, laughed at her reaction.[37]

FRIENDS OF FRANKLIN AND WILKINS have described each as shy and prone to melancholia when things did not go their way; it was just that their shy, sad manners manifested themselves in ways that confounded the other. Wilkins responded to strong personalities with withdrawal and silence, refusing to communicate with those he deemed rude, insulting, or threatening. Franklin tended to withdraw from those who slighted her, but more often countered her adversaries with a brash front that frightened off a meek fellow like Wilkins. Unlike Wilkins, who avoided eye contact with others, Franklin looked directly into the eyes of those she spoke to, aiming her human beam of vision as unerringly as she pointed her X-ray machine at a DNA specimen.[38] When describing the "hot and heavy form of disputation which many scientists enjoy," Anne Sayre observed that "Rosalind enjoyed it, and found it useful. Wilkins disliked it very much."[39] Raymond Gosling agreed: "Rosalind had passion. Rosalind had temperament, and nobody was expected to have temperament in King's . . . while Maurice was always very careful not to show any emotion whatsoever, and therefore they couldn't have been more different."[40]

Some have suggested that Franklin's vigorous verbal attacks on Wilkins were a remnant of her time in Paris, where laboratory disputes took on the contours of grand opera, or that they might have been a function of their divergent ethnic backgrounds—the *sturm and drang* of the Franklin family's evening discourses clashing with the stiff-upper-lip conduct of the Wilkins clan. In 2018, Franklin's sister, Jenifer Glynn, offered still another reason for her stormy relationship with Wilkins: "Rosalind had great patience in explaining something to someone who didn't know about a particular topic. But she became quite cross when having to explain something to someone she felt ought to know about it."[41] Sayre was even more blunt: Franklin's respect "was never, all things considered, very difficult to achieve, but it was also not difficult to forfeit."[42]

Mary Fraser, a biophysicist at King's College and the wife of biophysicist Bruce Fraser, captured the Franklin–Wilkins problem in a 1978 letter:

> Now Rosalind Franklin arrived and I suppose we assumed she would fit into the casual role of relaxing amidst the beakers, balances, centrifuges, and Petri dishes—but she didn't. Rosalind didn't seem to want to mix—no one was particularly worried, after all it was her decision and everyone respected an individual's right to be different. If one was unkind about Rosalind Franklin, they would say she hated her fellow human beings (male and female). Her manner and speech was rather brusque and everyone automatically switched-off, clammed-up, and obviously never got to know

her. She couldn't be bothered with social chit-chat—it was a bore and a waste of time. Also, we poor mortals with feet of clay who loved it!!

Now why did Rosalind Franklin and Wilkins get on so badly? Here was Maurice Wilkins, tall, quiet, gentle, a brilliant experimentalist who normally would never quarrel although he could be stubborn. Here was a man with X-ray pictures taken of his prepared specimens of DNA and he knew they were brilliant. He looked at the X-ray photographs and he knew all the evidence was there and he wanted to know the structure of DNA and yet he felt utterly frustrated by his own lack of knowledge of the mathematical skills to work out that structure.

Here was Rosalind Franklin, a dedicated scientist, already she had done good work, and she looked at the photographs and knew the answer would not come easily (alas no help from computer analysis in those days)—ahead lay months of dreary mathematics and model building. Perhaps she felt afraid she would be pressured by Wilkins' impatience and she fought for the right to do things in her own time, and in her own way. In more ways than one, perhaps she felt trapped in a nightmare of spiraling DNA molecules and she spontaneously erupted every time she and Wilkins tried to discuss the problem. Rosalind was too obsessive and took everything too personally—if she had suggested to Wilkins that help was needed with the problem it would have been alright but she didn't want any help.

Wilkins was in a dilemma and he sought help from other fellow scientists. Can we blame him? Dedicated people like Rosalind Franklin (the sex is irrelevant), great artists, scientists, writers, mountaineers, sportsmen, or the Florence Nightingales have an obsession and their fellow human beings are secondary to their overriding passion. They are impossible people to live with but nevertheless make the pages of history of the human race. Rosalind Franklin just failed to make the pages of history in scientific achievement but she made a lot of ripples in her short life.[43]

Marjorie M'Ewen, another physicist working in the lab, went a step further: "I fear Rosalind had a personality problem. I have never known
anyone who was so totally devoid of humour, had this not been so I
think a number of minor [disagreements] could have been very easily
resolved." Still, M'Ewen expressed regret over how Rosalind's memory was defamed and deformed in both *The Double Helix* and *Rosalind
Franklin and DNA*. She closed by cursing "a plague upon the Watson's
and the Sayre's of this world!"[44]

Eventually, Wilkins took to asking Raymond Gosling what he might
do to improve his poor relations with Franklin. A sympathetic Gosling
recalled, "She and Maurice were temperamentally very far apart, and so
they remained far apart, and things went from bad to worse in a personal
sort of way . . . Rosalind did not like discussing the work with Wilkins,
but I think, to give him his due, that he tried awfully hard to discuss
with her."[45] Wilkins even took Gosling's juvenile advice to present Franklin with a peace offering in the form of a box of chocolates, which she
spurned as a tired gesture from a "middle-class" bloke.[46]

FRANKLIN'S DISDAIN REACHED a boiling point in early May 1951. Wilkins
was having difficulty hydrating the Signer DNA fibers in order to mimic
the conditions of DNA in living organisms. When removed from cells,
DNA tends not to take up water easily, and merely soaking the specimens in a water bath does little to hydrate them. For months on end,
Wilkins could not swell the fibers "beyond about 20 or 30%"—a short-
coming, as he later explained, that led him "in the wrong direction . . .
and I began to push the idea that DNA consisted of only one helical
chain (which would have been sufficient to contain all the genes)."[47]

One morning, Franklin entered the lab armed for battle.[48] As Wilkins
hunched over his bench, carefully spinning out a long, fibrous specimen
with a glass rod, she stood over him shaking her head in disapproval.
She examined the water bath he used to hydrate his DNA fibers and
offered a solution: bubble hydrogen gas into the camera chamber and
make certain that the gas "was thoroughly humidified by salt solutions."

Wilkins objected, fearing that a salty "spray might be carried over to the fibers" and thus cause artifactual, rather than biological, changes to the DNA. With few words and less patience, Franklin deftly demonstrated how to avoid such contamination.[49] In her subsequent work, she perfected a means of controlling the hydration of a DNA specimen by placing it over specific drying agents, which would pull the water out of it, and then reversing that process by increasing the humidity. This technique had the added advantage of allowing her to use the same specimen many times.[50]

Instead of thanking Franklin for her advice, Wilkins fumed over being shown up in front of the entire laboratory by a woman, no less. For nearly twenty years, he griped that there was "nothing original, nothing inventive, in Rosalind's suggestion."[51] Worse, to his mind, "she was very superior about this. Her attitude was always superior. But it was simply an accident that it worked, and it was this, this accidental thing, which represented her contribution."[52] In 2003, he finally admitted that her discovery was hardly accidental and that Franklin, who was accomplished in physical chemistry techniques, "knew the best salts to use for various humidities, and in spite of difficulties from [the] spray she was right to urge [the] use of salts."[53] Regardless of the version he told, Wilkins forfeited Franklin's respect that morning—and he knew it.

THE HOSTILITIES INTENSIFIED in July 1951, at a "protein seminar" organized by Max Perutz in Cambridge (some three months before Watson arrived to work there). Delighted to be lecturing in the hall where Ernest Rutherford had professed so many of his discoveries, Wilkins recounted his work at the Naples Zoological Station and reported that his latest X-ray patterns included an easily discernible central "X" or "cross-ways."[54] Neither Wilkins nor those attending the seminar that summer day were able to interpret his research findings, but all were beguiled nonetheless—except Franklin, who became increasingly irritated. Once the lecture was over, she quickly exited and waited outside the hall to tell him, in no uncertain terms, that the X-ray work was *her* domain: "Go back to your

microscopes!" (Wilkins and Willy Seeds were using ultraviolet microscopy, an alternative method for analyzing DNA samples, which did not
involve X-ray diffraction.)[55] In 2003, Wilkins recalled this humiliating
moment as if it had happened only a day earlier. He was "shocked and
bewildered" by what he perceived to be an outrageous command just
when he "reported encouraging progress. Why should she want me to
stop? What right had she to tell me what I should do? Could she not see
that the new progress would contribute to her work as well as mine?" To
Wilkins's credit, he excused himself from the verbal skirmish, hoping
that "the crisis might simply blow away and things might become normal again. But that was not to be."[56] Yet Franklin's reaction is understandable in light of John Randall's letter accepting her at King's—which
clearly gave her the assignment of leading the X-ray diffraction work on
nucleic acids.[57] That said, Randall allegedly failed to make this assignment clear to Wilkins.

A few hours after the confrontation, Wilkins was almost literally pole-
axed, when Geoffrey and Angela Brown invited him to join them punting on the Cam River. Franklin and a few others were in a separate boat.
At one point, Wilkins, who was in a recumbent position, saw the other

Punting on the Cam River, alongside King's and Clare colleges, Cambridge.

boat approaching at a rapid clip and Franklin holding her punt pole high in the air, "bearing down on them in what he saw as a menacing way." Wilkins exclaimed, "Now she's trying to drown me!" Everyone laughed at the ridiculous situation—everyone, that is, except for Wilkins and the person holding the punt pole.[58]

Once back in London, Wilkins confessed his distress to a "most helpful" Jungian psychotherapist, who suggested a peace accord. By this time their "routine Saturday lunches at the Strand Palace Hotel had petered out" and the therapist encouraged Wilkins to invite Franklin to a reconciliatory dinner. After looking for her without much success, he finally found her sprawled out on the floor of her lab dressed in a smudged lab coat, "busy fitting together the electrical wiring for the Ehrenburg fine-focus X-ray tube." After dousing her X-ray setup—with benzene to rid the pumps of the sludge and vacuum grease they collected in use—Franklin gave Wilkins her full attention. Wilkins recalled that she "seemed quite willing to talk." But because the work was so hard, the ambient temperature so hot, and the atmosphere in the lab "so close," Wilkins could not help but be repelled by her body odor. He irrationally objected to the very idea of "sitting down to dinner" with Franklin in her present smelly state. Although he respected her hands-on approach to her research, he was unable to muster the will to enjoy a cordial supper with her and he simply "drifted away."[59] From the distance of more than a half century, it is hard to believe that this most consequential of scientific squabbles might have been smoothed over if only Wilkins had a spare stick of Old Spice.

This malodorous interaction did not prevent Wilkins from continuing to attempt to find a détente of sorts. Shortly after the Cambridge seminar, he wrote to Franklin suggesting a number of different approaches to the helical problem, the use of Patterson equations, and some "promising" new developments. He ended his note with a most cordial "Hope you had a good holiday. M.W."[60] Sadly, none of his ploys for peace worked, and the row between them festered.

Nearly two decades later, in a 1970 interview with Anne Sayre, Wilkins tried to write off the problem as merely "an incompatibility of personality." Unable to leave it at that, however, he soon descended into a sorrowful tone. "I don't see why it isn't possible to have civil, civilized conversations. I don't see that this is too much to ask," he said plaintively. Franklin was "very fierce" and "simply denounced people's ideas, and this made it quite impossible as far as I was concerned to have a civil conversation." He had little choice but "to walk away" and put aside his DNA work until she left the laboratory for good. "You may say it was [Prime Minister Neville] Chamberlain all over, peace at any price, but there are things I don't think I have to face. All that fierceness, scowling—and she could be very abrupt. There was no possibility whatever of discussion."[61]

Sayre believed that Wilkins hated Franklin "passionately, with [a] corked up passion . . . characteristic of his inhibited nature. He hates her as much as if she were alive today, and working next door to him daily, and frustrating him at every turn . . . few hatreds linger with this much force for more than a decade beyond the grave."[62] Hate is a strong word and a complex emotion, especially if intermingled with amorous feelings gone awry. What is obvious is that Franklin never really left Wilkins's tortured mind. For the rest of his life, he endured the angry words of critics over how badly he treated her. As a man of decency, albeit one with blind spots when it came to Franklin and to women in general, the animosity distressed and diminished him in every way possible. He was haunted by her.

BY OCTOBER 1951, they had stopped speaking to each other, and Randall was forced to negotiate a treaty of sorts. Franklin was to continue her X-ray analyses of DNA, while Wilkins tackled his own DNA work using microscopic methods. The manner in which Randall handed down his edict made Wilkins "feel like a naughty child."[63] He even had a Freudian psychosexual dream over the situation: "In the nightmare I was a fish on a fishmonger's slab: 'Would you like a nice filet Madam? Or would you like it on the bone?' Rosalind could be terrifying."[64]

Franklin's mother claimed that even after the Randall accord Wilkins

"made her life miserable at King's."[65] A letter Franklin wrote to Adrienne
Weill on October 21, 1951, provides support for her mother's allegation:

> Dear Adrienne,
> I'm sorry I have been so silent for so long. I came back from my holiday
> to the blackest of crises in the lab, which lasted weeks and weeks, took
> all my energy and left me too fed up to write to anybody. Things have
> calmed down a little now, but I still want to get out as soon as possible,
> and think seriously of coming back to Paris if Paris will have me . . .
> I wish I could think of some way of getting back to work there [Paris]
> before next October, but I'm afraid it's impossible.
> Love from
> Rosalind[66]

The Lecture

By choice she did not emphasize her feminine qualities. Though her features were strong, she was not unattractive and might have been quite stunning had she taken even a mild interest in clothes. This she did not. There was never lipstick to contrast with her straight black hair, while at the age of thirty-one her dresses showed all the imagination of English bluestocking adolescents. So, it was quite easy to imagine her the product of an unsatisfied mother who unduly stressed the desirability of professional careers that could save bright girls from marriages to dull men. But this was not the case. Her dedicated, austere life could not be thus explained—she was the daughter of a solidly comfortable, erudite banking family. Clearly Rosy had to go or be put in her place. The former was obviously preferable because, given her belligerent moods, it would be very difficult for Maurice to maintain a dominant position that would allow him to think unhindered about DNA.

—JAMES D. WATSON[1]

Rosalind . . . had this great zest for living. She lived with such intensity . . . Whatever she was doing she put her whole soul into it—And she took immense trouble over her clothes, and she always wore lipstick. You can well say that a sentence from the book [by Watson] sets its tone; there are many such.

—MURIEL FRANKLIN[2]

Wednesday, November 21, 1951, began like many mornings for Dr. Rosalind Franklin. She awoke early in her one-bedroom flat, situated on the fourth floor of Donavan Court, a red-brick, sandstone-trimmed, eight-story structure erected in 1930 at 107 Drayton Gardens. It was (and is) a quiet street of smart three- and four-story terraced houses and mansion blocks running north–south between Old Brompton Road and Fulham Road in the South Kensington area of London.

The apartment was "modestly" furnished with taste and flair. But to her friends who were used to living in tiny flats and sharing a bathroom with their flatmates, Franklin's new place looked "luxurious."[3] She had hesitated to sign the lease, fearing that the rent was too steep for her salary. She was eventually convinced that such false economies were silly when she had the money to spend based on "the private income [an inheritance from her grandfather's estate] she had always previously scorned."[4]

Her mother fully approved of her choice: "she furnished it to no set, conventional pattern, but with immense care." The flat had a spacious dining and living room ("with a roll-out couch"), a bedroom with a picture window, a full bathroom, and a galley kitchen. She even made her own curtains. It "[was] pretty and attractive in an entirely personal way, full of small treasures brought home from her journeys abroad. It was never empty, for during any long absence, it was invariably lent to friends." Her need for everything in her life to be just so occasionally led to "sparring" with her landlord. Given that it was Rosalind Franklin involved in such matches, she rarely lost.[5]

Franklin's domestic star shone most brightly in the kitchen. Her French-inspired specialties included rabbit or pigeon braised in red wine, roast artichokes with crispy breadcrumbs, and new potatoes simmered in butter rather than British water. All these dishes were accented with lots of olive oil, fresh herbs, shreds of hard, aged Parmesan cheese, basil, and garlic. When it came to that pungent bulb, she secreted it into the roast beef and Yorkshire pudding she cooked up for her father, who professed a hatred of garlic but never was able to detect it in his daughter's delicious Sunday dinners. On the contrary, Ellis Franklin rarely turned down a second helping.

To the many men and women who befriended her outside of the lab, she was never the brusque, harsh, intense Dr. Franklin. Instead, they universally agreed that she was the essence of grace and fun. She carefully placed her guests around the dinner table to ensure the best conversations, and she put little gifts at their table settings in order to emphasize that each guest was a special one. Her home life was a perfect example of how well she compartmentalized the varied parts of her life: the dutiful daughter, the enchanting hostess, and the cold, determined scientist. Most remarkable is how rarely her friends and family saw the complete tableau of the many roles she played so well.[6]

The morning of November 21 was chilly (49° F), blustery, and rainy during an unusually wet month; there were only eight days without measurable precipitation.[7] Franklin prepared her usual light breakfast of black tea whitened with a spot of milk and a single digestive biscuit. Then she slipped on a conservative dark skirt and buttoned up a snowy white blouse. Her graduate student Raymond Gosling recalled that she was always dressed "sensibly" rather than "attractively"—a wise choice considering that her daily work required hours of X-raying DNA, often requiring her to lie on the floor and wrestle with fussy equipment, in a musty basement lab. Even so, he thought her appearance was "charming and feminine."[8] Franklin, her mother insisted, "took a great care over her clothes, many of which she made herself. [The hems of her] skirts went up and down according to the fashion—and she was always well dressed and elegant and up to date in the style of clothes she wore."[9]

Before leaving her flat, Franklin carefully made up her face and applied lipstick. Long after James Watson published his mean and lipstick-less description of Franklin, Muriel Franklin made certain to remind others of her daughter's routine habit of presentation to the world.[10] Grabbing an umbrella from a brass stand in the hallway and putting on her raincoat, she locked the door behind her, descended in the lift, and walked down a wide hallway to the street. With each deter-mined step, one could hear the clickety-clack of her squared high heels hitting the pavement. Her claustrophobia dictated avoiding the twelve-minute trip by Underground so, instead, she took a thirty-or-so-minute bus ride to the corner of Aldwych and Drury Lane. From there, she walked several more minutes, careful to avoid both the commerce of the Strand and the legal morass of the Inns of Court, into the presumed clarity of King's College.

Entering through the imposing iron gates on the Strand, Franklin made her way into the quadrangle abutting the Thames. Giving wide berth to the bomb crater in its center, she veered left to climb the steps of the King's Building, an imposing, eight-floor Georgian pile of granite, balustrades, and arched windows. Like so many King's students and staff, she rarely, if ever, looked above the main doors to where two allegorical figures stood guard: one holding a cross and the other a book. Drawn from the college's coat of arms, they represent its motto, *Sancte et Sapienter* ("With Holiness and Wisdom"). Clickety-clack, her heels echoed on the slick, wet marble floor of the main foyer, dominated by a huge staircase and larger-than-life marble statues of the Greek play-

The main lobby and stairwell of King's College, London.

wright Sophocles and Sappho, the Greek lyric poet—each stationed without collegiate edict or signage to indicate their purpose there.[11]

She walked toward a side door, descended one flight of stairs, and strode through a pair of heavy fireproof doors into the biophysics lab. Once at her workspace, she took off her raincoat and covered her tasteful outfit with a newly-starched white lab coat, which was cut for a male frame and was, as a result, too big and bulky. This last accoutrement was essential in her physical and psychic transformation from the delightful Rosalind so beloved by her friends into the peremptory and often difficult Dr. Franklin.[12]

ON THIS NOVEMBER DAY, John Randall asked his staff to present a colloquium on nucleic acids in the College Theatre at 3 p.m. There were three speakers. Wilkins led off with an updated version of the lecture he gave in Naples in May and reprised in Cambridge in July, presenting the evidence of DNA's "'cross-ways' X-ray pattern indicating [a] . . . helical structure." He also discussed the extensibility of nucleic acid fibers (a stretching process he called "necking"), the optical properties of DNA, and his work on extracting DNA from the nuclei of *Sepia* octopi sperm cells. He "had little new to say" during this talk, but years later claimed that he "probably mentioned [Erwin] Chargaff's important results [on the equal amounts of adenine and thymine, and guanine and cytosine in each species' DNA]." No one else attending the colloquium recalled him presenting anything related to Chargaff's work.[13]

Alec Stokes spoke next, on his new helical theory. He had worked out a mathematical formula (independently of Crick, who had done the same in Cambridge a week or so earlier) that explained the diffraction of repeating structures in two-dimensional lattices or gridlike structures, known as the Fourier transform of a helix. Stokes proudly concluded that his work represented a new method for interpreting the diffraction patterns of DNA.[14] When interviewed twenty-three years after the event, he recalled, "I gave my talk on helical diffraction theory—one thing I do remember is remarking in the introduction to my talk that I had heard that Crick and Cochran had worked out the same—but that my talk was based entirely on my own work. The trouble is that I have a very dim memory of what was said [with respect to Franklin's talk that followed]—I had just given a paper myself, you see."[15]

Franklin was scheduled to speak last—the worst spot in an afternoon symposium. More than two hours into the proceedings, the sleepy audience members had already begun to focus on their post-symposium pints of ale. Glancing at the men who would judge her work, she braced herself and stepped up to the stage. Placing a sheaf of notes on the oak lectern, she cleared her throat and, in her Queen's English accent, delivered the details of her highly technical research. Seated in the front row of hard wooden benches, directly in her eyeline, was Maurice Wilkins. A row or two behind him sat a cross-legged Jim Watson, armed with a few newspapers in case the lecture failed to be of interest. This was the first time Franklin and Watson saw each other in person.

François Jacob, the 1965 Nobel laureate in Physiology or Medicine, recalled young Watson's behavior at meetings around this time. He was striking not only because of his "tall, gawky, scraggly" physique, but also

The lecture theatre, King's College, London.

because he exhibited a style no one else could approach, let alone imitate. He tended to enter lecture halls "cocking his head like a rooster looking for the finest hen, to locate the most important scientists present," and sit near them. His manner of dress was purposefully odd, with "shirttails flying, knees in the air, socks down around his ankles." And then there was "his bewildered manner, his mannerisms, his eyes always bulging, his mouth always open, he uttered short, choppy sentences punctuated by 'Ah! Ah!'" All combined, these qualities, Jacob estimated, added up to "a surprising mixture of awkwardness and shrewdness. Of childishness in the things of life and maturity in those of science."[16]

SOME FIFTEEN PHYSICISTS ATTENDED the November 21 colloquium. Yet there exist only three written versions of what Rosalind Franklin presented during her lecture: Watson's 1968 memoir, Wilkins's 2003 memoir (annotated with the occasional interviews he granted over the years), and Franklin's 1951 lecture notes prepared before the event. Watson's version of events is the most commonly cited account, which is unfortunate because he is often unreliable when describing events or actions involving Franklin. As Horace Judson noted, Watson's account is "brief, evasive, and devoted to his failure to take notes, his failure to understand what Franklin was talking about, his failure to remember correctly the little he said he did grasp. Watson's failures were, for his story, all that he found interesting."[17] That said, we shall begin with Watson's interpretation.

On Tuesday evening, November 20, Watson attended a sherry party at Bragg's house in Cambridge. Before retiring for the night, the more than slightly tipsy Watson wrote to his parents, "tomorrow I will go into London to hear a lecture on Nucleic Acids at King's . . . then on Friday, I will probably go to Oxford for a visit since others from the lab are going."[18]

Watson was the wrong man to attend the lecture, for far more important reasons than his inherent weirdness. As Crick recalled years later, Watson was "only a new boy in crystallography."[19] Although he told his parents that he was quickly comprehending this arcane topic, Watson had only a few weeks of study under his belt and understood very little of the complex mathematics, the interpretation of diffraction patterns, or even the critical differences in terminology, such as a crystal's unit cell as opposed to an asymmetric unit—a point which baffled him during Franklin's lecture. Add his "childishness in the things of life" to the equation and it is hardly surprising that Watson's recollection of Franklin's talk and her appearance still has the power to offend:

> She spoke to an audience of about fifteen in a quick, nervous style that suited the unornamented old lecture hall in which we were seated. There was not a trace of warmth or frivolity in her words. And yet I could not regard her as totally uninteresting. Momentarily I wondered how she would look if she took off her glasses and did something novel with her hair. Then, however, my main concern was her description of the crystalline X-ray diffraction pattern.[20]

Watson went on to describe Franklin's insistence that "the only way to establish the DNA structure was by pure crystallographic approaches." He saw her approach as weak because it did not include model building, except as "a last resort." She did not even "mention Pauling's triumph over the α-helix," let alone wish "to ape his mannerisms."[21] She was not wrong in her assertion; the slow acquisition of accurate X-ray data was essential to solving DNA's structure. But Watson had no intention of wasting his precious time in such a pursuit; hence, he mocked her as a recipient of "a rigid Cambridge education only to be so foolish as to misuse it."[22]

Watson's misinterpretation of Franklin's lecture—the false details which he began spreading within hours and repeated for decades— was that "Rosy did not give a hoot about the creation of the helical theory . . . since to her mind there was not a shred of evidence that DNA was helical."[23] After the talk, only Wilkins asked a few questions, but these were of "a technical nature," and the paucity of discussion left Watson disappointed. The rest of the audience, he claimed, remained silent, looking at their feet, fearful of one more of Franklin's withering rebukes. "Certainly a bad way to go into the foulness of a heavy, foggy November night," Watson insisted, "was to be told by a woman to refrain from venturing an opinion about a subject for which you were not trained. It was a sure way of bringing back unpleasant memories of lower school."[24]

WILKINS'S ACCOUNT IS more complex and enigmatic. Late in his life, he begrudgingly admitted that Franklin's lecture was "a first-class account of various likely aspects of DNA structure. She presented clearly reasons why the phosphate groups should be on the outside of the molecule, and the importance of understanding the role of water in DNA structures A and B."[25] Yet, during the fifty years preceding this admission, Wilkins serially repeated Watson's tall tale about Franklin's anti-helical belligerence.[26] In 1970, he insisted to Anne Sayre that "if [Franklin] hadn't had this anti-helical attitude there's no doubt the solution would have been there ahead of Crick and Watson. . . . She was very bloody-minded about it. What could I do? I couldn't even talk to her."[27]

Two years later, in 1972, when Wilkins was shown the written version of Franklin's November 21 lecture presentation—which definitely does discuss the possibility of a helical DNA structure—Wilkins dissembled: "my own memory is that Franklin said nothing about helices, but I cannot be certain. I think it would have been out of character to present in a talk the speculations in the notes."[28] And in 1976, his story turned accusatory: "there was very incomplete communication during that period. I think it unfortunate that Franklin did not keep us informed about her progress on aspects other than the evidence that DNA could not be helical."[29]

This oddly comforting myth—for Wilkins, that is—eventually landed in the pages of his 2003 memoir: "I certainly do not remember her discussing helical structures; and Jim Watson's recollection was the same." He postulated, without proof, that she spoke the anti-helical gospel to avoid a Wilkins–Stokes–Franklin collaboration. "To ensure her independence," Maurice suggested, "it is not surprising that she dropped the helical parts in her lecture notes. I think she wanted to work in well-established ways."[30]

There exists no film recording of her lecture, and no member of the audience was able to precisely recall what she said that afternoon. Yet it seems historically safe to assume that—just as every academic scientist who presents their research at an important conference—Rosalind Franklin used her prepared lecture notes as the source for her lecture. In 1976, her colleague Alexander Stokes agreed that despite his unclear memory of that afternoon, "it was consistent with what he knows that she gave what her notes said."[31] What possible motivation would a scientist as devoted to facts as Rosalind Franklin have to omit key, hard-won experimental results—especially when speaking in front of her boss and men she perceived to be hostile? So what was in Franklin's lecture that, when read aloud, created such confusion?

Fortunately, the archivists at Churchill College, Cambridge, have saved and digitized all of Rosalind Franklin's red-bound Century schoolbooks and scraps of paper, which document each day's scientific experiments, speculations, analyses, and calculations while she was working at King's College. Included within is a double set of eight yellowing pages titled "Colloquium Nov. 1951," bundled together with six more pages of sketches and mathematical formulae to herself. These scribbles—along with Franklin's progress report for 1951–52—unravel an entirely different yarn than the anti-helical narrative concocted and perpetuated by Watson and Wilkins.

Much of these comments appear in shorthand. Some serve as brief prompts, meant to remind her of the results she had long committed to memory, as she glanced down nervously at the pages before making planned pauses of eye contact with her audience. Scattered throughout are classic Franklin entreaties for caution and the need to acquire ever more data. She emphasized to her audience the need for better pictures, better DNA fibers, and better dexterity in handling the camera and the DNA specimens. There is even a direction to "Show Photos" when she displayed her images on the screen behind her.[32]

The caveat Watson elected to hear loudest that afternoon was that

"Rosy regarded her talk as a preliminary report"—an admission that gave him leave to discount her data as unconfirmed and unreliable. That her data was preliminary was an obvious point. She had been conducting her DNA research for only about nine months and was still working out the kinks in her experimental methods. It would have been foolish, incautious, and unscientific to state otherwise—three traits no one would ever accuse Franklin of displaying.[33]

For decades after Franklin's November 1951 lecture, Watson, Wilkins, and Crick each falsely claimed that Franklin had vehemently nixed the notion of a DNA molecule's helical shape. One eyewitness, Raymond Gosling, has refuted this notion on several occasions: "This is what Watson says, which I think is unfair—that Rosalind was simply anti-helical, and that she did all she could to fight it. This is simply not true."[34] Especially in the years after Watson, Wilkins, and Crick won the Nobel Prize, Gosling's objections were simply no match for these scientific titans.

In fact, the notes for Franklin's November 1951 lecture fail to support the anti-helical charge. Franklin took great pains to describe how the better she became at hydrating the DNA fibers, the more they resembled the *in vivo* (living) form of DNA, especially in the way they stretched and uncoiled—another point Wilkins erroneously claimed she missed making. The addition of water to the DNA fibers, she wrote, generated "a complete change in picture towards much simpler."[35] She next delved into her systematic study on the use of salt and water solutions to discover three "more or less well-defined states:" a) a wet state; b) a crystalline state; and c) a dry state. This preliminary finding of three forms of DNA was probably the result of her not yet having perfected the experimental conversion into what a few months later she would describe as only two distinct states: the *dry crystalline* (A) form and the 75–90% *wet paracrystalline* (B) hydrated form.

Specifically, she found that the dry, crystalized A form was easier to X-ray but ultimately more difficult to interpret because, as it turned out, the X-rays bounced off the atoms in the molecule in a manner that produced artifacts. The wet B form was initially harder to photograph but, as she later discerned, proved easier to visualize a helical structure in its diffraction pattern. Between 1951 and 1952, Franklin kept an open mind and went through both "anti-helical" and "pro-helical" phases while analyzing the conflicting data she was generating.[36] More than a quarter of a century after the fact, protein crystallographer C. Harry Carlisle of Birkbeck College explained the scientific rationale of her cautiousness: "I am convinced from Rosalind's excellent X-ray studies on both the A and B forms of DNA that she was not in the least anti-helical." The reason she focused on the A-form, Carlisle observed, was that the X-ray diffraction data for it was the strongest in resolution and, she hoped, more likely to produce hard, reproducible results as opposed to theories and conjectures. It was a result of the way she was trained to conduct rigorous scientific inquiries.[37] Thus, a more accurate description of Franklin's November 1951 lecture was that she did not yet have enough data to *definitively* state that DNA was helical, and that there was some evidence (which ultimately proved to be artifactual) to the contrary. Her data was preliminary because, given the technology she had on hand, mapping out every atom and bond on an organic molecule as complex as DNA took many months to years. Time, patience, increasingly better X-ray technique, and non-interference by her hungry audience were the ways forward, she advised.

Franklin did make several other crucial experimental points that afternoon, which completely sailed above Watson's head. For example, she had enough data to hypothesize how water molecules tended to surround the phosphate groups located on the outer portion of the DNA molecule.[38] She also described how the diffraction patterns of the wet DNA revealed a 3.4 Ångstrom arc on the meridian and two "oblique smears at about 40 degrees to it." On the equator of that pattern, she found a "sharp intense spot" suggesting "high order." For the dry form, she reported a gradual diminution of the equatorial spots, leaving only the 3.4 Ångstrom meridional arc and two side arcs. With respect to the "crystalline" pattern, she found a "27 Ångstrom spot" (just as Astbury had found in 1938), which was "much too marked to result merely from diff. betw diff nucleotides, & must mean nucleotides in equivalent positions occur only at intervals of 27 Å. Suggests 27 Å is length of turn of spiral." The word "spiral" was then a commonly used synonym for a helix. As she calculated density estimates, the data pointed to there being more than one chain in each molecule, and as the crystalline form made its transition to the wet form, she observed a "large length change," indicating that the wet "helix has not got same structure as in Xtal—Xtalline form involves some strain of helix . . . cf. Pauling."[39] This last comment suggests that she did mention both helices and Linus Pauling's protein work, contrary to Watson's claim.

Her lecture notes go on explain, "near-hexagonal packing suggests that there is only one helix (containing possibly more than 1 chain) per lattice point. Density measurements (24 residues/27 Ångstroms) suggest more than 1 chain."[40] The carefully written conclusion section is even more definitive in destroying the anti-helical mythology: "Big helix or several chains, phosphates on outside, phosphate-phosphate inter-helical bonds, disrupted by water. Phosphate links available to proteins."[41] In the next line, she describes her search for "evidence for spiral structure" and suggests that is the case because a straight untwisted chain would be "highly improbable" (i.e., such a structure would prove unbalanced and unstable in nature).[42]

A few weeks later, Franklin typed up her findings in an interim progress report on her work from January 1, 1951 to January 1, 1952, which both Randall and Wilkins reviewed. In this five-page, double-spaced report, the words "helical" or "helix" appear five times. It is highly doubtful that between late November, when she gave her lecture, and the Christmas holidays of 1951, when she began writing up the progress report, she had accomplished any additional discoveries.

> The results suggest a helical structure (which must be closely
> packed) containing probably 2, 3, or 4 co-axial nucleic acid chains
> per helical unit, and having the phosphate groups near the outside.
> It is the phosphate groups which would be capable of absorbing
> water in large quantities and of forming strong inter-helical bonds,

in the presences [sic] of considerable quantities of water, thus giving the substance a 3-dimensional crystalline structure. These bonds would be disrupted in the presence of excessive quantities of water (leading first to the "wet" structure of independent helices with parallel axes, and ultimately to the solution of DNA in water) and would remain strong in the absence of water, thus explaining the cementing effect of strong drying. The <u>dry</u> structure is distorted and strained due to holes left by the removal of water, but contains intact the skeleton of the crystalline structure giving the substance a three-dimensional crystalline structure.[43]

THERE WAS ONE OTHER major finding—mentioned in both her notes for the lecture of November 21, 1951, and the 1951–52 interim progress report—which ultimately proved essential to unlocking the structure of DNA. More than a year before Watson and Crick published their famous paper, Rosalind Franklin determined that the crystalline state, or A form, of DNA could be classified as a monoclinic, face-centered, or C2, unit cell.[44] This arcane crystallographic observation requires some explanation before we conclude that there would be no way to ascertain the double helix structure without it (even though that is entirely true). In February 1975, the *New Yorker* reporter Horace Judson approached Max Perutz at his Medical Research Council laboratory, newly opened on the south side of Cambridge, nearly five miles from the dusty halls of the old Cavendish, and requested an explanation that he could relay to his readers. Perutz's response remains the best elucidation of the vagaries of crystallography this historian has seen.[45]

The physicist began by explaining how crystals display symmetry in a variety of ways; more precisely, they do so in a matter of degrees, some being more symmetrical than others. The crystals sharing the least amount of symmetry are called "triclinic, in which the three axes, or planes, of the crystal are all oblique to each other, none of the angles at the corners being right angles." On the other end of the spectrum are orthorhombic crystals "where all three planes intersect at right angles." Perutz then proceeded to discuss Franklin's vital finding. In between the two ends of the crystal spectrum is the monoclinic category, "where two of the three angles are right angles, but the third can take any angle." The "minimum symmetry" for a monoclinic crystal is twofold, meaning that when you rotate "the crystals by half a revolution, they come back into congruence again." In Perutz's estimation, Franklin stopped short soon after discovering that crystalline DNA had "monoclinic symmetry and the axis of symmetry wasn't parallel to the fibers, but perpendicular to them, to the chains." Unfortunately, this finding was of "a peculiar geometric consequence, the crucial consequence she failed to recognize."[46]

Perutz had offered this explanation to countless students over the years and quickly sensed the reporter's confusion. So, he took two pencils out of his jacket pocket and placed them on a nearby table, side by side, lead point by lead point to the north, and eraser by eraser to the south. In an encouraging manner, he elaborated, "If I rotate this pair together, in the plane of the table, they don't return to symmetry with their original position until they've gone around the full three hundred sixty degrees." To illustrate his words, Perutz turned the two pencils to the west, south, east and north positions. He then changed the orientation of one of the pencils so that the lead points and the erasers were at opposite ends. "Lying head to tail, the two chains return to symmetry after a half rotation. You see. They exchange directions." Perutz next rearranged the pencils so that they each rotated by one-half of a complete rotation. "If DNA was monoclinic from its X-ray diagram and the axis of symmetry was perpendicular to the chain"—he could barely contain his excitement—"then it immediately followed that there must be one chain running up and the other down. That in physical fact there was a dyad, [i.e., a two-stranded chromosome with] one chain upside down to the other."[47]

Perutz then added another layer of complexity by explaining the meaning of a crystal's space groups and unit cells. He used a metaphor of "fancy wallpaper with a repeating pattern. Then the unit cell simply means the smallest repeat of the pattern, identical in size, shape and contents. Except that in a crystal it's three-dimensional."

Drawing a simple, three-dimensional box of atoms with one of the pencils, Perutz circled the eight corners of the box, denoting an atom at each corner. He then pointed out how the box repeats itself like "a wallpaper pattern and forms a three-dimensional lattice of atoms. With only the corners occupied, the lattice is called P, for primitive, and if it has monoclinic, twofold symmetry, it's called a $P2$ space group. But in some other substance you could well have another atom or molecule at the center of a face of each box, as well." Penciling in a second box with a ball in the middle of each end, Max continued, "this face is called, by convention, the C face, and if this box also possesses twofold symmetry then the space group is $C2$, or face-centered monoclinic. Each corner of the unit cell, and each occupied face, is also called a lattice point." Such abstract terms were developed by mathematicians seeking to define "the orderly arrangement of the components of a crystal." In nature, there exist 230 geometrically different—or nonequivalent—space groups, Perutz explained. "They were all worked out in the nineteenth century by classical crystallographers calculating all the possible ways to arrange a lattice that will repeat in three dimensions, long before X-ray crystallography began."[48]

When it came to Rosalind Franklin's work, Perutz has a valid critique about "the crucial consequence she failed to recognize." What Franklin did not grasp—probably because she was a physical chemist used to working with inorganic compounds such as coal rather than a physicist who had long studied biological molecules—was the import of her $C2$ finding with respect to its explaining a mechanism of cellular replication and the function of a double-stranded helix of DNA. Watson, too, missed the implication of a face-centered, monoclinic space group $C2$ crystal. He further failed to report this information to Crick the following day. His gap in understanding was not in biology but, instead, due to the fact that he was not a seasoned crystallographer and, at the time of the lecture, did not even comprehend the difference between the terms "unit cell" and "unit symmetry." Thus, Max Perutz retrospectively chided Franklin for failing to discern the implications of her data while giving Watson a pass for the same scientific sin.[49]

THE ONLY RECORD OF what happened directly after the November colloquium appears in the pages of *The Double Helix*. In Watson's telling, the colloquium ended with Wilkins tensely making conversation with Franklin. The gangly American hung back, waiting for his opportunity to invite Wilkins to take a walk along the Strand and seek out supper at Choy's restaurant on Frith Street in Soho.[50] In 1958, *Fodor's Guide to Britain and Ireland* praised Choy's for serving "the best Chinese food in Britain" and for its "authentic Oriental atmosphere." Unlike most of the Chinese restaurants in England at that time, Choy's was licensed to serve alcohol, had a very good wine cellar, and was open until 11 p.m. (including Sundays)—no small issue in the not-so-swinging London of the 1950s.[51]

At Choy's, Watson was surprised to find that the man seated across the table was not the aloof, retiring, stiff physicist he had encountered in Naples the previous summer. Wilkins was now eager to chat about the laboratory, his research, and his difficulties with Rosalind Franklin.[52] Over platters of chop suey, chicken curry, chips, and fried rice, all lubricated with black tea and cheap red wine, Watson and Wilkins engaged in the old-school bonding of men as they conspired to exclude Rosalind Franklin from the DNA hunt. Wilkins painted an ugly landscape of how little headway she had made during her brief tenure at King's. To be sure, she was adept at making far clearer and sharper X-ray photographs than he was able to obtain, but, to his mind, she had done nothing to explain what she was actually photographing.[53] These complaints were but a soft sigh in the whispering campaign he was mounting to ice her out of the King's lab. Only this time, he spread his venom to an outsider.

Wilkins raised serious doubts about Franklin's calculated estimates for the "water content of her DNA samples," which she had announced an hour earlier in her lecture.[54] The precise measurement of DNA's water

content was essential in determining its structure, but difficult to ascertain at this point. Based on the approximate water and salt density numbers Franklin was able to calculate, it was unclear if there were two or
three nucleotide chains per molecule. Her numbers indicated a value in
between the choice of two or three (and possibly even four) chains, but
were not yet definitive enough to determine which one was correct. She
needed better data to make a better assessment. Once that value was
ascertained, however, figuring out the exact number of nucleotide chains
would be achieved by using either a simple equation or from observations
of the X-ray patterns. The paradox of Wilkins's carping is that, at least,
Franklin was on the right track, while he was puttering around without
making much progress. Moreover, Wilkins, the helical man, also failed
to recognize the importance of her discovery of the C2 symmetry of
DNA's unit cell. If this observation was valid—and it was—the number
of helical chains was likely be an even number (i.e., two, or possibly four)
and not three, as Wilkins was insisting upon at the time.[55]

In his relaxed state, Wilkins confessed that his physicist colleagues
did not think highly of his segue into biological research. They were
polite at academic gatherings, but the insecure Wilkins was certain
that once he was out of earshot, they were belittling him for dropping
out of the feverish pace of post–Second World War physics research.
He found even less support from the English biologists, all of whom
were old-fashioned botanists and zoologists. Discerning that Wilkins
needed some collegial comfort, Watson aimed his bulging eyes in the
older man's direction, listened, nodded, and encouraged. Watson told
Wilkins how the cutting-edge biologists of his peer group dismissed
such old fogies as collectors, categorizers, list makers, and taxonomists
who spun imaginary theorems on how life began, none of them based
on solid scientific data or even the acknowledgment that genes were
composed of DNA.[56] By the time their meal ended, Watson thought he
had successfully bucked up his new colleague, until Wilkins suddenly
reverted to ragging on Franklin and the moment was lost. Watson paid
the bill as he watched a dejected Wilkins skulk out of Choy's and into
the dark, foggy London night.[57]

The Dreaming Spires of Oxford

To-night from Oxford up your pathway strays!
. . . that sweet city with her dreaming spires,
She needs not June for beauty's heightening,
Lovely all times she lies, lovely to-night!

—MATTHEW ARNOLD, "THYRSIS"[1]

The morning after the King's College colloquium, Watson met Crick in London's sooty Paddington Station, not far from Rosalind Franklin's childhood home in Notting Hill. The cavernous train depot was still in the process of repair thanks to several German air sorties during the Second World War. The roof above the platform where they awaited their Great Western Railroad Express train had been only recently replaced, long after the old one was destroyed in 1944 by two Luftwaffe 230-kilogram flying bombs.

Still groggy after his late evening with Wilkins, Watson was eager to make his first trip to Oxford. If he took the time to explore Paddington before entering the station, he might have spied St. Mary's Hospital. There, in the fall of 1928, in a tiny laboratory high up in one of its red-brick towers, the Scottish microbiologist Alexander Fleming accidentally discovered penicillin, which he derived from some mold growing in a Petri dish he had set aside before going on summer holiday. By the end of the Second World War, penicillin was hailed as the wonder drug of the twentieth century and, in 1945, Fleming shared the Nobel Prize in Physiology or Medicine.[2] Neither Watson nor Crick leaves a recollection of penicillin's origins from that day, which also coincided with the American holiday of Thanksgiving. Nor did they connect the antibiotic dots leading directly to the woman they were meeting that afternoon:

Dorothy Hodgkin and Linus Pauling, 1957.

Dorothy Hodgkin, one of the most prominent X-ray crystallographers in the world. Among her many accomplishments was working out the molecular structure of penicillin.[3]

Crick was meeting with Hodgkin to discuss a new theory he had devised explaining the X-ray diffraction of helical organic molecules.[4] Watson described their excitement before boarding the train: "At the train gate Francis was in top form. The theory was much too elegant not to be told in person—individuals like Dorothy who were clever enough to understand its power immediately were much too rare."[5]

Dorothy Hodgkin was blessed with an ability to hide her brilliance behind a sweet disposition. As a result, she rarely threatened her male counterparts and climbed up the greasy pole of academia long before they had a chance to push her down. Like Rosalind Franklin, Hodgkin applied complex mathematical equations to the interpretation of two-dimensional X-ray diffraction patterns. Unlike Franklin, Hodgkin was not opposed to speculative model-building before nailing down each and every data point. Yet even the simplest molecule she studied—penicillin, which contained only twenty-seven atoms—took her several years to solve. Thornier molecules, such as insulin and vitamin B12, which contained hundreds of atoms, took far longer.

Crick's new theory had its origin a few weeks earlier, on the afternoon of All Hallow's Eve. Max Perutz had just received a letter from Vladimir Vand, a Czech crystallographer working at the University of Glasgow, proposing an explanation on how helical molecules diffracted X-rays.[6] Perutz passed it to Crick, who immediately found flaws in Vand's reasoning. Letter in hand, he bounded up to the third-floor office of William Cochran, a talented crystallographer who had to modulate his thick Scottish accent so that his English colleagues could understand him. Although Cochran was not involved in deciphering large biological molecules, he loved poking holes in Crick's flightiest ideas and served as an excellent sounding board for some of his better ones. The two agreed to work out a better series of formulae than those proposed by Vand. Absorbed all morning in the chalking up and erasing of equations on a scratched slate blackboard, Crick's head ached by the time he adjourned to the Eagle for lunch. "Feeling a bit off-color," he left the pub to rest his weary head at "the Green Door," the Cricks' "tiny, inexpensive flat" on the top floor of a several-hundred-year-old house across Bridge Street from St. John's College.[7] Sitting by the gas fire in his cramped living room, he grew weary of doing nothing and picked up his pad of scribbled equations. Soon enough, he arrived at an answer.[8]

At dusk, he stopped work because he and Odile had plans to attend a wine tasting at the spirits shop of Matthew and Son on nearby Trinity Street. Watson claimed that Crick's morale was lifted by the invitation because it meant acceptance into the vaunted Cambridge social life, a change from the harsh treatment he endured daily at the Cavendish Laboratory.[9] The evening of wine and cheer proved less tasty and fun than the Cricks had hoped, and they left early. Watson guessed that this was because there were "no young women" present and the bulk of the guests were "college dons contentedly talking about the burdensome administrative problems with which they were so sadly afflicted."[10] Crick characterized Watson's description of the wine tasting as "absolute nonsense" and "typical of Jim to just get this sort of thing with a slant in wrongly."[11]

Crick did, however, admit to trying the 1949 vintage hocks and moselles served by Mr. Matthew.[12] Apparently, he used great restraint in his tasting and went home "unexpectedly sober," where he resumed his seat by the fireplace and reexamined his molecular calculations.[13]

The next morning, Crick sailed into the Cavendish armed with a series of clumsy mathematical equations, which, he told Perutz and Kendrew, could be used to predict the helical structure of proteins. Within a few minutes of this announcement, Cochran walked in with his own set of far more elegant equations explaining the same thing. By checking the Pauling α-helix against Perutz's X-ray protein photographs, they were able to confirm their new helical theory and Pauling's model. Crick and Cochran promptly wrote up their findings and sent the paper off to the offices of the journal *Nature*—an important step in understanding the molecular biology of helical molecules.[14] Many accounts of Crick's early days at Cambridge center on his bombast and sketchy performance in the laboratory. Yet if he had done nothing more there than publish his helical theory, he would have enjoyed some scientific distinction. On the pages of *The Double Helix*, however, Watson could not resist injecting a sexist and silly "slant" into the equation of Crick's success: "For once the absence of women had gone along with luck."[15]

SO, WHAT DID CRICK puzzle out that night? As Perutz later explained to the *New Yorker* writer Horace Judson, when X-rays pass through a helical molecule that is placed perpendicular to a sheet of photographic paper, the result is a zigzag pattern. The X-rays hit the film in a "striking arrangement of short horizontal smears that step out along the diagonals from the bull's eye, in a characteristic X or Maltese cross. The zigs cause one arm of the cross, the zags the other. The exact angle of the cross is caused by the angle of the zigs to the zags—that is the pitch of the helix."[16] Thus, Crick and Cochran (and Vand) demonstrated that something had not yet been mathematically realized: "if you follow out the layer lines—step by step along the arms of the cross—at a certain distance the spots begin to march back together to cross again. The cross doubles at top and bottom of the target and becomes two diamonds." Although it required the right setup of X-ray equipment and enough resolving power to show such a pattern, there existed a reciprocal relationship "between the diffraction pattern and the actual space between the atoms within the molecule," Perutz explained. "Big spacings between the repeated planes produce spots close to the target, but as you move out, you are reading planes with finer and finer spacings."[17]

Crick was not alone in dreaming about helices. During the summer of 1951, shortly after reading about Pauling's α-helical protein model, Alec Stokes and Maurice Wilkins were also working on "calculating X-ray diffraction from the structure." In 2003, Wilkins recalled how he discussed the same relationship with Stokes and the next day Stokes "produced a Bessel function calculation of diffraction from a helix." Stokes made these calculations on a sheet of paper during the hour-long train ride from his home in Welwyn Garden City to London. Passing by a travel poster for the popular seaside resort of Bexhill-on-Sea, he decided to call his explanation "Waves at Bessel on Sea."[18] This was the work that Stokes lectured on at the November 21 colloquium.

Around the same time, Franklin was zeroing in on the detection of first three, and ultimately two forms of DNA, the dry or crystalline form (A) and a wet form (B). Stokes had "a clear memory of Rosalind's B pattern" and applied his "plot of calculated intensities" to it, noting with mounting excitement how well they corresponded. Stokes and Wilkins could hardly contain their excitement and rushed into Franklin's room to inform her of their "important good news." She listened to their explanation before angrily cutting them off: "How dare you interpret my results!" As a result of this encounter, the two men discussed their ideas with Crick but never published them. Stokes had to settle for a mere acknowledgment in Crick's *Nature* paper.[19]

A FEW MONTHS BEFORE Crick and Watson made their pilgrimage to Dorothy Hodgkin's laboratory at Oxford, Rosalind Franklin made a similar visit, bringing along several of her X-ray photographs of DNA. Hodgkin told her that they were "the best she had ever seen." Unfortunately, a misunderstanding soon arose between the two women. According to Jack Dunitz, a crystallographer working in Hodgkin's laboratory, it may have stemmed from the fact that Franklin, an inorganic physical chemist by training, "lacked Dorothy's comprehensive grasp of the [organic] chemistry behind the structures that so fascinated her."[20] While looking intently at Franklin's pictures, Hodgkin observed that the quality was good enough to calculate the space group of the molecules simply by studying the diffraction patterns. Franklin agreed and told Hodgkin that she had already "narrowed the possibilities down to three." Hodgkin then exclaimed, with perhaps a bit too much zeal, "But Rosalind!" She went on to explain that two of the three structures Rosalind proposed were physically impossible. The issue had to do with "handedness"—the chiral orientation of the sugars in the DNA molecule. What Rosalind "failed to take into account" while analyzing her X-ray images was that *all* of the sugars in DNA are right-handed in their orientation, as are the helices they construct. "Two of the three possible structures suggested by Franklin," Dunitz recalled, "required both left- <u>and</u> right-handed forms" and were, thus, inconsistent with her data.[21] Hodgkin saw the error immediately, even though she was not working on DNA. Franklin did not.

The two scientists might have made a wonderful team of mentor and mentee. They were both graduates of Newnham College, Cambridge, experienced many of the same travails of working in a male-dominated field, and shared a passion for X-ray crystallography. Sadly, a warm, working relationship was not to be. In Dunitz's telling, Franklin was too easily embarrassed by her rudimentary knowledge of biological chemistry to take Hodgkin's advice. Without corroborating evidence, he believed that Franklin's pride was savagely wounded by Hodgkin's exclamation, "But Rosalind!" In 2018, James Watson added his own pungent interpretation: "She really probably needed to be friends with a good crystallographer . . . who knew the helical theory . . . [but Dorothy Hodgkin] figured out Rosalind didn't know anything in a few minutes." When asked to elaborate on a meeting he did not attend, Watson added, "I'm sure Rosalind was afraid of the meeting, because of Dorothy's reputation and wanting to impress her . . . she [Rosalind] quickly revealed herself as . . . not smart enough to even be in the same room."[22] One can only speculate whether this encounter was marred by a real or perceived criticism, competitive dueling, or something more benign, such as Franklin's insistence on doing all the DNA work by herself. The fact remains that Franklin lost a potential collaborator and colleague who might have helped her in the coming months.

ONCE THE CAVENDISH DUO SETTLED into the second-class carriage bound for Oxford, Crick grilled Watson about what he had learned at King's College the day before. Already aware of Wilkins's and Stokes's helical theory work, Crick was particularly interested in the data Watson culled from Franklin's talk.[23] Soon after the train pulled out of the station, however, he realized that Watson failed to fully understand Franklin's lecture. With each passing tie on the railroad, Crick grew more irritated by Watson's failure to complete the most basic of student tasks—taking good notes. Watson tried to allay Crick's anger by telling him about his debriefing dinner with Wilkins at Choy's. This failed to mollify Crick, who worried that Wilkins had not told Watson everything he knew. Watson countered snidely that Wilkins was incapable of such furtive behavior—even though both of them most definitely were.[24]

Instead of owning up to his inattention during Franklin's talk, Watson argued that Crick should have attended because that would have given them a clearer understanding of her findings. This, Watson believed, was the cost of being too sensitive to Wilkins's insecurities. Only a few days before, Crick had privately told Watson that he could not attend the lecture without Wilkins becoming suspicious or worse.[25] What Watson could not understand was Crick's insistence that it would have been "grossly unfair" for both Watson and himself to hear Franklin's research results at the same time. English manners dictated that "Maurice should have the first chance to come to grips with the problem." Watson was not as intellectually threatening to Wilkins as Crick was, so Crick used the excuse of his meeting with Hodgkin to absent himself from the seminar. Nary a sentence in Watson's 1968 memoir goes by, however, before Watson rationalizes ignoring such intellectual protocols. To wit, he rationalized that it was acceptable for him and Crick to begin building a DNA model because at their Sunday dinner in Cambridge a few weeks earlier, and again after the Franklin talk, Wilkins gave "no indication that he thought the answer would come from playing with molecular models."[26]

Before the second stop on their train ride, Crick began sketching out diagrammatic sketches on the blank sides of a manuscript he pulled from his valise. The calculations he made were flawed, though he did not yet know it, because Watson had grossly undervalued the water content of the DNA samples Franklin reported. Watson admitted that he could not understand Crick's mental machinations and, instead, turned to reading *The Times*. Soon enough, the ideas sparking in Crick's brain were powerful enough to make them both ignore the world around them and lose themselves in the molecules they hoped to decipher. Crick explained that there were only a few structural answers that corresponded to both his (and Bill Cochran's) helical theory and Watson's misremembered version of Franklin's data. Although the mathematics completely confounded Watson, he was able to follow the major points of Crick's proposed plan of attack. Crick thought the X-ray data suggested a double, triple, or quadruple helix, made up of strands of nucleotide chains. They still needed to determine "the angle and radii at which the DNA strands twisted about the central axis," however, which would require accessing more X-ray diffraction data from their competitors at King's College.[27]

About an hour after leaving Paddington, Watson and Crick arrived at the comparatively tiny Oxford station. They walked the half mile into the city center, several hours before their scheduled meeting at Hodgkin's laboratory in the University Museum of Natural History. The excitement Crick generated must have been palpable. How inspiring it must have been for Watson that rainy afternoon when Crick declared that they should have an answer soon enough; all told, he predicted a week or so of playing and shaping the molecular models to come up with the correct solution.[28]

FOR THE ANTIC ADVENTURES he later recounted with such success in *The Double Helix*, it was essential that Watson create an outright villain in Rosalind Franklin. But his inclination to deprecate her scientific abilities conjured up such a poor competitor that he needed to concoct another rival hot to pursue DNA. That character was Linus Pauling—even if Pauling did not yet know he was in the battle, let alone with two nobodies like Watson and Crick. Despite Watson's Chicken Little proclamations in print that Pauling might have solved the riddle of DNA first, Pauling was in fact the straw man in a drama that his son Peter later dubbed the "race that never was." Peter Pauling insisted that "the only person who could conceivably be racing was Jim Watson. Maurice Wilkins has never raced anyone anywhere." Francis Crick, Peter explained, simply enjoyed pitching "his brain against difficult problems." For his father, on the other hand, "nucleic acids were interesting chemicals, just as sodium chloride is an interesting chemical, and both presented interesting structural problems . . . as a geneticist, the gene was the only thing in [Jim Watson's] life worth bothering about and the structure of DNA the only real problem worth tackling."[29]

Inviting Pauling to the party is an excellent example of Watson's skill at playing people off against one another. Jim understood that Pauling had gotten into the collective heads of the Cavendish staff when he beat Bragg, Kendrew, and Perutz the year before with his α-helix structure. The humiliation felt by the Cambridge group was no childish booboo; it was an international gaffe of epic proportions printed in black and white for all to see, forever, on the time-honored pages of the *Proceedings of the Royal Society of London* and the *Proceedings of the National Academy of Sciences*.[30]

Watson had by now observed Crick's personality closely enough to know how to manipulate him. Crick may have been a newcomer to the Cavendish when Bragg et al. concocted their disastrous paper on the configuration of polypeptide chains, but he was enough of a team player to feel the sting of their "fundamental blunder." Although there had been a boisterous group discussion over the Bragg–Perutz–Kendrew hypothesis before sending the paper off for publication, it was one of the few instances when Crick kept his mouth shut and "said nothing useful." Crick being Crick, he long regretted this rare moment of silence and wished he had been able to steer the old man away from such a mess. Watson understood that scientists are a competitive lot, overachievers who recall their failures with far greater clarity than their successes. Here, now, Watson told his older but junior-ranked colleague, was a golden opportunity to show up the great Linus Pauling and demonstrate that he was not the only scientist brilliant enough to divine the structure of complex biological molecules.[31]

As the two men ambled along the streets of Oxford, they argued over the possible configurations of the DNA molecule. Passers-by of both town and gown variety could not help but notice how they were shouting at each other, hurling uncouth epithets and jargon-laden declarative sentences to top whatever was last uttered. Nor could an objective observer fail to miss the exhilaration generated by their beautiful minds. For them, the hours whizzed by as they erroneously imagined models with a centrally-placed sugar–phosphate backbone. They still had no idea of how to space the poorly-fitting nucleotide bases their model left openly exposed to the outside. They chose to ignore this issue, assuming the structural dilemma would magically vanish once they figured out the proper internal arrangement and solved the chemical question of how to neutralize the negatively charged phosphate groups on the backbone. Unfortunately, these conjectures contradicted the data Rosalind Franklin had presented only a day earlier. As Franklin had correctly announced—and, for the time being, Watson misled Crick by dismissing Franklin's finding: "Big helix or several chains, phosphates on <u>outside</u>, phosphate–phosphate inter-helical bonds, disrupted by water. Phosphate links <u>available</u> to proteins."[32]

Crick and Watson took a big bite into a red herring rather than solving the problem in front of them: the three-dimensional arrangement of

DNA's constituent atoms and the negatively charged phosphate groups surrounding them. The only break they took was for a sandwich at a cheap spot near the High Street. They skipped the temptation of coffee, instead making the rounds of the many Oxford bookstores until they finally entered the great mother of all booksellers, Blackwell's—directly across from the university's Sheldonian Theatre. Somewhere in the cluttered stacks arranged higgledy-piggledy in its several old, connected buildings, they located the store's only copy of Pauling's *The Nature of the Chemical Bond*. They plunked down the few quid needed to take the book out of the store and stood in the middle of the well-named Broad Street, pulling at either end of the volume, as if in a scholarly game of tug-of-war, as they flipped the pages searching for the correct measurements of the "candidate inorganic ions." For once, Pauling failed to elucidate; Watson and Crick could not find anything within its pages to help "push the problem over the top."[33]

Their manic giddiness had abated by the time they walked into the Museum of Natural History and down the hall to Dorothy Hodgkin's mineralogy and crystallography laboratory. At its entrance was a brass plaque noting that this was the room where, during an 1860 meeting of the British Association for the Advancement of Science, T. H. Huxley defended Charles Darwin's recently published theory of evolution against the Bishop of Oxford, the Right Reverend Samuel Wilberforce.[34] The long room was lit by huge Gothic-style windows, the upper portions of which were blacked out to accommodate a loftlike darkroom suspended from the high ceiling. In the center of the room was a large oak table on which the X-ray diffraction photographs developed by Hodgkin's assistants were spread out for her review.[35]

The conversation, Watson reported, initially focused on Hodgkin's work on insulin. In the few minutes remaining, Crick quickly "ran through" his helical theory and, far more briefly, their "progress with DNA," but as dusk approached "there seemed no point in wasting more of her time." The Cavendish men proceeded on to Magdalen College, one of the wealthiest constituents in the Oxford University firmament, where they had tea and cakes with Crick's friend the immunologist Avrion Mitchison and a chemist named Leslie Orgel, both of whom were fellows of the college. Mitchison was the son of a wealthy Labour member of Parliament (Lord Mitchison of Carradale) and a glamorous, best-selling novelist (Naomi Mitchison). He was also the nephew of the distinguished geneticist and evolutionary biologist J. B. S. Haldane and grandson of the even more eminent physiologist John S. Haldane.[36] As Crick and Mitchison gossiped about mutual friends, Watson sipped his tea and fantasized about living the life of a Magdalen don.[37]

This mildly caffeinated repast at Magdalen was followed by dinner with Crick's close friend George Kreisel, a mathematical logician and Ludwig Wittgenstein's favorite student. They dined at the six-hundred-year-old Mitre Hotel restaurant. Bottles of claret helped mute their complaints about the overcooked, bland food. Kreisel dominated the conversation in his heavy Austrian accent, boasting about making "a financial killing" by buying and selling postwar European currencies. Such talk bored Watson, and his mood picked up only after Avrion Mitchison rejoined them. Watson and Mitchison excused themselves and walked along the medieval street to Watson's lodgings. Halfway there, Watson remarked that he had no real plans for Christmas, hoping to be invited to the Mitchison family's fabulous home on the southwestern tip of the Kintyre peninsula in the Scottish county of Argyll. He even inquired about a possible invitation for his sister, who was being chased by a Danish actor of whom Watson jealously disapproved. "By then," Watson recalled, "I was pleasantly drunk and spoke at length of what we could do when we had DNA."[38]

A few weeks later, on December 9, 1951, Watson wrote to Max Delbrück about Crick, the scientific contacts he was making, and the joys of eating in the Senior Common Room of Magdalen College, Oxford ("where no one can speak at breakfast"). He went on to note that "drinking port after dinner at High Table is an experience very difficult to describe but immensely interesting to participate in," and remarked on his futile search for female companionship: "As you can no doubt guess women at Cambridge and Oxford are very rare and so, much ingenuity must be spent in finding lively and pretty girls for one's parties." Before signing off, he described the real meat of the meal: "About my scientific work I will write later when I have some results. We believe the structure of DNA may crack very soon. Time will tell. At present we are quite optimistic. Our method is to completely ignore the X-ray evidence."[39]

Mr. Crick and Dr. Watson Build Their Dream Model

As always, [it was] very interesting [for] me to go and see Crick in action. I have a tremendous amount of respect for Rosalind and for Wilkins and for Crick, but Crick is the fireworks character, he's show biz, he's terrific to watch in a meeting, he's terrific, even better in discussion, he's got a quicksilver mind that darts about—it was a great pleasure to go there. And, ah, it was a bit of a difficult meeting because as I say, Rosalind didn't like to show her hand until every button was sewn on every last dead soldier, and you could parade them all, the living and the dead, and look at it. And here was this guy leaping about the place with a model . . . and, so, she didn't expect it to be right.

—RAYMOND GOSLING[1]

Watson and Crick returned to Cambridge in the early afternoon of Sunday, November 25. They needed that day to recuperate from their alcoholic revelry in Oxford. On Monday morning, Watson stumbled down the narrow staircase connecting his dismal bedroom to John and Elizabeth Kendrew's kitchen. After the usual morning greetings, he described his and Crick's "scoop about DNA."[2] John Kendrew glanced passively at him over the black cellulose acetate plastic rims of his National Health Service–issued spectacles. Watson immediately understood the reason behind his indifference: Kendrew admired Crick but was well acquainted with his spectacular flights of fancy and numerous scientific near-misses. Kendrew turned back to *The Times*'s reportage on the new Tory government[3] and calmly dabbed his face for any crusty yellow remains of the fried egg he had just consumed—perhaps symbolic of the cliché he imagined Crick would soon be wiping off of his face—then slurped a final sip of tepid tea and excused himself to retreat to his medieval rooms at Peterhouse College.

Elizabeth Kendrew, long used to playing the role of cheerleader for her scientist spouse, clapped her hands in delight and encouraged Watson to elaborate on his "unanticipated luck." He soon grew bored with Elizabeth, who could do little more than nod her head at his rapidly expelled scientific explanations. With haste, Watson excused himself to dash off to the Cavendish so that he might play with the molecular models and whittle down the many structural possibilities within.[4]

ROSALIND FRANKLIN'S BIOGRAPHER Anne Sayre described the so-called race to discover DNA's structure as one not only between personalities but also between "two different methods of structural determination."[5] The X-ray diffraction studies preferred by Franklin required an enormous amount of hard work and the investment of time. Molecular model building, as developed by Pauling and mimicked by Watson and Crick, involved a great many inductive leaps and risks.

In simple molecules, such as penicillin, the X-ray diffraction effects were sharper, and it was an easier task to identify the locations of the individual atoms comprising them. Visualization was far more difficult for complex organic molecules because their diffraction patterns tended to be weaker; specifically, the atoms faded from focus and appeared blurred. "From fiber structures, such as hair or DNA," Dorothy Hodgkin explained, "the diffraction effects are more limited still and direct methods of structure analysis are impossible to apply."[6] In other words, Rosalind Franklin faced a Himalayan challenge in using X-ray crystallography to decipher the complex structure of DNA fibers, and yet, this is precisely what John Randall hired her to do.

Building molecular models, on the other hand, was an entirely more complicated affair than the model automobiles and airplanes children build. Not only are there no instructions included, but each bond angle, bond length, and atomic or molecular doppelgänger must be custom-designed to scale and properly placed in order to approach a structure invisible to the human eye. There is, however, an irritating rub: before "proving" a suitable model, an enormous amount of X-ray data must be acquired. Anne Sayre explained this dilemma succinctly: "Quite plainly, if nothing is known about a substance, a model cannot be built at all; if too little is known, then any model produced will be vague, uncertain, composed chiefly of hope, and insufficiently verifiable."[7]

So, which method does one choose? The scientifically obvious, albeit time-consuming method was not to make a choice but, instead, to painstakingly gather data in the form of hundreds or thousands of X-ray diffraction images, then to apply mathematical formulae, and *then* to construct a three-dimensional model. Franklin was comfortable settling into such labor-intensive chores; Watson and Crick were not. Dorothy Hodgkin thought that Franklin's troubles had much to do with the fact that "she was necessarily involved in collecting the accurate data on DNA, first, within the framework of which the model should be built. It was natural for her to postpone model building until her data collection was complete, and until she had extracted all the information she could from the data that would limit the kind of model she should build." Had Hodgkin been there to advise Franklin, she would have likely told her that, more than a year before Watson and Crick solved the double helix, "there were enough pieces of general information available about the geometrical form of the bases, sugar and phosphate groups, to make model building a reasonable course to pursue separately."[8]

WITHOUT ACCESS TO Franklin's X-ray data, Watson and Crick ran headlong into their own set of roadblocks. The first barrier was not having enough satisfactory, precision-cut metal model "atoms" lying about the Cavendish Laboratory. In the not-so-distant past, all good chemistry and physics laboratories maintained a well-equipped workshop. Populated by a staff of glassblowers, metalsmiths, tool and die machinists, and other skilled craftsmen, these workers fabricated the apparatus the scientists designed in order to conduct their experiments. About a year and a half earlier, Kendrew had attempted some model work on a polypeptide amino acid chain, but while he still had plenty of carbon, nitrogen, and hydrogen pieces to share, "there were no accurate representations of the groups of atoms unique to DNA. Neither the phosphorus atoms nor the purines and pyrimidine bases were on hand." Constructing these model pieces to scale could take weeks since there was "no time for Max [Perutz] to give a rush order."[9]

The delay forced Watson to begin his Monday morning "adding bits of copper wire to some of our carbon atom models, thereby changing them into the larger sized phosphorus atoms."[10] He next tried fashioning some other pieces to represent the inorganic ions that he thought might attach to the molecule, but with less success because he had no clear idea of how to account for their bond angles. The conclusion from his soon-aborted thought experiment was one Franklin had emphasized in her lecture: "we had to know the correct DNA structure before the right models could be made."[11]

An idle Watson awaited Crick's arrival filled with the hope that he would pull a metaphorical rabbit out of his hat and deliver the answer to their problem.[12] When Crick finally did arrive, at "ten-ish," he confessed that he, too, was stumped. He barely tried to figure things out and, instead, devoted most of his Sunday off to reading *A Perch in Paradise*, a somewhat salacious and entirely forgotten novel about the sexual exploits and peccadillos of Cambridge dons.[13] The only intellectual puzzle on Crick's mind was which character in the novel represented which of his many friends, acquaintances, and colleagues.[14]

Over cups of coffee, they mulled over a few possible configurations, played with Watson's makeshift pieces, and pondered the (incorrect) X-ray diffraction data Watson vaguely remembered from Franklin's talk. Watson still expected the solution would suddenly present itself, in epiphanic fashion, simply by his "concentrating on the prettiest way for a polynucleotide chain to fold up."[15]

Finding no suitable answers and with stomachs growling, they left the

Cavendish for their habitual lunch at the Eagle. Crick silently chewed on the problem he had just left behind. The logical next step, of course, was a phone call to Rosalind Franklin, inviting her to tea and offering to combine their respective talents in solving the structure of DNA. They had had little contact with her, and their impressions of her, drawn exclusively from Wilkins, were of a fierce and condescending woman. Despite her impeccable credentials, these two young men elected to treat her as an interloper who did not belong in the macho endeavor of physics. As Crick bluntly admitted, "I'm afraid we always used to adopt—let's say a patronizing attitude towards her."[16]

While tucking into his lunch, Francis decreed that the model-building would begin in earnest as soon as they returned to the lab. First, they had to decide whether the structure contained one, two, three, or four strands of nucleotides somehow joined together in a helical configuration. They then guesstimated a three-stranded structure joined by "salt bridges in which divalent cations like Mg++ held together two or more phosphate groups."[17] Crick posited that calcium ions might possibly hold together the sugar–phosphate backbone. This turned out to be a purely cationic fantasy; at no point did Franklin, Wilkins, or Alec Stokes ever suggest the presence of divalent cations forming bonds or molecular bridges. Watson admitted that in taking such a tack they "might be sticking our necks out." But instead of figuring out another solution to the problem, he blamed the King's College team for their lack of imagination with respect to models and for failing to identify which salt was present (in fact, they did identify it as being sodium). As they gobbled down slices of gooseberry pie, Watson and Crick hoped that the ad hoc addition of magnesium or calcium ions to the sugar–phosphate backbone "would quickly generate an elegant structure, the correctness of which would not be debatable."[18]

Unlike Linus Pauling's "pretty" model of the α-protein, Watson and Crick's triple-stranded monstrosity was anything but lovely to look at. The three chains of model molecules were braided, repeating every 27 Ångstroms along the helical axis. The pincers holding the little pieces of metal and wires were awkwardly bound to a ring stand pilfered from a neighboring lab. Several of the proposed atomic contacts were too close for chemical comfort. Although Watson and Crick could see that their model was a poor fit, they remained deliberately blind to the fact that they were hurtling down the wrong track.

Breaking for dinner at Crick's home, the duo tried to explain their day to Odile. She bubbled with joy, assuming that their discovery was so important that it might bring some income into the Cricks' depleted bank account and allow them to buy a new car or move into a larger house. Watson later made light of Odile's convent upbringing, the "artsy-craftsy world" in which she moved, her wayward ways in handling money, and her poor understanding of the concept of gravity (she insisted that it ended three miles above the earth). He concluded that her weak grasp of science made any effort to discuss such matters with her a complete waste of time.[19]

The following morning, Watson and Crick fiddled with their model a bit more, satisfying themselves that it "fit the seminar facts [Watson] had conveyed to Francis." Thanks to the comfortable perch of hindsight, it is easy to conclude that their model was hopelessly incomplete and incorrect. Yet, as Crick explained, while conducting research on the vanguard, the scientist is "always in a fog."[20] He was likely thinking of this episode when he wrote that perfect phrase.

Fortunately, just as we have a blueprint of Franklin's thoughts about DNA on November 21, 1951, we also have a first draft of Crick and Watson's ideas during the final week of November 1951. Crick used a broad-nib fountain pen to write an eighteen-page manuscript by "Crick and Watson"—an authorial order that would soon change forever. The words burst off the pages, giving the reader the impression that the royal blue ink could hardly wait to get out of his pen in a virtual storm of ideas. They were, Crick wrote, "stimulated by the results by the workers at King's College, London, at a colloquium on 21st Nov. 1951," and had "attempted to see if we can find any general principles on which the structure of D.N.A. might be based. We have tried, in this approach, to incorporate the <u>minimum</u> number of experimental facts, although certain results have suggested ideas to us."[21]

On November 26, after completing the final page and, literally, bending the copper wires of their model to his will, Crick picked up the telephone and asked the Cavendish operator to place a call to Maurice Wilkins in London. Confidently, he told Wilkins that he and Watson had modeled the double helix of DNA. Before Wilkins could stammer out an answer, Crick invited him to come to Cambridge as soon as possible. Later that day, John Kendrew came into their office and diplomatically inquired how Wilkins had taken the news. Crick was so thrilled by his potential discovery that he had not a clue and, according to Watson, declared that "it was almost as if Wilkins were indifferent to what we were doing."[22] In retrospect, it is impossible to believe this statement. It was hard enough for Wilkins to deal with Franklin working at odds with him in his own laboratory. He must have quaked and quivered with anger over potentially being scooped by the Cambridge interlopers. It would hardly be the last time the taste of bile rose into his throat upon hearing the latest news from the Cavendish.

Watson's claim that Wilkins had no interest in building a DNA model was inaccurate and self-serving. Wilkins was in fact pursuing a triple helix model at the same time Watson and Crick built their model. The day after Franklin's lecture, an Australian physicist named Bruce Fraser, one of the youngest members of the King's College team, popped his head into Wilkins's office "with a mysterious smile." He beckoned Wilkins to come into his lab, next door. Fraser and his superior, a biophysicist named William Price, were using infrared spectroscopy to analyze the chemical bonds of DNA. Wilkins could not help making an indirect dig at Franklin by adding, when he recounted this event in his memoir, that "the interaction between Price's group and ours was another good example of the co-operative spirit in the lab."[23]

With the flourish of a sculptor unveiling his latest creation, Fraser displayed a DNA model with three helical chains demonstrating "the right pitch, diameter and angle, and . . . linked together by hydrogen bonds between the flat bases which were stacked on each other in the middle of the model."[24] Fraser built his model based on the data Franklin had presented the day before along with "the general state of thinking in our lab." Fraser (and Wilkins) also borrowed from J. M. Gulland's work at Nottingham University, which demonstrated how the DNA chains "were joined together by inter-base hydrogen bonds," and from the work of a Norwegian physicist named Sven Furberg.[25] In his 1949 doctoral dissertation under the direction of J. D. Bernal at Birkbeck College, London, Furberg hypothesized a single-stranded (and ultimately unstable) DNA model known as the "zig-zag chain."[26] Although he was incorrect in this assumption, Furberg did establish the important fact that the plane of the purine or pyrimidine base "is almost perpendicular to the plane in which most of the sugar atoms lie."[27]

Fraser's molecular tower had its own set of structural problems. Most glaring were the three chains he built. They were equally spaced, in a manner that failed to match either the X-ray diffraction patterns or a precise 1:1 ratio of purines and pyrimidines recently detected by the New York–based biochemist Erwin Chargaff. Fraser and Wilkins stared at the model for several days, but nothing came to either man's mind as to what to do "with the three helices. We found ourselves completely stuck."[28]

Allowing his animosity for Franklin to once again get the better of him, Wilkins blamed this mistake on her so-called anti-helical views and for suggesting that DNA was configured in three chains.[29] Yet in her lecture notes and interim report, she actually posited "2, 3, or 4 co-axial nucleic acid chains per helical unit."[30] Ultimately, Wilkins was forced to accept Franklin's insistence that there was "no use at all in trying to build more models" until they had more data to guide them. What he could not do was stop scapegoating her for leading him astray: "Thinking that there were three chains," Wilkins whined, "had completely stopped us in our tracks. Our main mistake was to pay too much attention to [Franklin's] experimental evidence."[31]

MINUTES AFTER CRICK'S PHONE CALL, Wilkins "rushed round the labs" to tell everyone about the invitation to Cambridge. After calling his troops to arms, he informed Crick that he would be on the 10:10 train the following morning, Wednesday, November 27. He paused for a moment—

King's Cross station.

perhaps for drama, but more likely the result of his plodding manner of speaking—before telling Crick that "Rosy, together with her student R. G. Gosling, would be on the same train," as would Fraser and Willy Seeds. Watson could not help but snarl over their collective enthusiasm: "Apparently they were still interested in the answer."[32]

The next morning, the five King's College scientists met in the vast King's Cross train station to make the reverse trip that Watson had made six days earlier. On the way north, Wilkins offered clumsy bits of conversation but was ignored by Franklin, who stared out the window at the countryside dotted with farms, long stretches of pasture, cylindrically rolled bales of hay, and sleepy cows. The silence in the compartment was in direct contrast to the internal apprehension felt by all five of these young adults. "We knew that Francis and Jim were very bright," Wilkins said, describing the hour-long journey, "and we wondered what they had come up with."[33]

Upon arrival in Cambridge, the immediate decision facing the King's group was how to travel the final 3.4-mile distance to the Cavendish. Wilkins suggested they split the fare for a cab. Ever the contrarian, Franklin insisted on taking the bus. One can only imagine the level of discomfort had she joined the four men, knees touching and shoulders huddled together, in the back of a black taxi.

They regrouped in Free School Lane. Following Wilkins as ducklings follow their "papa duck," the physicists made their way into the Austin Wing of the Cavendish Laboratory. Wilkins kept up his inane banter, hoping to shore up the morale of his team who, for all they knew, were about to be bested by a couple of non-experimentalists. Ignoring his cheerful demeanor, Franklin focused her attention on Gosling, who was increasingly uncomfortable at being placed in the middle of a spat between his two senior colleagues.[34]

According to Watson, Wilkins "poked his head" into Room 103 to announce their arrival, thinking "that a few minutes without science was the way to proceed. Rosy, however, had not come here to throw out foolish words, but quickly wanted to know where things stood."[35] Watson went on to elaborate a longer narrative of the visit's progress, beginning with Max Perutz and John Kendrew welcoming Wilkins and the others to the Cavendish, then begging off so that Crick, ever the showman, might have his day in the sun. Watson and Crick had planned to begin with Crick teaching the King's scientists about "the advantages of the helical theory" and how "Bessel functions gave neat answers." The two would then outline the assumptions and facts of their model, followed by a friendly lunch at the Eagle and a return to the lab for an "afternoon free to discuss how we could all proceed with the final phases of the problem."[36] Although the day did begin with Crick introducing his helical theory, he was soon interrupted by Wilkins, who informed him that "without all this fanfare, Stokes has solved the problem in the train while going home one evening and had produced the theory on a small sheet of paper the next morning."[37]

The oft-told version of what happened next has Franklin inspecting the model and, "tickled pink," laughing at Watson and Crick's sorry effort, then proceeding to shoot their work down with the accuracy of a sniper. Some accounts even have her squealing with delight, "Oh look, you've got it inside out!"[38] This was the way Watson and Crick recalled her critique. Gosling was more nuanced in his recollection:

[O]nce in their lab and in front of their model, our relief must have been palpable. Rosalind let rip in her best pedagogic style "you're wrong for the following reasons . . ." which she proceeded to enumerate as she demolished their proposal . . . It also confirmed Rosalind's view that one could build atomic models "until the cows came home" but it would be impossible to say which were nearer the truth. If Maurice would back off and let us (she and I) get on with the measurement of the diffracted intensities and the admittedly slow and painstaking calculations, then ultimately, "the data would speak for itself."[39]

Watson angrily countered that Franklin "did not give a hoot about the priority of the creation of helical theory and, as Francis prattled on, she displayed increasing irritation. The sermon was unnecessary, since to her mind there was not a shred of evidence that DNA was helical. Inspection of the model itself only increased her disdain. Nothing in Francis's argument justified all this fuss." In his memory, she became "positively aggressive" when they segued to the topic of those pesky magnesium divalent ions, which Crick and Watson fantasized were the glue that held the strands of their triple helix together. The "Mg++ ions" would be "surrounded by tight shells of water molecules," she insisted, and could not possibly act as the "kingpins of a tight structure."[40] Franklin probably stated her criticisms calmly, confidently, and forcefully, as was her manner, hardly meriting his "positively aggressive" description. To Rosalind Franklin's data-driven mind, looking at Watson and Crick's deeply flawed model was akin to a skilled musician being forced to listen to a symphony riddled with wrong notes.

Regardless of the tone of her voice, once Franklin had pointed out the impossibilities of the triple helix model, the air escaped from Room 103. As Watson reported, Crick's "mood was no longer that of a confident master lecturing hapless colonial children who until then had never experienced a first-rate intellect." Despite a request by Crick and Watson to join forces, there would be no collaboration with the King's group.[41] Historian Robert Olby characterized the offer starkly: "Franklin and Gosling very understandably would have nothing to do with such a suggestion. They had witnessed two clowns up to pranks. Why should they condone such behavior by joining forces with them?"[42] Watson and Crick lost this tussle; the clear victor was Franklin.[43] As Crick plainly put it, "We made asses of ourselves."[44]

The rest of that day was uncomfortable for all the participants, but especially for Watson, who finally accepted that Franklin's annoying "objections were not mere perversity." By this point in the discussion, "the embarrassing fact came out that [Watson's] recollection of the water content of Rosy's DNA samples could not be right." The model underestimated the water content by a factor of ten, all because he had not listened carefully to Franklin's lecture. "There was no escaping the conclusion," he admitted, "that our argument was soft. As soon as the possibility arose that much more water was involved, the number of potential DNA models alarmingly increased."[45]

After lunch, the two teams took a walk along the backs and through the Great Court of Trinity College, but no amount of cajoling by Crick could convert the King's biophysicists. As Watson described it, "Rosy and Gosling were pugnaciously assertive: their future course of action would be unaffected by their fifty-mile excursion into adolescent blather." Wilkins and Seeds were more reasonable, but this may have simply been a function of their "desire not to agree with Rosy." The conversation slid further downhill when the young scientists returned to the Cavendish. Wilkins broke the silence with the wan observation that "if they moved with haste, the bus might enable them to get the 3:40 train to Liverpool Street Station." All that was left to say was a perfunctory goodbye.[46]

Many have tried to pinpoint the precise time when Watson began his decades-long hate affair with Rosalind Franklin. As point of privilege, this historian offers the moment in Room 103 when Franklin exclaimed something that Watson insisted sounded like, "Oh look, you've got it inside out!" More than fifty years later, Franklin's disdain—regardless of her precise words—still rang loudly in Watson's ears. In 2018, sitting in his Cold Spring Harbor office, with a bitter tone as if the event had happened that morning, he recalled, "She was never very nice to us, to me especially . . . Rosy always had to let you know that her brain was better than yours—even if that wasn't the case. She wasn't modest enough to know what she didn't know."[47]

WHEN BRAGG HEARD what had transpired one floor below his head-of-department office, he was livid. Perutz tried to calm him down, but Bragg wanted heads to roll. DNA was strictly in the King's College domain. Who was this Watson fellow to interfere with another Medical Research Council unit's work? And what the devil was Crick doing, taking time off from his own work and barging into the research of a colleague at an outside institution? Bragg thundered on that at this rate Crick would never complete his doctorate, a result he simply would not tolerate. As Watson observed, "Now when he should be enjoying the rewards accorded the most prestigious chair in science, he had to be responsible for the outrageous antics of an unsuccessful genius."[48]

The telephone wires between Bragg's book-lined study in Cambridge and Randall's basement warren in London must have burned hot with abuse and apologetic words. Randall had already heard about the expedition from Wilkins and was, understandably, furious.[49] Although there is no record of what Randall and Bragg discussed that afternoon, we have superb indirect evidence thanks to the recent discovery of "the lost correspondence of Francis Crick." These letters were wedged into boxes containing the papers of Sydney Brenner, the 2002 Nobel Prize laureate for Physiology or Medicine, who shared a Cambridge University office with Crick from 1956 to 1977. In a quirk of historical record-keeping, these letters were not discovered until 2010, when Brenner donated his (and nine more boxes of Crick's) papers to the Cold Spring Harbor Laboratory archives.[50]

In a typed letter dated December 11, 1951, Wilkins formalized a peace treaty of sorts with Crick. It began with a warm "My Dear Francis," and an apology for "rush[ing] off on Saturday without seeing you again." Little else in the letter is dear or warm, because it was copied to (and probably dictated by) John Randall:

> I am afraid the average vote of opinion here, most reluctantly and with many regrets, is against your proposal to continue the work on n.a. [nucleic acids] in Cambridge. An argument here is put forward to show that your ideas are derived directly from statements made in the colloquium and this seems to me as convincing as your own argument that your approach is quite out of the blue . . . I think it most important that an understanding be reached such that all members of our laboratory can feel in future, as in the past, free to discuss their work and interchange ideas with you and your laboratory. We are two M.R.C. Units and two Physics Departments with many connections. I personally feel that I have much to gain by discussing my own work with you and after your attitude on Saturday begin to have very slight uneasy feelings in this respect. Whatever the precise rights or wrongs of the case I think it most important to preserve good inter-lab relations. If you and Jim were working in a laboratory remote from ours our attitude would be that you should go right ahead. I think it best to abide by the view taken by the majority of the structure people here and your unit as a whole. If your unit thinks our suggestion selfish or contrary to interests as a while of scientific advance, please let us know. I suggest you show this letter to Max for his information, and having discussed the matter with Randall I am, at his request, letting him have a copy.[51]

A few hours later, Wilkins sent off a handwritten note to Crick, without John Randall's hot breath bearing down on his neck. The second letter far better reveals the nature of their long friendship and Wilkins's advice for ameliorating the awkward situation:

This is just to say how bloody browned off I am entirely & how rotten I feel about it all & how entirely friendly I am (though it may possibly appear differently). We are really between forces which may grind all of us into little pieces. So far as your interests are concerned I do very much suggest it is best to make some sacrifices of credit for ideas in this connection. You can see how the wind is blowing when I say that I had to restrain Randall from writing to Bragg complaining about your behaviour. Needless to say I <u>did</u> restrain him, but so far as your security with Bragg is concerned it is probably much more important to pipe down & build up the idea of a quiet steady worker who never creates "situations" than to collect all the credit for your excellent ideas at the expense of good will. And you see it <u>does</u> make me a bit confused about our discussions if you get too interested in everything which is important; where I say confused I mean confused, I am now largely incapable of any logical thinking in relation to polynucleotide chains or anything. And poor Jim—may I shed a crocodile & very confused tear? & send him my best wishes & regards & friendly greetings to both of you & if you <u>should</u> have any ill feeling about the part I have played I hope you will tell me! Regards to John too![52]

Two days later, on December 13, Crick hand-wrote a characteristically charming letter to Wilkins:

Just a brief note to thank you for the letters and to try to cheer you up. We think the best thing to get things straight is for us to send you a letter setting out in a mild manner our point of view. This will take a day or so to do, so we hope you'll excuse the delay. Please don't worry about it, because we've all agreed that we must come to an amicable arrangement. Meanwhile, may we point out what a fortunate position you are in? It is extremely probable that in a short space of time you and your unit will have solved decisively one of the key problems in biomolecular structure. By doing so you will have opened the door to many of the really crucial biological problems. [So] cheer up and take it from us that even if we kicked you in the pants it was between friends. We hope our burglary will at least produce a united front in your group![53]

With DNA safely back in his domain, Wilkins bungled still another "fortunate position" to solve the double helix structure first. After the trip to Cambridge, he was sitting in his office, "slightly depressed," when Franklin entered "to discuss a new idea about helical DNA." Wilkins was surprised, because they had not spoken directly since "Stokes and I had fled from her outburst." Making sure he was not dreaming, he offered her a chair, and she proceeded to explain her thoughts about the new B pattern she had discerned. This was the same form that Stokes and Wilkins had already concluded was helical. To Wilkins's surprise, Franklin "had a very sensible thought." Specifically, she told him that "the relative intensities of the layer lines seemed to indicate that, in the DNA molecule, there were two concentrations of matter separated by three-eighths of the repeat distance along the length." Wilkins looked at the picture, which depicted a "helical molecule containing two clumps of matter separated by the three-eighths distance." Unfortunately, both he and Franklin were baffled by its meaning. Here, again, was a sterling opportunity for them to collaborate that passed almost as quickly as it appeared. For the next half century, Wilkins struggled to explain how "a mental block prevented Rosalind and me from seeing that the two concentrations of matter, separated by three-eighths of the repeat distance along the fiber, were the helical chains of two-chain DNA."[54] This may have been his way of elevating his thinking at the time to suggest that they were, together or separately, far closer to solving the DNA puzzle than they actually were in late 1951. At that moment, however, Wilkins was still entangled by his wishful thinking for a "more stable" three-chain structure. He had no idea how to proceed even when the answer was staring him in the face.[55] Francis Crick soundly derided Wilkins's subsequent claims of how close he had been to solving DNA: "He had as much information as we had and now he says he picked up the point in Chargaff's article, but he's talking through his hat. He may have been mooning about, but he didn't <u>see</u> it, and that's all there is to it."[56]

THE DNA MORATORIUM at Cambridge was ordered from on high, from Bragg to Perutz to John Kendrew and finally to Watson and Crick. Crick was told to stick to his last and finish up his doctoral thesis, while Watson was directed to work on tobacco mosaic virus (TMV), a single-stranded RNA virus named for the green and yellow speckles it creates on infected tobacco leaves.[57] The virus was much studied in the early days of virology and molecular biology, not only for the damage it brought upon the tobacco industry but also because it contained "ruthless Trojan horse techniques" to insert itself into the host cell and take over its reproductive mechanisms.[58]

After enduring Franklin's molecular censure, Watson was forced to admit that his triple helix model "smelled bad." To make the freeze even deeper, Bragg instructed Watson to send the molds, jigs, and other parts they had used to build their failed model to Wilkins at King's. Meanwhile, the squabbling between Wilkins and Franklin only intensified after their visit to Cambridge. "Rather than build models at Maurice's command," Watson joked, "she might twist the copper-wire models around his neck."[59] For nearly six months, the jigs and molds reposed in a battered cardboard box in a remote corner of the King's College biophysics empire. In June 1952, Wilkins asked Watson and Crick if they would like him to return the molds. The Cavendish duo said yes, "half implying that more carbon atoms were needed to make models showing how polypeptide chains turned corners."[60]

The ever-ready Jim Watson had no

Watson and Crick. Detail from a group portrait of the staff of the Cavendish Laboratory, 1952.

intention of abandoning his life's quest. Fortunately, "at no time did [John Kendrew] try to reinterest me in myoglobin." Kendrew sensed that Bragg's "moratorium on working on DNA did not extend to thinking about it."[61] At the lab, Watson gave the appearance of tinkering with TMV, which he called "the perfect front to mask my continued interest in DNA."[62] Surreptitiously, he spent the "dark and chilly" Cambridge winter learning more theoretical chemistry and perusing the genetics journals in the hope of finding "a forgotten clue to DNA."[63]

As his first term at Cambridge drew to a close and Christmas approached, Crick presented him with a gift: a second copy of the book they had scrambled to find at Blackwell's in Oxford a few weeks earlier, Linus Pauling's *The Nature of the Chemical Bond*. The volume was special because of Francis's inscription on the flyleaf: "To Jim from Francis–Christmas '51." Thumbing through Pauling's chemical masterpiece, Watson hoped he might still find some hints to unlocking DNA. The avowed atheist from Chicago recalled with irony his appreciation of Crick's gift: "The remnants of Christianity were indeed useful."[64]

CHRISTMAS 1951 also marked the point when Watson and Crick's wings were nearly clipped. Bragg remained so incensed about the triple helix fiasco that over the holidays he actively conspired to show Crick the door. On January 18, 1952, Bragg wrote a confidential letter to A. V. Hill, the muscle physiologist who had recruited Crick to Cambridge a few years earlier. The purpose of the missive was to banish Crick from the Cavendish:

> There is a young man working here, in Perutz's team, who I believe at one time was a protégé of yours and advised by you to take up biophysics. This is Crick. I am worried about him and if you take more than a passing interest in him, I should like to consult you about him. He is working for a Ph.D. here, though he is 35, because the war stopped him trying before. My worry is that it is almost impossible to get him to settle down to any steady job and I doubt

whether he has put enough material for his Ph.D. which should be taken this year. Yet he is determined to do nothing but research and is very keen to hang on here. With a wife and family, he ought to be looking out for a job. I think he overrates his research ability, and that he ought not to count on getting a job with no other commitments. Are you interested in his career enough to want to discuss it? I should like some help in deciding what line to take with him.[65]

Fortunately, Hill got Bragg to calm down, cease, and desist. The personality problems Crick had with Bragg and his gate-crashing on the King's College turf were, however, career-threatening. We now understand that no matter how irritating the ebullient Crick might have been, his scientific insights often proved invaluable. His grasp of biology, from theory down to the molecular level, was truly breathtaking. At this particular time, however, he had yet to demonstrate those unique talents to anyone but himself—and James Watson. For the moment, at least, he was safe at Cambridge—even if he did not know how close he had just come to being sacked.

PART IV

MORATORIUM, 1952

*Incidentally, a lot of those stories of old Jim Watson's are pure
imagination, about fights with Crick and all the rest of it . . .
he wasn't always accurate. That is not the book of a mature
man at all; it's really an almost literal transcript of his letters
to his father and mother when he was 25, so you've got to
remember that. A rather brash young man coming to Europe
for the first time [and] his violent reaction to it. . . . [it's] a
novelist's interpretation.*

—SIR WILLIAM LAWRENCE BRAGG[1]

Dr. Pauling's Predicament[1]

20 June 1952

To Whom it May Concern:
I am not a Communist.
I have never been a Communist.
I have never been involved with the Communist Party.

—LINUS PAULING[2]

L inus Pauling, like many highly accomplished men and women, attracted his share of enemies. His critics disparaged him as an academic exhibitionist. Those who admired his work worried that he often behaved like a self-appointed guru of all things scientific. The chemist only aggravated matters with his insatiable love for seeing his name in print in the scientific journals—where he published at an astounding rate—and on the pages of the daily newspapers, which loved to cover his exploits. Even his clothes drew attention. Instead of dressing in the standard American professor's issue of Harris tweed jackets, white Oxford cloth shirts, gray flannel pleated-trousers, dark knit ties, and heavy cordovan shoes, Pauling wore brightly printed sports shirts, baggy khaki pants held up by colorful suspenders, open-toed sandals, and jaunty berets. The tangents of his long, wispy gray-white hair approached that of the world's most tonsorially challenged scientist, Albert Einstein. He was the archetypical academic, who both loved the decorum of university life and the freedom to taunt it. As he explained, "there were two qualities of my personality pushing in opposite directions: the one to conform, and the other to rely on my own assessment of the situation."[3]

In the early 1950s, Linus Pauling waged battle on two fronts. The first involved conquering new vistas of molecular biology and chemistry. The second was in his political activism, and it captured the full attention of the U.S. government. Thick in the middle of the McCarthy era, when careers and lives were ruined by the mere pointing of a finger, Pauling had a red bull's eye painted on his back. With each successive appearance he made in the political arena, the circles on that target only widened, alerting more foes to shoot him down.[4]

Many liberals of this period felt betrayed by President Harry Truman's political drift to the right, one that only became "righter" with the rise of the Communist "Red Scare," government-required loyalty oaths, and the United States's incursion into war on the Korean peninsula. Pauling grew particularly cross over the shifting winds and articulated his views on the radio, in newspapers, and at protest marches, criticizing the very people who had the power to make his life miserable. Pauling's politics were actually aligned with those of Franklin D. Roosevelt's old New Deal Democrat coalition, but his independence of thought, charisma, fame, and vociferous support for ultra-leftist causes aroused a great deal of public attention. At worst, he might have been what the FBI director J. Edgar Hoover characterized as a "fellow traveler" of the Communist Party. In reality, he was more of a "peacenik," an activist opposed to war in all forms. During the Second World War, Pauling refused to apply for a U.S. government security clearance, nor did he contribute to the building of the atomic bomb. From the 1940s to the end of his life, he was a leading figure in the anti-nuclear weapon and antiwar movements, which won him the 1962 Nobel Peace Prize.

As Senator Joseph R. McCarthy's influence infected the nation, Pauling made himself especially vulnerable to attack by joining groups with known (or perceived) ties to the Communist Party of America, including the Progressive Citizens of America, the Independent Citizen's Committee of the Arts, Sciences and Professions, and the American Association of Scientific Workers (an offshoot of an international organization run by the Nobel Prize–winning nuclear physicist and founder of the French Communist Party, Jean Frédéric Joliot-Curie). Pauling grabbed the lapels of the Red-baiters even more tightly by serving as parole advisor to Dalton Trumbo, who wrote the screenplays for such Hollywood classics as *Roman Holiday, Exodus,* and *Spartacus* as well as the 1939 National Book Award–winning antiwar novel *Johnny Got His Gun.* Trumbo was one of the Hollywood Ten, a group of prominent screenwriters, producers, and directors investigated by the federal government for their Communist activities. In 1950, Trumbo served eleven months in federal prison for refusing to name names to the U.S. House of Representatives Un-American Activities Committee (HUAC). Pauling was also a vociferous defender of Julius and Ethel Rosenberg, who in 1950 were arrested for spying activities on behalf of the Soviet Union. He made several public pleas for their clemency before they were executed in 1953.[5] All the while he was embroiled in these controversial activities, the entire nation, as Jim Watson described it, was stuck in an "awkward cold war thought up by the American paranoids, who should be back in the law offices of middlewestern towns."[6]

As a result of these doings, Pauling endured a series of career-threatening investigations by HUAC, the FBI, the U.S. State Department, and the California Institute of Technology. Not a single inquiry managed to prove his membership in a Communist cell, nor did years of surveillance by FBI agents haunting his classroom and public lectures find flaws in his allegiance to his country. Nonetheless, the merest hint of Communist sympathies was enough to make one a social pariah in the early 1950s. When Pauling walked through the Caltech campus, his colleagues crossed the street to avoid greeting him. For a man who craved the attention of his peers, such Cold-War shoulders were torture. In 1950, after Senator McCarthy leveled false charges against him, a teary-eyed Ava Helen Pauling told one of his former

Pauling testifying on the denial of his passport, c. 1953–55.

students, "I don't know how my husband can hold up much longer."[7] But he *did* hold up and it was his science that saved him.

IN THE FALL OF 1951, Pauling received an invitation to speak before the Royal Society, the British equivalent of the United States's National Academy of Sciences and one of the most distinguished scientific bodies in the world. Set to take place on May 1, 1952, this was no mere lecture. Pauling was asked to present his research on the molecular structure of proteins, much as an erudite lawyer is called to plead an important case before the Supreme Court. Many of the world's savviest chemists, biologists, and physicists—all of whom had questions about the α-protein helix—would be in the audience, armed with well-thought-out comments and critiques. As he prepared his lecture, Pauling considered including something about nucleic acids. He was long acquainted with Oswald Avery's work on the transformative substance observed in pneumococcus, but initially found the work to be unimportant: "I didn't accept it. I was so pleased with proteins, you know, that I thought that proteins probably are the heredity material, rather than nucleic acid, but of course that nucleic acid played a part."[8]

After receiving the lecture invitation, Pauling wrote to Maurice Wilkins and, subsequently, to John Randall asking to see Wilkins's "good fiber pictures of nucleic acid." Both requests were denied.[9] Wilkins, who often hectored others on the importance of openness in science, tended to make exceptions to that rule when it came to sharing his own unpublished work. In 1997, Wilkins recalled his rejection of Pauling's request in a characteristically muddled fashion: "I said, 'No, thank you very much for asking,' or something, that 'we need more time,' we wanted to spend more time looking at them ourselves. I don't feel ashamed of saying 'we'd like more time, if you don't mind.'"[10]

Unfazed by the rejections from Randall and Wilkins, Pauling put DNA aside and concentrated on preparing his Royal Society lecture. For the next several months, he and Robert Corey "tested, refined, and rethought their [protein] structures."[11] A more mundane task was to renew his U.S. passport. In the section of the application form asking the reason for his travel, Pauling listed, "scientific purposes—to take part in a discussion meeting on the structure of proteins that had been arranged by the Royal Society of London for May 1, 1952, to give lectures on scientific subjects before universities, to discuss scientific questions, especially the structure of proteins with foreign investigators, and to receive an honorary doctorate (*Docteur de l'Université*) from the University of Toulouse."[12] On a crisp fall morning in Washington, DC, his application found its way to the desk of Mrs. Ruth Bielaski Shipley, chief of the U.S. State Department's Passport Division from 1928 to 1955.

Mrs. Shipley wore a daily uniform of prim, dark woolen or linen suits, topped by hats resembling collapsed soufflés.[13] Beneath such millinery disasters was a tightly knotted bun of blue-gray hair. Her shark-like eyes were obscured by old-fashioned pince-nez glasses anchored by a black ribbon pinned to her dress. The corners of her mouth curled downward as if locked in a perpetual frown. Beneath that stern visage was the intense pride she took in reviewing every passport request that came to Washington, even though she had a staff of over two hundred people to handle the daily avalanche of applications.

Taking the Subversive Activities Control Act of 1950 as her God-given commandment, Mrs. Shipley numbered its principal sponsor, Senator Patrick McCarran, secretaries of state Cordell Hull, Dean Acheson, and John Foster Dulles, FBI director J. Edgar Hoover, and Senator Joseph McCarthy's legal counsel Roy Cohn as friends and admirers. President Franklin D. Roosevelt cautiously praised her as a "wonderful ogre."[14] *Time* magazine described her as "the most invulnerable, most unfirable, most feared and most admired career woman in Government."[15] *The Reader's Digest* called Shipley "the State Department's watchdog" and informed its 40 million readers that "no American can go abroad without her authorization. She decides whether the applicant is entitled to a passport and also whether he would be a hazard to Uncle Sam's security or create prejudice against the United States by unbecoming conduct."[16]

In retrospect, it is astounding to think that one woman, neither elected nor congressionally confirmed, had the full and singular author-

Ruth Shipley at the beginning of her career at the U.S. State Department, 1920.

ity "to comply with or to deny applicants."[17] Technically, Shipley was supposed to submit the more complex cases to "a board of advisers who constitute a supreme court of arbitration on the matter."[18] More times than not, whenever she detected the faintest scent of Communist sympathy, Mrs. Shipley "did her duty as she saw it" and reached for a large rubber stamp spelling out "REJECT," pressed it onto a bright red inkpad, and resolutely banged it on the application.[19] Among her most famous rejections were for the playwrights Arthur Miller and Lillian Hellman, the singer, actor, civil rights activist, and pro-Stalinist Paul Robeson, the sociologist, scholar, and civil rights activist W. E. B. DuBois, the Manhattan Project physicist Martin D. Kamen, and Jim Watson's PhD advisor, Salvador Luria of Indiana University.[20]

By January 24, 1952, Pauling was worried that he had not heard anything back from Shipley's office, wrote to her about his passport renewal. Three weeks later, on February 14, Mrs. Shipley sent him a typewritten letter that was impossible to confuse with a Valentine's Day card:

> My dear Dr. Pauling:
> In reply to your letter of January 24, you are informed that your request for a passport has been carefully considered by the Department. However, a passport of this Government is not being issued to you since the Department is of the opinion that your proposed travel would not be in the best interests of the United States. The passport fee of $9.00 which accompanied the application you executed on October 17, 1951 will be returned to you at a later date.
> Sincerely,
> R. B. Shipley
> Chief, Passport Division[21]

This was hardly a snap decision. Mrs. Shipley had been following Pauling's activities for at least four months. In October 1951, she asked for and received a State Department investigation document on Pauling, which consisted of a close reading of his FBI file. In it, an anonymous source called the chemist "a professional do-gooder" whose wife goaded him into the political arena. The source went on to describe Ava Helen Pauling as a "complete fool with regard to politics" who "assures her husband daily on the hour that he has one of the three greatest minds in the world today, and that he should not deny the uninformed and ignorant of his leadership and ability."[22] For Shipley, this report provided "good reason to believe that Dr. Pauling was a Communist."[23]

She picked on the wrong man. Pauling saw the denial of his passport as the perfect opportunity to raise awareness of governmental capriciousness. On February 29, he fired off a letter to President Harry Truman, who only four years earlier had awarded Pauling the Medal of Merit for "exceptionally meritorious conduct in the performance of outstanding services" during the Second World War.[24] Pauling beseeched the commander in chief to "rectify this action, and to arrange the issuance of a passport to me. I am a loyal and conscientious citizen of the United States. I have never been guilty of any unpatriotic or criminal act."[25] Yet even the president of the United States was unwilling to check Shipley's absolute power. Doing his best imitation of Pontius Pilate, the commander in chief replied that it was a Passport Division issue. Shipley rejected the appeal and Truman remained silent.[26]

There were protests from colleagues, a plea from the head of the National Academy of Sciences, and, finally, a visit by Pauling to Shipley's office in Washington. Seated across the martinet's gunmetal gray, Steelcase of Grand Rapids desk, the professor explained the importance of his trip. He voluntarily added, under oath, that he was not and never had been a member of the Communist Party. Mrs. Shipley was unmoved. On April 28, hours before the last plane that could get him to London in time for his lecture left the gate, he received a telegram from Foggy Bottom. The bottom line was: no passport.

ON MAY 1, 1952, Robert Corey wobbled and hitched his crutches to the lecturer's bench of the stately, semicircular lecture hall of the Royal Institution, London. Standing directly under a portrait of King Charles II, he read Pauling's lecture—but his delivery was dry, hesitant, and uninspiring. The Caltech crystallographer Edward Hughes, who also spoke on Pauling's behalf, fumed at the response. "For the rest of the day," Hughes remarked, "the Englishmen sat there, telling us what was wrong."[27]

Right, wrong, or left, Europe wanted Pauling. The international scientific community made him a bigger star than he already was by writing editorials, offering glib quotes of condemnation in newspapers around the globe, and organizing protests against the actions of the U.S. government.[28] The Nobel Prize–winning chemist Sir Robert Robinson wrote a letter to *The Times*, published on May 2, chastising the U.S. government's "deplorable" action. A State Department official stationed at the U.S. embassy in Grosvenor Square, London, sent the clipping, via diplomatic pouch, directly to Secretary of State Dean Acheson with a cover letter stating that "this one case is resulting in a definite and important prejudice to the American national interest."[29] For days, Pauling's political quarantine dominated the front pages of the London newspapers. Across the English Channel, French scientists heaped even more scorn on the U.S. State Department. To make their outrage louder, they named Pauling honorary president of the Second International Congress of Biochemistry, to be held in Paris that July.[30]

In Washington, the heat index matched the public furor over the government's grounding of Linus Pauling. So many prominent scientists and citizens wrote to their congressmen that several members of the U.S. House and Senate—including senators Henry Cabot Lodge, Jr., and Richard Nixon, hardly pro-Communists—demanded that the State Department explain the passport refusal. As Pauling grumbled to a reporter for the California Institute of Technology student newspaper, *Tech*, "This whole incident, to be blunt, stinks."[31]

Mrs. Shipley refused to recant her position. In a memorandum dated the next day, she scoffed, "As I had to defer to the scientists on scientific matters on which they were expert, they would have to defer to the Department in so technical a matter as the refusal of a passport."[32] Reportedly, Secretary of State Acheson was surprised to learn that U.S. citizens whose applications were rejected by Shipley did not have an appeals process available to them. To stanch the flow of State Department blood, Acheson quietly decreed that Pauling could have a limited passport to conduct his academic business in England and France, as long as he confirmed what was already well known: he was not a member of the Communist Party.[33] There was no public announcement or apology, nor was Acheson's name listed anywhere on the corrective memoranda to show that Mrs. Shipley had been overruled by her boss. On July 11, Pauling presented himself at the Los Angeles Federal Building where, once again, he signed an affidavit attesting to the fact he was not and never had been a Communist. His "limited passport" was granted three days later, on July 14. He was in the air on the 16th, to New York, thence to London on the 18th and Paris on the 19th.[34]

Although Pauling—and the cause of international scientific cooperation—won this round, in the bigger picture he took a severe licking. In fact, Mrs. Shipley's denial of his passport played a crucial role in blocking his path toward solving the structure of DNA.[35] Had Pauling been allowed to go to London earlier, he undoubtedly would have visited King's College; and had he done so, Rosalind Franklin would likely have shown him her latest X-ray images. By May 1952, she had developed a sharp new picture of the B, or wet, form of DNA, which eventually ruled out the three-stranded structure that so distracted Watson, Crick, Wilkins, and Pauling and demonstrated "the cross-like reflections" of the double helix. As Wilkins later confessed to a BBC reporter, if Pauling had merely presented himself at the King's College laboratory unannounced that spring, "I'm sure I couldn't have resisted showing him every damn thing we had. Because [he] was this sort of god-like presence. It would be such an honor to show these things."[36]

Chargaff's Rules

[In 1944] there appeared a publication by Avery and collaborators on the mechanism of the so-called Griffith phenomenon, the transformation of one pneumococcal type into another ... This discovery, almost abruptly, appeared to foreshadow a chemistry of heredity and, moreover, made probable the nucleic acid character of the gene. It certainly made an impression on a few, not on many, but probably nobody on a more profound one than on me. For I saw before me in dark contours the beginning of a grammar of biology. Just as Cardinal Newman in the title of a celebrated book, The Grammar of Assent, *spoke of the grammar of belief, I use this word as a description of the main elements and principles of a science. Avery gave us the first text of a new language, or rather he showed us where to look for it. I resolved to look for this text.*

—ERWIN CHARGAFF[1]

Maurice Wilkins long stewed over being the "third man of the double helix," but the truly forgotten man in this endeavor was the Austrian émigré Erwin Chargaff. Born in 1905, Chargaff was the scion of a middle-class Jewish family who moved from Czernowitz to culturally rich Vienna while he was still a child.[2] By his early teens, he was fluent in five languages (Greek, Latin, French, German, and English) and well schooled in history, mathematics, literature, music, "a little physics, and a ridiculous quantity of 'natural history' [*Naturphilosphie*]."[3]

Every weekday morning, on his way to the 9th District Maximilian Gymnasium (the Austro-Hungarian equivalent of an elite U.S. high school), Chargaff "passed the house on Berggasse where, at the entrance door, a plaque announced the office of 'Dr. S. Freud.' This meant nothing to me," he later recalled. "I had not heard the name of the man who discovered entire continents of the soul that, arguably, might better have been left undiscovered."[4] When he matriculated into the University of Vienna in 1923, at age eighteen, Chargaff pursued chemistry instead of the humanities, on the grounds of job security and earning power. A self-thwarted litterateur, Chargaff dotted his cluttered prose and Mitteleuropean-accented conversation with obscure references to books, music, and art—many of them requiring a well-stocked library to decode.

Five years later, having earned his doctorate, Chargaff traveled to America for a fellowship in chemistry at Yale. As a Jew living in an affluent Protestant, New England town, Chargaff resented being at the bottom of New Haven's "caste-conscious" hierarchy. While visiting home in 1929, he married Vera Broida, whose family had emigrated from Vilnius, Lithuania, to Vienna. In 1931, they moved to Berlin, where Chargaff won an appointment as a "chemical assistant" in the Institute of Hygiene at the University of Berlin. He toiled at his laboratory bench for three years until "the tramp of marching boots" motivated him to leave Hitler's Germany for a two-year stint at the Institut Pasteur in Paris under Albert Calmette, a direct disciple of the great Louis Pasteur and creator of a tuberculosis vaccine.[5] Chargaff quickly realized that even cosmopolitan Paris would not provide a haven from the poisonous spread of Nazism,[6] so in 1935, the couple moved again, this time to New York, where Chargaff had been offered a position in the biochemistry department at Columbia University's College of Physicians and Surgeons. He stayed at Columbia for the remainder of his academic career, taking the C-train subway every morning from his thirteenth-floor apartment on Central Park West and Ninety-Sixth Street to his cluttered laboratory at the Columbia–Presbyterian Medical Center in Washington Heights.[7] Despite decades spent in the United States, the displacement from the land of his youth—combined with the horrific deaths of so many relatives during the Holocaust—left him feeling "rootless," devoid of (as he perversely employed Hitler's infamous expression) "blood and soil."[8]

Erwin Chargaff.

Chargaff had spent nearly a decade studying the chemistry of the human blood clot-
ting system when he first read Oswald Avery's
landmark 1944 paper on the transformation
factor of DNA. Avery's work so captivated
him that he "abruptly" turned "the rudder"
guiding his research.[9] In addition to Avery's
work, Chargaff was also "deeply impressed
by a little book written by the great Austrian
physicist Erwin Schrödinger which carried
the modest title *What Is Life?*"—the same
volume that brought Watson, Wilkins, and
Crick to the genetic quarry.[10] Chargaff spent the rest of his career exploring the cell nucleus, "which was also known as the seat of the—at that
period—still mythical units of heredity, the genes."[11]

CHARGAFF'S WORK WAS VITAL to unlocking the double helical structure
of DNA. Between 1944 and 1950, his laboratory developed the methods of partition chromatography and ultraviolet spectrophotometry to
determine the "differences in content and order of the DNA [nucleotide]
bases, i.e., the purines and pyrimidines."[12] He published a dense thicket
of findings that later became synopsized as Chargaff's Rules. Specifically, he demonstrated that while each species contains its own specific proportion of nucleotide bases, the molar (molecular) ratios of the
purine-to-pyrimidine nucleotide bases "approached" 1:1; in other words,
the amount of adenine "closely approximated" thymine, as did guanine
in relation to cytosine.[13] Unfortunately, Chargaff's cautious nature led
him to write in 1950 "whether [the 1:1 ratio] is more than accidental, cannot yet be said."[14] As it turns out, the ratio is *exactly* 1:1, or A=T and G=C,
and not at all accidental. But as Chargaff later bemoaned, "my greatest
defect as a scientist—and one of the explanations of my lack of success—
is probably my reluctance to simplify. In contrast to many others, I am
a 'terrible complexifier.'"[15]

THE 1:1 RATIO EVENTUALLY SERVED as Watson and Crick's "open ses-
ame" in discovering the structure and function of DNA. How, then, was
Chargaff not able to take things further and make sense of the genetic
implications of his findings?[16] One explanation is that he worked from
the nineteenth-century Germanic scientific premise that "there must be
one level on which all of life is chemical." As such, he relied almost
exclusively upon the titration, purification, and distillation methods of a
bench biochemist. Unlike Linus Pauling and Watson and Crick, Chargaff failed to appreciate the essential three-dimensional structure of the
atoms and molecules composing DNA. He had no clue how to use or
interpret X-ray crystallography images and derided molecular biology as
"essentially biochemistry without a license."[17]

DURING THE FIRST SIX MONTHS of the moratorium on DNA research
that Bragg had decreed at the Cavendish, Jim Watson took hundreds of
photographs of tobacco mosaic virus (TMV) specimens "with a powerful rotating anode X-ray tube which had just been assembled."[18] Unable
to contain his work to regular hours, he often returned to the Cavendish after 10 p.m., when the big heavy doors on Free School Lane were
slammed shut for the night. To gain entry, he either bothered the porter,
asleep in his flat next door, or borrowed the only other key from muscle
physiologist Hugh Huxley. Luckily, the Kendrews imposed no curfew,
unlike his former landlady on Jesus Green, and he could stay out as late
as he liked. By late spring, he had enough evidence to detect a helical
pattern in TMV, but he also concluded that "the way to DNA was not
through TMV."[19]

One spring evening, Watson read Chargaff's papers about "the curious regularities in DNA chemistry" and told Crick about them the next
morning, but "they did not ring a bell, and he went on thinking about
other matters."[20] Crick's bell was not rung until a few weeks later, while
he was enjoying a pint of beer at the Bun Shop with John Griffith, a
theoretical chemist with an interest in biochemical genetics.[21] The two
men had just attended a lecture given by the astronomer Thomas Gold

on his "steady state model," a long-since discarded alternative to the Big Bang theory. Describing it as "the perfect cosmological principle," Gold hypothesized that the universe is always expanding at the same density and rate, making such changes unobservable. More poetically, there is no beginning or end to the universe and, on the grand scale, the universe always appears the same.[22] Gold had a talent for "making a far-out idea seem plausible" and his "perfect cosmological principle" set Crick to wondering if there was a "perfect biological principle"—specifically, "the ability of the gene to be exactly copied when the chromosome doubles during cell division."[23]

Reviewing the various molecular permutations in his head, Crick "had the feeling that DNA replication involved attractive forces between the flat surfaces of the [nucleotide] bases."[24] Building on this hunch, he asked Griffith to make the necessary calculations to demonstrate either a mechanism of complementary or direct-copying DNA. Several days later, when the two men "bumped into each other in the Cavendish tea queue," Griffith told Crick "that a semi-rigorous argument hinted that adenine and thymine should stick together by their flat surfaces. A similar argument could be put forward for attractive forces between guanine and cytosine." Griffith's equations, which he did not yet want "to defend too strongly," essentially verified the same "odd results of Chargaff's" which Watson "had recently muttered" to Crick.[25]

In late May 1952, Chargaff came to Cambridge for dinner and drinks at Peterhouse College with John Kendrew,[26] during his first return visit to Europe since the war. He had just been promoted to full professor at Columbia and was scheduled to give several lectures across the continent and in Israel, as well as to deliver an important paper on DNA at the Second International Congress of Biochemistry in Paris that June.[27] Before saying their good-nights, Kendrew asked Chargaff if he would be willing to speak with "two people in the Cavendish Laboratory who were trying to do something with the nucleic acids. What they were trying to do was not clear to him; he did not sound very promising."[28]

Chargaff's recollection of the meeting was characteristically snarky: "This intrinsically unmemorable event has so often been painted—'Caesar Falling into the Rubicon'—repainted, touched up, or varnished in the several auto- and allo-hagiographies that even I, with my good memory for comic incidents and great admiration for the Marx Brothers films found it difficult to scrape off the entire legendary overlay."[29] The older man and the two younger ones hated each other at first sight. Crick and Watson found Chargaff to be arrogant and insufferable, which he likely was. In turn, Chargaff was unimpressed by Crick's nonstop chattering, not to mention Watson's Greek chorus of eye-bulging and snorting. He ridiculed Watson's Midwestern accent and, later on, took to referring to the two molecular biologists as "pygmies."[30] Watson recalled how the conversation rapidly degenerated when Kendrew let "loose only the possibility that Francis and I were going to solve the DNA structure by model building. Chargaff, as one of the world's experts on DNA, was at first not amused by dark horses trying to win the race."[31]

In 1978, having ruminated on this encounter for more than twenty-five years, Chargaff admitted, "my diagnosis was certainly rapid and possibly wrong. The impression: one, thirty-five years old, the looks of a fading racing tout, something out of Hogarth ("The Rake's Progress"); Cruikshank, Daumier; an incessant falsetto, with occasional nuggets glittering in the turbid stream of prattle. The other, quite underdeveloped at twenty-three, a grin, more sly than sheepish; saying nothing of consequence." Chargaff resented that Watson and Crick were overly influenced by Pauling's α-helix model of proteins and not enough by his "attempts to explain the complementarity relationships" of adenine and thymine, and cytosine and guanine. He was "baffled" by their "enormous ambition and aggressiveness, coupled with an almost complete ignorance for chemistry, that most real of exact sciences." Nonetheless, Chargaff long insisted that it was this very conversation that led Watson and Crick to their "double-stranded model of DNA."[32] Whether it was pride, generational miscommunication, or his failure to understand their questions to him about the "pitch," or angle, of the helix—an essential calculation in Crick's helical theory but a topic on which the biochemist was clueless—Chargaff sarcastically dismissed them as "two pitchmen in search of a helix."[33]

Crick conceded that the meeting did inspire a critical connection he never forgot. The moment of recognition occurred shortly after he slighted Chargaff by asking, "Well, what has all this work on nucleic acid led to; it has not told us anything we want to know." The oversensitive Chargaff replied, "Well, of course, there is the 1:1 ratios." Crick made the error of asking, "What is that?" To which Chargaff spluttered, "Well, it is all published!" Crick flippantly replied that he had not seen Chargaff's work because he never read the literature, and further inflamed Chargaff's scorn by admitting "that he did not remember the chemical differences among the four bases."[34] Sometime after Chargaff explained the 1:1 chemical ratio, however, Crick had an epiphany of sorts: "the effect was electric. That is why I remember it. I suddenly thought, 'Why, my God, if you have complementary pairing, you are bound to have a one to one ratio.' "[35]

An account of this episode would be incomplete without its comical conclusion. Crick made an impromptu visit to John Griffith's rooms at Trinity College the same afternoon he met Chargaff; having forgotten the details of Griffith's complementarity ratios and "quantum-mechanical arguments," he realized he needed to hear them again. Upon opening the door, he found Griffith in a passionate tangle with a young woman; undeterred, he corroborated Griffith's calculations, scribbled the formulae on the back of an envelope, and beat a hasty retreat. Of the intrusion, which marks Griffith's exit from the DNA story, Watson sourly observed: "It was all too clear that the presence of *popsies* [British slang for attractive young women] does not inevitably lead to a scientific future."[36]

CHARGAFF'S FAILURE TO ESTABLISH a productive working relationship with Watson and Crick was compounded by the fact that he was betting on the wrong horse. What Chargaff did not tell Watson, Crick, or Kendrew in the spring of 1952 was that for the past year he had been supplying Maurice Wilkins with DNA specimens. Chargaff and the unthreatening Wilkins had bonded the summer before at the Gordon Research Conference on nucleic acids and proteins in New Hampshire, where they were among the minority insisting that DNA was the central player in heredity.

In October 1951, Randall had "divide[d] the living child in two." Rosalind Franklin's half included the far superior Signer DNA specimens, much to Wilkins's chagrin.[37] As a result, he experimented with his preparations from *Sepia* octopi sperm heads obtained in Naples. By December 1951, however, Chargaff was sending Wilkins packages of DNA extracted from calf thymus and *B. coli* cultures by express airmail from his lab in New York. In return, Wilkins sent him monthly progress reports.[38] Still, the Chargaff samples were a far cry from the superb Signer DNA. They tended to degrade soon after extraction, making them unusable for extended X-ray analysis, and they failed to transition well from the A form to the B form, despite adequate hydration methods.[39]

On January 6, 1952, only a few weeks after Bragg ordered Watson and Crick to drop their modeling research on DNA, Wilkins sent Chargaff several X-ray images he had taken which he deemed "rather better than the best Astbury picture of calf thymus." In an accompanying letter on his King's College Biophysics Research Unit stationery, he drew the now iconic Maltese cross pattern to reflect "a helical array of pennies with 27 Å pitch of helix and 3.4 Å spacing between pennies." This finding occurred more than a year before Watson and Crick published their double helix model.[40] On the second page of the letter, he drew a cylindrical structure, with the phosphate and sugar portion of the molecule serving as the external backbone, in the shape of a spiral, or helix, and with the nucleotides, marked "N," in the spiral's center. Thus, the combination of the chemistry Chargaff so arduously divined and Wilkins's 1952 X-ray pictures and crude schematic drawings was surprisingly, but not definitively, close to the final answer discovered by Watson and Crick a year later, in 1953.

Bubbling with excitement, Wilkins asked Chargaff for strict confidentiality:

If you will excuse my enthusiasm I think we have this problem on the run and hope in the next six months to <u>prove</u> the details and show that the same nucleoprotein helical micelle exists in living cells such as thymocytes, etc. & not only in low water content inactive sperm. Could you treat the pictures and information as confidential for the time being please?

P.S. The reason I suggest you keep the information to yourself & your coworkers is that we have been embarrassed by the exceedingly great interest shown by some people over here in some of these results and a tendency to nerk off [act foolishly or recklessly] & try to work out the implications before we have ourselves. I do not think that it is impeding the progress of science to keep one's ideas to oneself for a short period of time, say 3–6 months, during gestation (if that is the right word). Most of the points I have mentioned are 1–2 months old. I [sic] like <u>you</u> to know the very latest results and ideas as so much of it depends on your work and you supply the material.[41]

Had they succeeded in fully interpreting these findings, the team of Chargaff and Wilkins might well have been the names we utter when referring to the double helix of DNA. Yet despite having most of the data sitting on their desks, a full year before Watson and Crick figured it out, they failed to crack the conundrum. Chargaff and Wilkins simply did not possess the intuitive brilliance that enabled Watson and Crick to sprint past their competitors and ultimately win. Chargaff's too quick dismissal of the Cambridge men proved to be the biggest mistake of his long and distinguished career. Though in his memoir Chargaff dismissed the analogy of Julius Caesar crossing the Rubicon and declaring "alea iacta est" ("the die is cast"), for the rest of his days he knew all too well that by crossing Watson and Crick, he had passed the point of no return. The biochemist's bitterness increased exponentially after Watson, Crick, and Wilkins won the Nobel Prize in 1962.[42] Angry at Stockholm's oversight of his own work, he "wrote to scientists all over the world about his exclusion."[43] In 1978, when asked why he hadn't come up with the double helix model himself, Chargaff replied in a manner every bit as hagiographic as the tales told by and about Watson and Crick. He was "too dumb," he said, to answer the riddle, but "if Rosalind Franklin and I could have collaborated, we might have come up with something of the sort in one or two years."[44]

Paris and Royaumont

I tried to rescue Maurice's morale by bringing him out to the Abbaye at Royaumont for the week-long meeting on phage following the biochemical congress . . . Later, I kept expecting Maurice to search me out, and when he missed dinner I went up to his room. There I found him lying flat on his stomach, hiding his face from the dim light I had turned on. Something eaten in Paris had not gone down properly, but he told me not to be bothered. The following morning I was given a note saying that he had recovered but had to catch the early train to Paris and apologizing for the trouble he had given me.

—JAMES D. WATSON[1]

The Second International Congress of Biochemistry in Paris attracted more than 2,200 chemists, physicists, biologists, and physicians. The Sorbonne's stately Amphithéâtre had barely enough seats to accommodate the crowd.[2] Called to order by Pierre-Olivier Lapie, the novelist, essayist, lawyer, and French minister for education, the seven-day congress featured lectures on a host of biochemical topics and climaxed with a black-tie evening of ballet at the Théâtre National de l'Opéra. For the adventurous, bored spouses in attendance, there were also day trips to the lace-making workshops in Chantilly and the forest of Compiègne, where in the Glade of the Armistice, two famous truces had been signed—one ending the First World War on November 11, 1918, and the other formalizing Hitler's occupation of France on June 22, 1940.[3]

Between sessions, Erwin Chargaff and James Watson passed each other in the Sorbonne's central courtyard. Extending his hand in greeting, Watson was rebuffed with only the "trace of a sardonic smile" from the elder scientist. At least, that is how Watson recalled the meeting.[4] Chargaff described the event quite differently: "I felt far from 'sardonic.' I was looking for a toilet; but whatever door I opened, there was a lecture room and the same large portrait of Cardinal Richelieu."[5] Insulting, indifferent, or merely heeding the call of nature, the forty-seven-year-old Chargaff succeeded at intimidating the young Watson—at least for the moment.

§

THE CONFERENCE'S MAIN ATTRACTION was the session on protein structure and biogenesis held on July 26. Although the keynote speaker was J. S. Fruton, an enzyme chemist from Yale University, it was Linus Pauling's hastily arranged lecture that brought the overcrowded hall to its feet. Compiled from the notes he prepared for his aborted Royal Society talk in May, his speech inspired a thunder of applause rarely heard in academic forums. The overwhelming response to both his science and his courageous defiance of an oppressive government policy did little to cheer a sullen Jim Watson sitting in the back of the auditorium. He panned Pauling's lecture as "only a humorous rehash of published ideas." He already knew Pauling's "recent papers backward and forward. No new fireworks went off, nor was there any indication given about what now occupied his mind."[6]

Watson's opinion was in the minority. After Ava Helen and Linus Pauling returned to their hotel room at Le Trianon in Saint-Germain-des-Prés, their suite overflowed with well-wishers and colleagues eager to congratulate the Congress's "honorary president." A few hours later, they sat like a king and queen at the head table of an ornately decorated banquet room for a ceremonial dinner. The cover of the menu featured a cartoon of prepubescent nymphs on a scaffold building a wall—each brick labeled as a different amino acid.[7] The pages inside announced a sumptuous meal: minestrone soup, lobster mayonnaise, roast leg of lamb, green salad, an assortment of cheeses, and peach Melba, accompanied by vintage Pouilly Fuissé, Pommard, and Champagne frappé brut, as well as coffee and assorted liqueurs.

In the middle of this academic feast, Wilkins quietly entered the banquet hall. The physicist was still in the queasy throes of dyspepsia from eating too much rich French food the day before. He purposefully sat next to "a remote man, who [he] thought would not expect [him] to make polite conversation. But [the man] soon began a very exciting account of his new research which showed that when a virus infected a bacterium in order to reproduce, nothing but DNA entered the bacterium."[8] At first, Wilkins thought the scientist was merely recounting Oswald Avery's pneumococcus experiments. The following morning, he realized that his dinner companion was Alfred D. Hershey, the Cold Spring Harbor geneticist who was to deliver the keynote address at the International Phage Conference at Royaumont Abbey.

THE ABBAYE DE ROYAUMONT is a thirty-kilometer train ride due north of Paris. King Louis IX (later Saint Louis) ordered his architect to build it during the years 1228 and 1235. The Abbaye is arranged in a lopsided quadrangle, with an odd spoke of a building jutting out here and there. The interior chamber is filled with an almost overwhelming wave of Gothic arches, elaborate rows of clustered columns, ribbed vaulting, and stunning stained-glass windows. Outside, a lush garden features a reflecting pond in the shape of a cross. Originally a Cistercian monastery, since its earliest days the Abbaye has hosted countless intellectuals, artists, and scientists for important meetings, performances, and lectures.

The Phage Group, informally run by Max Delbrück and Salvador Luria, managed to book the Abbaye for a weeklong summer meeting following the Paris Biochemistry Congress.[9] Watson was excited to reconnect with colleagues he had not seen in more than a year and, after running into Wilkins in Paris, invited him to come along. Wilkins

The Abbaye de Royaumont.

accepted, happy to meet a new cadre of scientists who were breaking ground on the understanding of genetics.[10]

Everyone attending the Royaumont conference had heard and wanted to learn more about Alfred Hershey's experiments on the genetics of bacteriophages. Tall, thin, and plagued by insomnia, Hershey worked for many years with a single assistant named Martha Chase.[11] He was a loner by nature and a man of few words. Once, when a visitor asked to tour his Cold Spring Harbor lab for interesting equipment, Hershey offered up a characteristically blunt response: "No, we work with our heads."[12]

In 1952, Hershey and Chase published a seminal study that came to be called the Waring blender experiment, because they separated the protein and nucleic acid components of bacteria with the same machine used to blend milk shakes and malteds at soda fountains. Hershey's scientific objective was to settle, once and for all, the debate over which was the true genetic material: protein, DNA, or some combination thereof. Their method was to radiolabel the bacteriophage by replacing the sulfur found *only* in proteins with radioactive sulfur, and the phosphorus found *only* in DNA with radioactive phosphorus. They then infected bacteria with the radioactive phage samples to see if, upon cellular replication, viral DNA or viral protein entered the next generation of cells. After centrifuging the mix in a blender to separate out the lighter phage particles from the heavier bacterial cells, they found that bacteria infected with phage containing radiolabeled-phosphorus DNA produced subsequent generations of bacteria with radiolabeled-phosphorus DNA. The bacteria infected with sulfur-labeled phage—which incorporated the radioactive protein—produced radioactive-free progeny in successive generations. The results were clear to all who listened: DNA definitely directed cell replication, and the proteins did nothing in that arena.[13] For his work, Hershey won the 1969 Nobel Prize in Physiology or Medicine.[14] In 1998, Jim Watson wrote an obituary of sorts for Hershey in the *New York Times Magazine*, recalling how "the Hershey–Chase experiment had a much broader impact than most confirmatory announcements and made me ever more certain that finding the three-dimensional structure of DNA was biology's next important objective."[15]

Linus Pauling was impressed, too. Immediately after Hershey's lecture ended, he got up and admitted the error of his ways. To the riveted audience, he boldly stated that DNA was "the genetic master molecule, the one that directed the making of proteins."[16] In a real sense, Pauling

Alfred D. Hershey.

announced his formal entry into the DNA race directly after Hershey's lecture, even if he still appeared to be walking rather than careening toward the finish line. Although he had not yet seen Franklin's or Wilkins's X-ray images, his associate Robert Corey did see them briefly when he was in London two months previously to deliver Pauling's lecture to the Royal Society. Corey informed Pauling that while the images were quite good, there was "no indication that either of them [Franklin or Wilkins] knew enough chemistry to be a serious threat." Most likely for ethical reasons, given that it was not his data to share, Corey did not provide Pauling with precise diagrams of Franklin's images. But he did confide to his boss that the King's College group was an utter bedlam of infighting and name-calling. No positive results could possibly emerge from such an unsettled environment, he said. As to the Cavendish, there was no indication that Bragg, Perutz, or Kendrew were interested in DNA. Pauling had not yet met Crick and was unimpressed by Watson, who only a few years earlier had been rejected by the Caltech doctoral program. Instead, Pauling comforted himself that time was on his side, the King's group posed no threat, and the men of Cambridge had not bested him in their previous scientific jousting matches.

ONLY A FEW HOURS BEFORE Hershey's lecture, Jim Watson was chatting with André Lwoff, a Cambridge-trained French microbiologist at the Institut Pasteur. Over croissants and coffee, Lwoff mentioned that Pauling and his wife were expected at Royaumont any moment. Watson scurried over to the auditorium to claim a good seat for Hershey's lecture, and watched with envy as Pauling made his entrance under the wing of the U.S. embassy's scientific attaché, Jeffries Wyman, a Harvard molecular biologist with a Brahmin pedigree who had segued into a diplomatic career.

"Immediately," Watson reminisced, "I began to think of ways that would allow me to sit next to him at lunch."[17] There was never a doubt of his success in the matter. After a full morning of presentations, lunch was served on the lawn of the medieval monastery. There, the grand poohbah of chemistry and Watson exchanged pleasantries and small talk about virus and X-ray diffraction research. Watson made sure to share the information that Max Delbrück was recruiting him for a post-doctoral fellowship at Caltech the following year.

Watson and Delbrück had corresponded regularly in the weeks lead-ing up to the phage conference so that Watson would be well prepared when he met Pauling. On May 20, Watson sent Delbrück a long dis-course on his TMV work, spiced with gossip from Cambridge, his wor-ries about being drafted into the U.S. Army, and the news that he and Crick "temporarily stopped [the DNA modeling] for political reasons of not working on the problem of a close friend. If, however, the King's group persists in doing nothing, we shall again try our luck."[18] Delbrück wrote back on June 4 to tell Watson that Pauling "has $10,000 from the National Foundation for Infantile Paralysis for DNA structure work which is lying idle for lack of manpower" and that "there will be a pro-tein meeting here at Cal Tech March '53 with most of your Cambridge friends invited; you might make that your point of return. Or, summer '53 when the Cold Spring Harbor Symposium will be on viruses."[19]

Watson's chat with Pauling did not go as well as he hoped. They briefly discussed the possibility of Watson going to Caltech the following year to study viruses. Watson brought up the new X-ray images being developed at King's College. Pauling countered that the "very accurate X-ray work of the type done by his associates on amino acids was vital to our even-tual understanding of the nucleic acids." Watson walked away engulfed by a wave of discontent because "virtually no words went to DNA."[20]

"I got much further with Ava Helen," he chortled.[21] Watson knew that the Paulings' second son, Peter, would be joining the Cavendish that fall as a research student. He also knew that had Peter's last name not been Pauling, he would have doubtless been rejected, just as he had been by all the other universities to which he applied. A self-admitted "sex maniac" and straight-C student at Caltech, Peter Pauling's spotty academic record went further south after he contracted mononucleosis during his junior year.[22] He and Watson first met at a party in the sum-mer of 1949, when Watson was working under Delbrück in Pasadena.

Thirty-four years later, Peter Pauling admitted he had no recollection of that event because he "had been concerned primarily with adolescent dreams of seducing [his] sibling's babysitter."[23]

Peter's mother was worried about her son, not only because of his excessive partying but also because he was entering a field where he would be constantly compared to and overshadowed by his father. She told Watson that Peter was "an exceptionally fine boy whom everybody would enjoy having around as much as she did." As she spoke, Watson fantasized about Peter's beautiful sister, Linda, and "remained silently unconvinced that Peter would add as much to our lab as Linda." On cue, Watson cooed to Ava Helen Pauling that he would be more than happy to mentor Peter and help him "adjust to the restricted life of the Cambridge research student."[24]

A WEEK AFTER the Royaumont phage conference, Watson was hiking through the Italian Alps. On August 11, at 1,600 meters above sea level, Watson sat down on a rock and wrote a long letter to Francis and Odile Crick about it. He reported that he gave his tobacco mosaic virus talk while wearing a carefully premeditated "not caring how I look" uniform of a baggy, untucked shirt, oversized jacket, far too short shorts, bunched up dark socks, and untied, beat-up, brown Oxford shoes.[25] The reason behind what became his usual garb at summer meetings was that his suitcase was "snatched from [his] train compartment" while he slept on the way from Paris.

Jim told the Cricks that the architecture of Royaumont reminded him of Cambridge, and that its atmosphere was far more conducive to great thoughts than that of Paris. He also amused them with an account of the formal garden party at Sans Souci, the Gouvieux–Chantilly country estate of the Baroness Édouard de Rothschild, when he had nibbled on smoked salmon and sipped flutes of chilled Champagne, properly served by a bevy of butlers, as he stared at the Rubens and Hals paintings hanging on the walnut-paneled walls. He wore a borrowed jacket and tie, he reported, and had put a large amount of "highly perfumed brilliantine" in his then longish hair to appear "Latin."[26] Only a few weeks earlier, when his mother visited him in Cambridge, she wrote to her husband that their son had "affected a sort of 'Einstein' cut—long and curley [sic]."[27] Apparently, Watson was pleased with the weird impression he made on the baroness and her guests: "The message of my first meeting with the aristocracy was clear. I would not be invited back if I acted like everyone else."[28]

Far more important than his distorted fashion sense or partygoing, Jim Watson told the Cricks about "Mrs. Pauling's small talk" and "gather[ed] that Peter has not yet calmed down and so we shall not have another quiet youth . . . in our midsts [sic]." To keep the boy on the straight and narrow, Watson "recommended to his mother a very small sum of money for him to live on and in this way may cause a tendency for the puritanical life," that thanks to his new grant funding, he was now "escaping from."[29]

Here, history intervened again to Watson and Crick's benefit. The coincidence of Peter Pauling coming to the Cavendish Laboratory only a few months before they returned to their DNA work was a huge factor tipping the horse race in their favor. In a matter of weeks, Watson and the younger Pauling became fast friends. Peter Pauling's first impression of Watson was of a "funny-looking fellow slightly my senior, with rather large ears and thin, rather wild hair."[30] Watson, on the other hand, fondly remembered Peter Pauling as "my most important friend in Cambridge . . . we were about the same age and he was great fun."[31] Seated next to each other in Room 103 of the Austin Wing, Peter Pauling was about to become Watson's bug on the wall of Linus Pauling's laboratory in Pasadena.

A Haphazard Summer

. . . the final remark I would make here is, of course, that you must remember I wasn't really working on the [DNA] problem, that's why the work was haphazard . . . I was working on a thesis on proteins. Anyhow, all I was saying was, the reason it was haphazard was that I personally wasn't working on the problem and I don't think Jim, you know, was working on it, as a strong interest, it wasn't a research program. So that's why it was haphazard.

—FRANCIS CRICK[1]

The summer of 1952 provided Maurice Wilkins with a respite from his troubles at home. In July, he set off on a long journey to Brazil, where he and several other British biomolecular scientists planned to "visit laboratories, hold a conference on important molecular advances, and generally liven up Brazilian science."[2] They were the guests of Carlos Chagas, the famed physician and bacteriologist who described what became known as Chagas disease, a parasitic infection caused by being bitten by insects infected with *Trypanosoma cruzi*.[3] A few months earlier, while on a train from Innsbruck to Zurich, Wilkins wrote to Crick, "Franklin barks often but doesn't succeed in biting me. Since I reorganized my time so that I can concentrate on the job, she no longer gets under my skin. I was in a bad way about it all when I last saw you."[4] Bold assertions aside, Wilkins was eager to get far away from Rosalind Franklin, whose effect on him was far more than skin-deep. His summer trip proved to be the perfect solution.

Wilkins enjoyed being feted as a distinguished scientist, sunned himself on Ipanema Beach, and shopped in the street markets of Rio de

Janeiro. Next, he traveled west to Lima, in search of "monster squid" from which to extract sperm heads and DNA. Finding none, he explored the Peruvian art scene, zigzagged along the Andes, and thrilled to the ancient cultures of Machu Picchu and Cuzco. Standing on a mountain peak, he looked out and meditated on "the beauty of Inca civilization and . . . the surreal ruins of its brutal destruction."⁵ He saw the rich and violent history of the Incas as an allegory for nuclear war. Seven years after President Harry S. Truman ordered atomic bombs dropped on Hiroshima and Nagasaki, Wilkins remained deeply troubled over the role he had played in their development, even though his "disillusion with the Bomb . . . led him to choose molecular biology." When it came to this massive weaponry, he wondered, "Where would it all end?" Enmeshed in a strange trance of "timelessness and detachment," he shelved the daily worries of King's College, stood back, and looked at the world anew. He saw "its past, present and future" and asked, "How did it all add up?" Characteristically he concluded, "there were no clear answers to those questions. All we could do, I realized, was to push on, exploring the world, while keeping the big questions clearly in our minds."⁶

Some type of psychological release was essential for this anxiety-ridden man. Only two months earlier, Wilkins had left England with "a great cloud hanging over our DNA work" because of his war with Franklin. He was also processing the aftereffects of breaking up with his girlfriend, Edel Lange. To his credit, while standing alone on that remote Andean mountaintop, Wilkins determined to "go back to our lab bench again and struggle to find the DNA structure." Where else could he go? Decades later, he reflected on that pivotal moment in his life: "If anyone told me that out of the gloom would emerge very soon one of the most important scientific advances of the century, I would not have been surprised except by how quickly it all took place."⁷

In early September, Wilkins took a long, indirect, propeller plane flight back home. He returned from "sunny Brazil" to his attic flat in Soho, only to find that London was "dark and cold" and he was "completely exhausted." Unpacking his suitcase, he took out all kinds of "beautiful objects from Peru" which he "would have loved to share" with Edel. But she had left his life and "would not come back." They had said their goodbyes six months earlier in the Alps. Standing alone in his flat, a lonely, loveless, and sleep-deprived Wilkins "exploded" and demolished all of the gifts Lange had given him. However, he recalled, "I did not smash the new things I had brought back from my trip—I knew my life had to go on."[8]

☊

AT KING'S COLLEGE, a desperately unhappy Rosalind Franklin spent the summer toiling over her X-ray images. Nearly twenty years later, her colleague Geoffrey Brown shook his head in sadness when describing how toxic the atmosphere in the biophysics laboratory had become. "Wilkins was not especially kind to Rosalind . . . especially towards the end and . . . the policy adopted, probably by Randall, but also probably at Wilkins suggestion, was simply to freeze Rosalind out."[9] On March 1, 1952, she wrote to David and Anne Sayre about her solitary confinement at work. Although she found her laboratory equipment and the facilities at King's to be "quite exceptionally good—in fact, scandalously good considering the shortage of money for such things," she was eager to escape as soon as she could. Sharply critiquing her colleagues, she said she liked the young people there who were "mostly thoroughly nice but none of them brilliant." A few of the senior people were merely "good and pleasant, but refrain from doing research so as to be able to keep outside the unpleasant atmosphere. And the other middle and senior people are positively repulsive and it is they who set the general tone . . . The other serious trouble is that there isn't a first-class or even a good brain among them—in fact, nobody with whom I particularly want to discuss any-thing, scientific or otherwise." Fortunately, she was able to wall herself in her little lab so as to have as little contact with them as possible, which alleviated the conflict but made for "distinctly boring" days.[10]

In the same letter, she described "a terrific crisis with Wilkins, which nearly resulted in my going back to Paris. Since then we've agreed to differ, and the work goes on—going on, in fact, quite well." Still, the

Rosalind Franklin on holiday in Tuscany, Italy, spring 1950.

strife was unbearable enough for her to make an appointment to meet with J. D. Bernal at Birkbeck College and ask if there was a position for her there. Precisely assessing her potential savior, she described him as condescending but pleasant, brilliant and inspiring. Bernal even gave her "some hopes of working in his biological group one day—I wouldn't at that stage make it clear that I wanted to go this year." She was careful to swear her friends to secrecy because "not a soul knows so far." Yet she knew that by leaving King's College for what was then mostly an extension and night school for working people, she would be going down a peg or two in prestige. "I suspect [Birkbeck is] more alive than other London colleges," she wrote to Anne and David Sayre. "It has only part-time evening students, and consequently they are all people who have come because they really want to learn, and to work. And they seem to collect a large proportion of foreigners in the staff, which is a good sign. King's has neither foreigners nor Jews."[11]

A few months later, on June 2, 1952, during a "wonderful trip" to Yugoslavia, Franklin found herself sailing "on the boat from Split to Rijeka." From the observation deck, she again wrote to Anne and David Sayre to say that little had changed in London but a plan of action had been made: "I still don't know anything about my future. I'll let you know when I do. I've seen Bernal, and he will take me if Randall agrees, but I've decided it would be bad politics to talk to Randall just before going away for a month, so that['s] a pleasure in store for when I get back."[12]

Sometime over the next four weeks, Franklin's fate at King's College was formally decided. There remains debate over whether Randall pushed her out of the laboratory, given the tense situation, or if the move was her decision. It was probably a combination of both. On June 19 she again contacted Bernal about the possibility of moving her work to Birkbeck and emphasized that Randall had no objection to such a move.[13] Randall himself may have eased the transition by speaking sotto voce with Bernal before Franklin approached him. More certain is that Randall had little incentive to extend her time at King's and did nothing to persuade her to stay. He must have been relieved to have found such an easy resolution to the vicious infighting in his laboratory.[14] As with all academic ventures, there was paperwork to be completed. On July 1, 1952, the Turner and Newall Fellowship Committee notified Professor Randall that Miss Franklin had asked to transfer the third year of her fellowship to Professor J. D. Bernal's crystallography laboratory at Birkbeck College so that she might apply her X-ray diffraction work to the study of tobacco mosaic virus.[15] The move did not actually transpire until March 1953.

The tragedy of the forced move to Birkbeck was that Franklin's work at King's College had never been better. Her experimental process required a multitude of camera angles and adjustments, often differing by a millimeter or less, gallons of sweat equity, and a shocking amount of dangerous X-ray exposure that would never be tolerated in a lab today. By spring, she and Gosling had gained proficiency in stretching out the sticky fibers, mounting them to the apparatus, and obtaining ever more precise X-ray photographs of both the "A," or dry, form and the B, or wet, form of DNA.

THE SEGUE, IN EARLY SUMMER, to the tedious but peaceful work of computing X-ray diffraction patterns was a welcome task for someone of

Franklin's mind. She applied the intricate, and often frustrating, Patterson equations to interpret the data because, as her Paris colleague Vittorio Luzatti may have advised, "it's what a crystallographer would do."[16] Patterson equations, created by the British X-ray crystallographer Arthur L. Patterson in 1935, allowed for the creation of a vector map, or "cylindrical Patterson," of interatomic distances for the molecule in question. Each map point was calculated from the intensities of the diffracted X-rays and from these data points the crystallographer endeavored to construct the molecule's dimensions and structure.[17]

Professional crystallographers like Luzzati and Franklin were devoted to this method of solving molecular structures, especially when confronted by a paucity of hard data. Patterson calculations provided subtle clues when exploring a molecule with a regular or repeating structure and helped define "its character so that the rest of the structure could be interpreted fully in three dimensions." The problem with Patterson equations, alternatively referred to as "beautiful" and "barbarous," was that they required an enormous amount of mathematical expertise to approximate a correct answer. As Horace Judson aptly described it, the result was a diagram that "looked like a geologist's contour map, all loops and meandering lines, of a particularly up-and-down square mile of the Dakota Badlands. To reason from this map back to the real structure was twistingly abstruse, and felt like putting one's mind through a sieve."[18] With far less eloquence, Max Perutz and John Kendrew declared the Patterson method as "infuriatingly elusive." By 1949, they abandoned it and moved on to other methods because "the physical meaning of this so-called Patterson synthesis is one of the most difficult conceptions in crystallography."[19] Francis Crick, too, found it "unreliable" in determining the structures of organic molecules.[20] As late as 2018, James Watson admitted that he never understood the Patterson methodology.[21]

Today, crystallographers can either code for or purchase the computer software to calculate Patterson equations, Fourier transforms, Bessel equations, and many other, newer complex mathematical modeling schemes. Such work is now done with the touch of a button on a computer, which produces results in a matter of minutes or less. In 1952, however, Franklin and Gosling used a cumbersome calculating device called Beevers–Lipson strips, "which assembled the values of the periodic functions all set out at appropriate intervals and arranged sequentially in a handsomely polished mahogany box." Nearly half a century after the fact, Raymond Gosling reported still having nightmares (especially when he had a bad hangover) about dropping a box of the strips on the floor and needing to put them back in the correct order. Yet as "tiresome" and "repetitive" as the computational work was, he found working on it with Franklin to be "great fun . . . since no-one had done it before. This worried me. But Rosalind was very professional and apparently confident that the task could be correctly completed."[22]

ON JULY 2, Franklin scribbled on a new page in one of her red Century laboratory notebooks:

> <u>Notes on first cylindrical Patterson</u>: There is <u>no</u> indication of a helix of diameter 11 Å. The central banana shaped peak fits curve calc[ulated] for helix of diameter 13.5 Å having two turns per unit cell. <u>If</u> a helix, there is only one strand. (2-strand would give [here, she draws two interlocking ovals]) . . . <u>if</u> a helix, it is v[ery] far from continuous uniform density.[23]

A few weeks later, on July 18, Franklin playfully drew up a black-bordered "memorial card" for the DNA helix. The card was drawn for her own entertainment (and disdain of Wilkins) rather than for general distribution.

> It is with great regret that we have to announce the death on Friday 18th July 1952, of D.N.A. Helix (crystalline). Death followed a protracted course of illness which an intensive course of Besselised injections had failed to relieve. A memorial service will be held next Monday or Tuesday. It is hoped that Dr. M.H.F. Wilkins will speak in memory of the late helix.
>
> (signed) R.E. Franklin. R.G. Gosling [24]

Wilkins was not amused by the faux funeral card, which he initially thought had been written by Gosling as an act of "friendly jocularity." When he later learned its true author (even though both Gosling and Franklin signed the card), he remained unforgiving as he exaggerated Franklin's desire to humiliate him and discounted the pranks his staff had pulled on her. Often overlooked is the fact that her note was referring to the crystalline, or A form, of DNA—which after months of Patterson analysis still gave off too many artifactual X-ray diffraction patterns to allow her to rule a helical structure in or out of contention. As Gosling often said, at no time did she think the B form was anything but helical.[25] All joking aside, Franklin wrote her own memorial card that day, one that all but marked the end of her tenure at King's College.

Ironically, it was Rosalind Franklin's conscientious, and stubborn, adherence to the painstakingly slow diffraction pattern methods that long kept her out of the history books. What she grossly underestimated was the jet-propelled speed of the model builders in Cambridge. At a fortieth anniversary symposium on the discovery of DNA's structure, Gosling bemoaned how before he and Franklin even had a chance to fully interpret the Patterson maps they so meticulously prepared, Watson and Crick announced their model. As Gosling sadly recalled, "Of course, once the cat is out of the bag you cannot put it back in, and so we looked again at our cylindrical Patterson function. We could see clearly the peaks representing the heavy phosphorus–oxygen groups lying on a double helix. One strand went up and one came down." When asked the obvious question of whether or not they would have found that answer on their own, he honestly answered, "I don't know. We might have done, but it is certainly very obvious when you are told that it is there."[26]

JIM WATSON'S 1952 SUMMER was busy, too. After his trips to Paris, Royaumont, and the Italian Alps, he wangled an invitation from Luca Cavalli-Sforza (formerly of Cambridge and then the University of

Parma) to attend the Second International Conference on Microbial Genetics. The three-day meeting was held in early September in the elegant town of Pallanza, overlooking Lago Maggiore, in the Piedmont region of northwest Italy.[27] Much to Crick's dismay, Watson pushed the pause button on his DNA fixation. Instead, he became "preoccupied with sex, but not of a type that needed encouragement." This was a poor gag that belied the importance of the meeting. The undisputed highlights of the conference were papers given by Cavalli-Sforza, William Hayes of the Hammersmith Hospital in London, and Joshua Lederberg of the University of Wisconsin establishing "the existence of two discrete bacterial sexes."[28]

In 1946, at the age of twenty-one, Lederberg had taken a leave from his medical studies at Columbia's College of Physicians and Surgeons to study for a PhD in microbial genetics under Edward Tatum at Yale. These two brilliant men worked on demonstrating genetic recombination, whereby bacteria entered a "sexual phase" of sharing and exchanging genetic material.[29] Instead of returning to Columbia in 1947 to complete his medical degree, Lederberg ventured west to become an assistant professor of genetics at the University of Wisconsin, Madison. Eleven years later, at age thirty-three, he shared the 1958 Nobel Prize in Physiology or Medicine with Tatum and George Beadle.

The wunderkind Watson was envious of ultra-wunderkind Lederberg's meteoric rise. In 1968, seven years after he won his own Nobel Prize at the age of thirty-nine, Watson carped that Lederberg "carried out such a prodigious number of pretty experiments that virtually no one except Cavalli dared to work in the same field. Hearing Joshua give Rabelaisian nonstop talks of three to five hours made it all too clear that he was an *enfant terrible*. Moreover, there was his godlike quality of each year expanding in size, perhaps eventually to fill the universe."[30] Referencing the microbiologist's father and maternal grandfather, who were both Orthodox rabbis, Watson added, "only Joshua took any enjoyment from the rabbinical complexity shrouding his recent papers."[31] Instead, Watson preferred William Hayes's "infinitely simpler" explanation of how "the discovery of the two sexes might soon make the genetic analysis of bacteria straightforward . . . that only a fraction of the male chromosomal material enters the female cell.”[32]

Upon returning to Cambridge in mid-September, Watson made a beeline for the university library and read every journal article by Lederberg he could find. Driven by his innate competitiveness, Watson hoped to find holes in Lederberg's experiments or clues that might advance the ball a few yards and “accomplish the unbelievable feat of beating [Lederberg] to the correct interpretation of his own experiments.” On October 27, he wrote to his sister about the experiments he was doing: “If it comes out, it will be very pretty, as it will solve a 5-year-old paradox and allow quite rapid progress in the field of bacterial genetics . . . it would be nice to beat Joshua Lederberg (Wisconsin) to the solution of his life's (still rather short—he is about 28) work.”[33]

Watson's “desire to clean up skeletons in Joshua's closet left Francis almost cold.”[34] After a long summer of slogging through the necessary work to complete his thesis, Crick was now getting ready to return to the problem of DNA. He worried that the more time Watson spent exploring the sex lives of bacteria, the less he would spend on solving DNA. Such distraction translated into the risk of losing their head start to Linus Pauling.[35] Now it was Crick's turn to get his partner back on track and into the wild thicket that was their scientific destiny.

CRICK HAD GOOD REASON to be concerned about Pauling. After the meetings in Paris and Royaumont, Pauling took the summer to visit his British colleagues who were working on the molecular biology of proteins. Rather than making the tour a victory lap for his α-helix model, Pauling was astute enough to engage those who had issues with his theories, listen carefully to their critiques and queries, and address them all, thus making his model stronger and better.[36]

His first stop was the Cavendish Laboratory. Surprisingly, at least to the Cavendish staff, it was neither Max Perutz nor John Kendrew whom Pauling most wanted to meet. He made a specific request that almost knocked Bragg out of his armchair: he wanted to spend as much time

as possible with this Crick fellow so that he might discuss his "mathematical formula for predicting how helixes would diffract X-rays."[37]
Not wanting to vent his dislike for Crick in public, Bragg begrudgingly made the necessary arrangements, hoping for the best and secretly
imagining that his obstructive graduate student might soon be irritating someone else for a change. Crick later disputed the conjectures that
Pauling's α-helix model inspired both his helical theory and the Watson
and Crick double helix. "Nothing could be farther from the truth,"
Crick stated in his inimitable manner. "Helices were in the air, and
you would have to be either obtuse or very obstinate not to think along
helical lines."[38]

One might easily imagine how a mere graduate student—even one
as brash, brilliant, and confident as Crick—would behave in the presence of the world's greatest chemist. As they drove along the streets of
Cambridge in a black taxi, Crick was agog in Pauling's presence. Pauling
loved the adulation and took it all in with his usual aplomb. For once in
Crick's thirty-six years—contrary to the famous opening line of Watson's *The Double Helix*—he was "in a modest mood."[39] With Pauling
seated right next to him, how could he be anything but humble?

During lunch, Crick avoided broaching the topic of DNA, but his
prudence was inspired by more than Bragg's moratorium. Crick did not
want to send Pauling down the path that he and Watson so fervently
wished to explore. He breathed a sigh of relief after learning that Pauling
had passed on visiting the King's College biophysics laboratory because
he was concentrating on finishing his protein work before tackling
DNA. Pauling informed Crick that Wilkins and Randall had refused to
share their data with him a few months earlier; he did not wish to make
the situation more uncomfortable than it already was.[40]

Instead of discussing DNA, Crick offered a theory to explain one of
the few holes in Pauling's α-helix: the absence of a 5.1 Ångstrom smudge
that was seen in most biological or natural substances. Crick was careful to engage the older professor without handing over the solution he
knew to be correct. Pauling told Crick that he had thought along similar
lines, and then fulfilled a dream of Crick's by inviting him to spend a

year working on it at Caltech. A thrilled Crick next asked if Pauling had ever considered whether or not the α-helixes might be coiled around one another. Pauling simply replied, "Yes, I have," and let the matter drop, proving that he was far cagier than his junior colleague.

Crick had been working on developing a mathematical equation predicting such protein coiling. Fearing that Pauling might have "stolen" his idea when he blurted it out during their cab ride, he hurriedly wrote a research note and sent it off to *Nature*. It arrived there in October, only a few days after Pauling and Robert Corey had submitted their own detailed paper on how an α-keratin protein could twist about itself, forming "coiled coils."[41] Because Crick's contribution was a brief "letter," rather than a full-length experimental investigation, his "Is α-Keratin a Coiled Coil?" was published six weeks before the appearance of Pauling and Corey's paper, on November 22, 1952.[42] This episode seemed uncomfortably similar to the debacle Crick caused a year earlier, when he accused Bragg, his superior at the Cavendish, of plagiarism, and it had the potential to devolve into a problem of international discord. The tense situation generated a rash of letters: from Peter Pauling to his father, explaining the matter; from Bragg to the editor of *Nature*, "telling him to get off the pot" and publish the Pauling paper; and between Crick and Pauling. On the advice of Perutz, Crick wisely chose to make peace with Pauling. The two scientists agreed to state that they had independently reached the same conclusion at the same time.[43]

Crick learned several things from this encounter: how fertile Pauling's mind was when it came to solving the structure of biological molecules, how formidable a rival he could be once he set his mind to a particular problem, and, perhaps most importantly—a lesson he took to Watson with a decided sense of urgency—that Pauling was close to completing his inquiry on proteins and ready to undertake the next big thing. As Pauling recounted in later years, "I always thought that sooner or later I would find the structure of DNA. It was just a matter of time."[44]

PART V

THE HOME STRETCH, NOVEMBER 1952– APRIL 1953

Young Watson played an enormous part in it. I don't think Crick would have ever done it apart from Watson, for a moment. Watson's enthusiasm was so enormous.

—SIR WILLIAM LAWRENCE BRAGG[1]

I couldn't have got anywhere without Francis. . . . It could have been Crick without Watson, but certainly not Watson without Crick.

—JAMES D. WATSON[2]

Linus Sings

And there among them a young boy plucked his lyre,
so clear it could break the heart with longing,
and what he sang was a dirge [or Linus song] for the dying year,
lovely . . . his fine voice rising and falling low
as the rest followed, all together, frisking, singing,
shouting, their dancing footsteps beating out the time.

—HOMER, *The Iliad*[1]

True to Crick's worst fear, Pauling was restless. Although he was busy teaching chemistry to fawning undergraduates, administering his chemical empire at Caltech, and preparing lectures and papers for much larger audiences, none of these activities were enough to fulfill his ravenous curiosity. He wanted a new scientific vista to conquer and the accompanying symphony of ever more accolades of success. Specifically, he wanted DNA.

On Tuesday afternoon, November 25, Pauling walked from his second-story office in the Gates and Crellin Chemistry Laboratory down the hall and into the seminar room of the Kerckhoff Laboratory of the Biological Sciences. There, he attended a lecture by Robley Williams, a microbiologist from the University of California at Berkeley. Working with the X-ray crystallographer Ralph Wyckoff, Williams had recently developed a new "metal shadowing" technique using the electron microscope. The method allowed for remarkably detailed three-dimensional images of bacteria. Pauling was mesmerized by the clarity and detail of the micrographs Williams projected onto the room's white screen. No slide impressed Pauling more than that of the salt of ribonucleic acid, or RNA.

Sitting in the front row of the darkened room, Pauling mentally compared Astbury's 1938 X-ray photographs of DNA with Williams's stunning images. The Astbury pictures depicted nucleic acids as "flat ribbons." The Williams pictures, on the other hand, presented RNA as cylinders or "long, skinny tubes." Pauling was well aware that he was looking at RNA rather than DNA, but for him, Williams's 1952 micrographs answered the query that was the subject of such heated concern in Cambridge and London: DNA had to be a helix.[2]

Long after he had gone home and finished supper, Pauling sat in his study pondering the molecular possibilities of DNA. The next day he holed up in his office "with a pencil, a sheaf of paper, and a slide rule."[3] At one point, he rummaged through a pile of scientific periodicals stacked high on his desk for a recent issue of the *Journal of the* (British) *Chemical Society* containing a paper on nucleotide chemistry by Daniel M. Brown and Alexander Todd from the Cambridge chemistry department. They demonstrated how "the inter-nucleotide links" in a strand of DNA were covalent "phosphodiester bonds joining sugar carbon atom #5 to sugar carbon atom #3 of the adjacent nucleotide."[4] This finding, in chemists' parlance, defines the complex bonds, or connections, of atoms comprising the helical phosphate–sugar backbone of DNA.

Pauling next reviewed his notes from the previous day's seminar and read the responses Williams gave when asked the diameter of the RNA molecule. "It was probably 15 Ångstroms," Williams replied twice to the same question, but he also admitted that it was difficult to measure precisely. Using Williams's density data to calculate the number of nucleotide chains per unit of DNA, Pauling wrote down in his notebook, "Perhaps we have a triple-chain structure!"[5] In 1974, Pauling recalled his surprise over this result because his calculations and his 1940 paper on complementarity suggested a two-chain, or double, structure. But Pauling incorrectly assumed there were artifacts in the existing data and stumbled down the wrong road. He later admitted, "I am now astonished that I began work on the triple helix structure, rather than the double helix."[6]

As Watson and Crick did a year earlier, Pauling deceived himself into thinking that the phosphate groups were on the inside of the helix. The nucleotide bases on the outside, he guessed, were packed tightly together in a helix and accounted for the observed volume and density of the molecule. As he calculated the bond angles, he posited that each strand consisted of "roughly three residues per turn. There are three chains closely intertwined and held together by hydrogen bonds between PO_4's [phosphate groups]."[7] His earliest drawings of the triple helix featured an extremely dense center without much space to fit in all of the atoms. Exhausted, he crawled into bed late on Wednesday night, November 26. The next day he celebrated the Thanksgiving holiday with his family.

Three days later, on Saturday, November 29, Pauling returned to his notebooks. The task before him was fitting his model to mesh with William Astbury's blurred X-ray images, Sven Furberg's even fuzzier ones, and Todd's chemical specifications. His attempt to "jam three chains' worth of phosphates into Astbury's space restrictions was like trying to fit the stepsisters' feet into Cinderella's glass slipper. No matter how he twisted and turned the phosphates, they wouldn't fit." A frustrated Pauling scribbled in his notes at the time, "Why are the PO_4 in a column so close together?" He played some more with the phosphate groups, stretching them out here and snipping their length there, but without much improvement, before calling a temporary halt to the brainstorming.[8]

On December 2, he asked an assistant to retrieve the latest crystallography literature from the library. Digesting these scientific papers— perhaps too quickly—Pauling wrote, "I have put the phosphates as closely together as possible and have distorted them as much as possible." Nonetheless, his model remained problematic. He was still cramming too many atoms into the center in a manner that did not reflect Mother Nature's usual methods. No matter, he fell in love with what he built: "an almost perfect octahedron, one of the most basic shapes in crystallography."[9]

Pauling was well aware that he was far from finished. Almost every morning of that December, the middle-aged chemist bounded down the stairs from his office to discuss his every nucleic thought from the evening before with a junior colleague named Verner Schomaker. He expounded his ideas with an enthusiasm far out of proportion to his evidence. Although he had no precise data on the bond angles or structure of the nucleotides, let alone those for the sugar–phosphate backbone, he talked himself into thinking he was correct. And he did not restrict his thinking to those he knew in Pasadena. On December 4, 1952, he wrote to E. Bright Wilson, a professor of chemistry at Harvard who was once his student, "I think now we have found the complete molecular structure of the nucleic acids."[10]

In late December, Corey inspected the model and offered an expert's opinion: the oxygen atoms were packed so closely together in the center of the model that they could not possibly accord with the known bond angle and lengths. And when the salt form of the molecule, sodium thymonucleate, was modeled, the center only became more crowded. There was simply no room for the sodium ion to fit, Corey argued. Undeterred, Pauling retreated into his office and by day's end emerged with a model of a phosphate-studded tetrahedron. Ignoring the biological imperative to explain how DNA passed on its genetic information during cell replication, he fixated only on the model, the model, the model. After all, his vaunted stochastic methods had never failed him in the past.

Pauling deluded himself so powerfully that on December 19, he wrote a long letter, laden with hints and asides, to Alexander Todd in Cambridge. He and Corey were "much disturbed that there has been no precise structure determination reported as yet for any nucleotide," he declared, but fear not: his laboratory was now committed to taking on this task. Acknowledging "that the Cavendish people are working in this field," Pauling demurred, "it is such a big field that it cannot be expected that they will do the whole job. On the other hand, we do not want to duplicate their investigations—it is more important that another nucleotide determination be made, in case that they carry out one of them." He then added coquettishly, "probably I shall write to Bragg or Cochran pretty soon and ask which ones they are working on. If there is no objection on their part, we should like to ask you to provide us with material, if you have some crystalline preparations of nucleotides or related substances that you feel would be especially worthwhile investigating . . .

The structure is really a beautiful one. It is hard to describe without some drawings, which have not yet been made. I shall keep you informed."[11] Here, Pauling was just fishing. He knew full well that the King's College lab was the primary group in Great Britain working on DNA structure, but was worried enough about Watson and Crick—probably from news relayed to him by his son Peter—to try and entice Todd to bite and tell him more.

DNA WAS NOT THE ONLY THING on Pauling's mind. Far more urgent were his troubles with the U.S. government, thanks to an informer named Louis Budenz. Once an active member of the central committee of the Communist Party USA and managing editor of the *Daily Worker*, Budenz abruptly recanted his political beliefs to HUAC in 1945

Louis Budenz (left) testifying before the Canwell Fact-Finding Committee on Un-American Activities in the State of Washington, January 27, 1948.

and, thereafter, earned a lucrative income as an informer. He logged over three thousand hours advising the FBI and naming names, and wrote popular books and articles on the Communist infiltration of nearly every level of American society.[12] For the cover story of the November 1951 issue of the right-wing *American Legion* magazine, Budenz asked "Do Colleges Have to Hire Red Professors?" One of the "red" academics prominently profiled in his long and reckless essay was Linus Pauling.[13]

On December 23, 1952, Budenz testified before a House of Representatives committee charged with investigating the politics of tax-exempt and charitable foundations. At the time, Pauling sat on the John Simon Guggenheim Foundation's advisory board, one of the organizations under scrutiny. Under oath, Budenz delivered a nasty Christmas gift by alleging that Pauling was "a member of the Communist Party under discipline. The Communist leaders expressed the highest admiration and confidence in Dr. Pauling." The same afternoon, Budenz named twenty-three other scholars who had won grants from the Guggenheim and other prominent foundations and three Guggenheim Foundation officials as Communists. All twenty-six individuals subsequently demonstrated they were not and never had been Communists, but by the time they were exonerated, it was too late for many of them. In addition to huge legal bills, several lost their research funding and some were dismissed from their jobs.[14]

In response, Pauling rebuked Budenz as "a professional liar. It is disgraceful that a committee of the U.S. Congress should permit and even aid such a scurvy, unconscionable person to cause trouble for respectable people. If Budenz is not prosecuted for perjury we must conclude that our courts and Congressional committees are not interested in learning and disclosing the truth."[15] Unfortunately, Budenz was granted immunity from prosecution for perjury by the very same congressional committee Pauling was attacking. The "rat" was allowed to scurry away in silence.

PAULING MANAGED TO compartmentalize Budenz's existential threat in a corner of his mind, so that he could concentrate on exploring "the first precisely described structure for the nucleic acids that has been suggested by any investigator."[16] He celebrated Christmas Day by inviting colleagues to his laboratory so they might admire his "extraordinarily" tightly-packed model. A prominent feature of the colorful array of balls and sticks, representing individual atoms and their chemical bonds, was that the nucleotide bases were placed on the exterior of the model, jutting out "like leaves on a stalk." This arrangement, Pauling declared, provided the needed space for the bases to be arranged in virtually any order, giving "maximum variability in the molecule and thus maximum specificity in the message."[17]

By noon, his lab was almost as tightly packed as his DNA model, filled with chemists, physicists, and biologists, all schooled never to contradict the master while he was giving a performance. Pauling told his legion of admirers that the model was only a first attempt and "probably capable of further refinement."[18] But he surprised the more skeptical men in the room when he announced that he was satisfied enough to submit a formal paper detailing his triple helix to the *Proceedings of the National Academy of Sciences*. He did precisely that, a week later, on December 31, 1952.

A few hours after the manuscript was sent off, Pauling wrote a long letter to John Randall at King's College, putting him, Wilkins, and Franklin on high alert:

> Professor Corey and I are especially happy during this holiday season. We have been attacking the problem of the structure of nucleic acid during recent months, and have discovered a structure which we think may be the structure of the nucleic acids—that is, we feel that the nucleic acid molecule may have one and only one stable structure. Our first paper on this subject has been submitted for publication. I regret to say that our X-ray photographs of sodium thymonucleate are not especially good; I have never seen the photographs made in your laboratory, but I understand they are much

better than those of Astbury and Bell, whereas ours are inferior to Astbury and Bell's. We are hoping to obtain better photographs, but fortunately the photographs we have are good enough to permit the derivation of our structure.[19]

If these declarations were not enough, he went one step further in planting his flag. On January 2, 1953, he and Corey dispatched a brief, twenty-four-line note to *Nature*, claiming priority for a model of DNA's structure. The notice was published in the February 21 issue and informed readers that Pauling and Corey's full description of DNA's structure would appear in the February 1953 issue of *PNAS*.[20] The point of this British preview was that because *Nature* was published in London, its Cambridge and King's College readers would be certain to learn of Pauling's news by the following day when they received their copies. Unlike his definitive exploration of protein structure—a venture that required years of careful and painstaking calculations—he crammed this job into a timespan of four weeks.[21] Linus sang, but he could not yet hear how off-key was his tune.

A Stomach Ache in Clare College

> *The Great American Evil—Indigestion: There can be no good*
> *health, or manly and muscular vigor to the system, without*
> *thorough and regular digestion. It is doubtless here that four-*
> *fifths of the weaknesses, breakings-down, and premature*
> *deaths, of American[s] begin . . . Do not depend on medicines*
> *to place your stomach in order; that is but casting out devils*
> *through Belzebub [sic], the prince of devils. . . . a great deal of*
> *the indigestion that prevails, is the result (we cannot too often*
> *recur to this) of a cause we have elsewhere alluded to, <u>excessive</u>*
> *<u>mental action.</u>*

> —WALT WHITMAN[1]

J im Watson loved the thrill of taking short cuts, skirting the rules, and telling white lies to get himself wherever he wished to go next. Skipping time-consuming, messy steps in experiments, slightly misrepresenting himself to his superiors, and surreptitiously appropriating the data of another scientist were among the many above-the-law but on-the-borderline behaviors he exhibited before he reached the age of twenty-five. This predilection even extended to arranging his living quarters in the fall of 1952. Having bunked in the back room of John and Elizabeth Kendrew's tiny house for almost a year, he hungered to live on his own and closer to the action that was Cambridge University.

In the late autumn, L. M. Harvey, the Assistant Registrary and Secretary of the Board of Research Studies at Cambridge, formally approved Watson as "a registered Research Student of this University working under the supervision of Dr. J. C. Kendrew," for the October term of 1952 through the Easter term of 1953.[2] This meant that Watson was eligible to live in college rooms. Like Darwin, Newton, Rutherford, and many other illustrious Cantabrigians, James Dewey Watson officially became a resident of the university.

For more than a year, Watson had been scouting the various colleges for potential housing. At first, it looked as if he might join Jesus College, because it had far fewer research students and offered a better chance for acceptance compared to "the large, more prestigious, and wealthy colleges like Trinity or King's."[3] That idea was scotched soon after Watson learned that Jesus was filled to capacity with boisterous undergraduates. Of the few research students who were coaxed to matriculate into Jesus, none were assigned rooms in which they could live. As Watson shrewdly concluded, "the only predictable consequences of being a Jesus man were bills for a Ph.D. that I would never acquire."[4]

In the fall of 1951, Max Perutz enlisted the help of Nicholas Hammond, a senior tutor at Clare College, an eminent classicist, and deco-

Clare Memorial Hall. Watson's rooms were on the ground floor, to the left of the central doorway.

rated Second World War hero. Through this connection, Perutz "slipped [Watson] into Clare as a research student." During his first year at Cambridge, Clare College had granted Watson dining privileges in hall. The problem was that these meals were brief affairs that allowed little time for socializing and, perhaps just as bad, the food consisted of nothing but "brown soup, stringy meat, and heavy pudding."[5]

One year later, he was assigned to live in a "double room," number 5 in R staircase of the newly-constructed Clare Memorial Court.[6] On October 8, 1952, he told his sister, "I now live in college and rather like it. My rooms are pleasantly large but rather dull. However, with Odile's help I hope to liven them up."[7] In *The Double Helix*, Watson admitted his collegiate dishonesty: "Working for another Ph.D. was nonsense, but only by using this dodge would I have the possibility of college rooms. Clare was an unexpectedly happy choice. Not only was it on the [river] Cam with a perfect garden but, as I was to learn later, it was especially considerate toward Americans."[8]

How could Watson not love his new address? His new digs were both affordable and prestigious. Especially attractive was his work commute:

Cambridge, Clare College.

Bridge across the Cam River to Clare College.

a mere ten-minute walk, across the Queen's Road and along a path between the Fellows' Garden and the King's backs, over the Clare Bridge across the river Cam. He then took the well-trod walkway between the Master's and the Scholars' Gardens of Clare College, through the Old Court, and down a narrow pathway abutting Gonville and Caius College and the Senate House. After three hard rights—the first onto stately King's Parade, then Bene't Street and, finally, Free School Lane—he sauntered into the Cavendish.

THE FOOD SERVED IN HALL continued to be unpalatable and, like many Americans unaccustomed to postwar British fare, Watson complained that it was overcooked. Even Clare's wood-paneled, coffer-ceilinged, crystal-chandeliered Great Hall could not alter the effects of the tired menus served within. On October 18, he wrote to Betty, "Clare food remains impossible so I eat many of my meals in the English Speaking Union. I also have food in my college rooms since I find I am quite hungry about 12 p.m. To my surprise, I find I can make myself tea."[9]

Most mornings, he sat at the battered counter of the Whim, an eating establishment on Trinity Street that opened on weekdays at 8 a.m. (and on Sundays at 10 a.m.) and served breakfast "much later than if [he] went to hall."[10] After plunking down three shillings and sixpence, he tucked into a full English breakfast (a fried egg, blood and pork sausage, back bacon, baked beans, toast, marmalade, and tea)—a meal both filling and cheap.[11] He bolted down these heavy repasts while reading *The Times*, avoiding the more conservative *Telegraph* and *News Chronicle*, so enjoyed by "flat-capped Trinity types," before making his way to the Cavendish.[12]

Lunch with Crick at the Eagle was costlier but essential to their collaboration. It was dinner that presented the most problems if he was to avoid the awful food at Clare. There were some nicer establishments such as the Arts and the Bath Hotel, but both were too expensive for everyday dining. After tiring of the fare at the English Speaking Union, he took to mooching dinner at Odile and Francis Crick's flat or, in a pinch, with the Kendrews, but these options, too, had their limits.[13] Eventually, he frequented the cheapest eating establishments in town—Indian curry joints and Cypriot–Greek greasy spoons.[14]

The discount suppers wreaked havoc on his tender gastrointestinal system. Watson's tender "stomach lasted until early November before violent pains hit [him] almost every evening." Home remedies, such as milk saturated with baking soda, did little to ameliorate his symptoms. Eventually, he sought the help of a Cambridge-educated doctor on Trinity Street. After signing into the "ice-cold" surgery, he was ushered into a small examination room where the medico proceeded to pound, percuss, and palpate his belly while asking him a series of embarrassing questions about his bowel movements and rate of flatulence. Eventually, the doctor handed him a "prescription for a large bottle of white fluid to be taken after meals." The white, chalky stuff hardly needed a doctor's signature for purchase: it was Phillips' Milk of Magnesia, an antacid and laxative consisting of magnesium hydroxide dissolved in water and flavored with sugar and peppermint oil. (Phillips' Milk of Magnesia remains a popular tonic for upset stomachs, indigestion, heartburn and that great bane of human civilization, constipation.)[15]

The milk of magnesia seemed to help but it could only do so much considering Watson's horrible diet. As soon as he ran out of the stuff, two weeks later, his symptoms returned with a vengeance. Like many a homesick student, Watson overdiagnosed himself—with a gastric ulcer, gallstones, and maladies far worse. He found little sympathy when he returned to the surgery. Barely looking at him, the doctor scrawled out another prescription for the milky laxative and told him to lay off the spicy tandoori chicken, greasy gyros, and cheesy spinach pie.

After leaving the doctor's office, Watson rode his bicycle over to the Cricks' newly purchased house on Portugal Place, a crooked cobblestone lane where stood a row of narrow houses, replete with uneven wooden floors and marble fireplace surrounds. Jim was "hoping that gossip with Odile would make [him] forget [his] stomach."[16] They began by gabbing about Peter Pauling, who was chasing Max Perutz's au pair, a young Danish woman named Nina. This line of discussion did little to quiet

Watson's bellyaching, so Odile suggested he turn his gaze several blocks south, to "a high-class boarding house" on Scroope Terrace owned by a French expatriate named Camille "Pop" Prior. Pop was widely known in town as the enterprising widow of a former professor of French language and an "indefatigable producer of every kind of dramatic and musical show."[17] To make ends meet, she boarded and taught English to a bevy of "foreign girls" who came to Cambridge hoping to improve their employment opportunities. Watson had little desire to learn French, but was intrigued by the chance to win "Pop's fancy" and gain entry to her famous sherry parties. Odile promised "to ring Pop to see if lessons could be arranged." The prospect of meeting more "popsies" cheered the ailing Watson immensely and off he went on his bicycle back to Clare College.[18]

Watson's indigestion did, however, force him to stay put in his rooms at Clare for days on end, stoking the coal fire, pulling his blanket over his head, and thinking deeply about DNA. The building was unusually frigid due to a brutal cold air system that blanketed much of England in early December, worsened by a thick, suffocating, sulfurous layer of smog—thanks to Britain's almost exclusive use of coal for heat.[19] At one point, he "huddled next to the fireplace, daydreaming about how several DNA chains could fold together in a pretty and hopefully scientific way." Such molecular musings were fortified by a stack of journals, reprints, and textbooks piled beside his bed, all containing theories on "the interrelations of DNA, RNA, and protein synthesis."[20]

It would be five more years before Francis Crick's famous articulation of "the central dogma" of gene function, describing how the information on a strand of DNA is copied by RNA, aided by specific editing and translational enzymes, in order to facilitate the synthesis of proteins by the ribosomes in the cellular cytoplasm.[21] Nevertheless, in early December 1952, Jim Watson scrawled down a rudimentary version of this formula: *DNA→RNA→protein.*[22]

Watson recalled thinking at that time that "virtually all the evidence then available made me believe that DNA was the template upon which RNA chains were made. In turn, RNA chains were the likely candidates for the templates for protein synthesis . . . the arrows did not signify chemical transformations, but instead expressed the transfer of genetic information from the sequences of nucleotides in DNA molecules to the sequences of amino acids in proteins." Watson taped the sheet of paper containing these scribbles onto the wall directly above his desk, as if he needed reminding that "the idea of the genes' being immortal smelled right." This dreamy realization helped him to fall asleep at night. When he awoke, the icy temperature of his bedroom "brought [him] back to the knowing truth that a slogan was no substitute for the DNA structure."[23]

AN ENTIRE YEAR HAD PASSED since Bragg told Watson and Crick to lay off DNA. Watson found the order to be arbitrary, if not silly, but was forced to present the appearance of obedience, given his tenuous status as a visiting fellow. Crick, whose position at the Cavendish was somewhat shakier, spent the year completing his doctoral thesis, calculating coiled coils of proteins, and deciphering the density of hemoglobin. Their lunches at the Eagle no longer centered exclusively on DNA but invariably, during their after-lunch strolls along the backs of King's and Trinity, "genes would creep in for a moment." When this did happen, their excitement carried them back to Room 103, where they "fiddled with the models . . . but almost immediately Francis saw that the reasoning which had momentarily given us hope led nowhere."[24]

Watson was by now bored with the tobacco mosaic virus and had an excess of intellectual energy to burn. While Crick stared at X-ray photographs of hemoglobin and filled pages of his laboratory notebook with calculations, Watson doodled DNA diagrams on the chalkboard. The irony was that, of the two scientists, Crick had far better spatial reasoning skills. He could visualize biological structures in three dimensions as if they were in his hand rather his head, a vital talent that Watson was still struggling to develop.

The frustration level in their tiny office steamed and frothed like a freshly made cappuccino, until one afternoon a smiling Peter Pauling entered, sat down, and carelessly parked his feet upon his battered desk.

It was a rare appearance, for his performance at the Cavendish had been less than exemplary. Bragg was close to terminating his position, regardless of his last name, because he spent more time socializing, chasing au pair girls, and "rowing the bow in the gentlemen's eight of the Peterhouse fifth boat" than "in the lab doing something."[25]

Watson expected Peter Pauling to start another monologue about his latest amorous conquest or the "comparative virtues of girls from England, the Continent, and California." He soon learned that "a fetching face, however, had nothing to do with the broad grin on Peter's face."[26] He told Crick and Watson that he had just come from lunch at Peterhouse College and after the meal, he had popped by the mailroom to retrieve a letter from his father. The letter was from his father.[27] In addition to talk of academic politics, family events, and other homebound topics, Linus Pauling let loose the news that Watson and Crick had "long feared": he was actively seeking the structure of DNA.[28] One can only imagine the shower of cortisol and adrenaline flooding their brains and bodies, setting them both up for a fight-or-flight response. Pauling was in the race! Maybe, Crick thought, if they could come up with a structure simultaneously, they might share credit of discovery with him. But how to do it? And how would they convince Peter Pauling to share some clues from his father so they might jump-start their efforts without giving away their own secrets?

Watson, Crick, and Pauling ran down the hall and up the stairs to the tea room, where Perutz and Kendrew were cooling their afternoon cuppas. As they passed the letter around, Bragg entered. Providence inserted itself into Watson's mouth, causing him to shut up: "neither of us wanted the perverse joy of informing him that the English labs were again about to be humiliated by the Americans. As we munched chocolate biscuits, John tried to cheer us up with the possibility of Linus' being wrong. After all, he had never seen Maurice and Rosy's pictures. Our hearts, however, told us otherwise."[29]

Peter and the Wolf

Many readers of Jim Watson's book The Double Helix *have told me that they interpret from it that my role was that of a double agent, spying on rival teams, using my special position to find out what one group was doing, and reporting to the other. This was in no way the case. I tried to understand as much as possible about what was going on in our laboratory and I wrote to my family and told them these things insofar as they affected me. My father has always kept me informed about what is interesting him and what work is going on.*

—PETER PAULING[1]

D uring the two weeks leading up to the Christmas holiday, all seemed quiet on the Pasadena front. That said, a voracious wolf in the form of Linus Pauling was hunting for fresh prey. During the holidays, Watson allowed himself the naïve assurance that if there was some terrific revelation about to break at Caltech, he would have heard of it by now. To make the most of his time off, he went skiing in Switzerland. Passing through London, he visited King's College and informed Wilkins that Pauling was scratching around the DNA molecule and might soon solve the puzzle. This confab occurred a week before Pauling sent off his official "DNA letter of intent" to Randall, on December 31. Wilkins appeared unconcerned, much to Watson's dismay. He was far more focused on getting rid of Rosalind Franklin. As joyous and free as one so neurotically bound up as Wilkins could be, he was now counting the days when with Franklin "at last out of his life, he would commence an all-out search for the structure."[2] All that was needed to complete this scene was one of those old-fashioned wall calendars, in which one ripped off each passing date to announce the present one.

※

UPON HIS RETURN TO CAMBRIDGE in mid-January 1953, the first person Watson sought out was Peter Pauling. Peter told him that he had received a letter from his father, dated December 31, 1952, that effectively lobbed a scientific grenade across a continent and an ocean: Pauling and Robert Corey had proposed a structure for DNA that would soon be published in the *Proceedings of the National Academy of Sciences*. In 1975, Peter recalled that his father wrote to him saying that a preprint was being prepared for Bragg and "he asked me if I, too, would like to have a copy of the manuscript. I knew that Bragg understood even less about DNA than I did, and that he would ignore the paper, so I answered yes, I would like a copy."[3]

The possibility of another one of Pauling's spectacular breakthroughs made Watson jumpy, to say the least. He quieted his nerves by spending the next several days working with Bill Hayes on their paper on bacterial sexuality and exchange—the study he hoped would topple Joshua Lederberg.

All such distractions were put on hold on the morning of January 28, when the mailman delivered two envelopes, both containing the Pauling DNA paper, to the Cavendish Laboratory, Free School Lane.[4] Bragg buried the paper in a huge pile of manuscripts sent to him by authors seek-

ing his blessing, rather than sending it down the chain of command to Max Perutz. He may have hoped to avoid the drama he knew would commence when Perutz showed the paper to Crick and that eccentric American, Watson. This was the last thing Bragg wanted, now that Crick was almost done with his thesis and, in, eight months' time, would be exiled for a year to the Brooklyn Polytechnic Institute.[5] Just as Wilkins was counting the days before getting

Peter Pauling, 1954.

Rosalind Franklin out of his daily view, Bragg had a similar metric for when he could again conduct his work free from Crick's "armor piercing" laughter.[6]

Peter Pauling recalled, "[I] had no idea of what a gene was . . . [and because my father's paper] meant nothing to me, I gave it to Jim and Francis to whom it seemed to be important."[7] In Watson's telling, the handoff was far more theatrical. As soon as Peter crossed the threshold of Room 103, Watson sensed that "something important" was about to happen and his "stomach sank in apprehension at learning that all was lost." Peter methodically informed Watson and Crick about his father's triple helix "with the sugar-phosphate backbone in the center"—a structure that sounded painfully similar to the model they had proposed a year earlier, which Rosalind Franklin had so definitively rejected. Watson's mood plummeted as he "wondered whether we might already have had the credit and glory of a great discovery if Bragg had not held us back." Breaking all rules of scholarly conduct, let alone a gentleman's manners, Watson did not even give Crick the opportunity to ask to see the manuscript; instead, he yanked it out of Peter Pauling's coat pocket.[8] His pupils dilated as he raced through the paper's introduction and explanation of methods, searching for the figures depicting Pauling's triple helix. His trained eyes were seeking the precise locations of the "essential atoms" making up the miraculous molecule that would soon make his own name immortal.

THERE ARE SEVERAL TYPES OF discovery in science. The most heralded kind is the discovery of the correct answer to a vexing problem of how things work—a terrific breakthrough that in the moment makes all in the world feel right. Almost as important, in the competition to explain things first, is discovering that your chief rival is wrong about *his* breakthrough. When it came to evaluating Pauling's DNA model, there were few people in the world who could reach such a conclusion quickly, just as there are few pianists dexterous enough to play the Rachmaninoff Third Piano Concerto. Jim Watson was one of them.[9]

Almost immediately, Watson recognized that Pauling's concerto contained too many faltering notes and clumsy cadenzas. Although Pauling and Corey declared that theirs was "the first precisely described structure for the nucleic acids that has been suggested by any investigator," the paper's prose hemmed, hawed, and confessed that "the structure cannot be considered to have been proved correct."[10] There was no new data presented and the X-ray photographs used for the model had been made by William Astbury of Leeds a decade and a half earlier. Pauling and Corey even admitted in print that the X-ray photographs they were able to produce in Pasadena were "somewhat inferior to those" developed in Leeds (not to mention the pictures Rosalind Franklin had produced, which Pauling had not yet examined). These flaws mattered little to Pauling because he was, well, Linus Pauling. In other words, thanks to his novel stochastic modeling methods, he and he alone had the knowledge and talent to absorb the extant literature and, like the chemical wizards who once called the discipline alchemy, conjure gold from lead.[11]

There were three colossal errors in the Pauling triple helix paper. The first had to do with how closely Pauling packed the atoms, with the sugar–phosphate backbone in the center and the nucleotides facing out, in three helical strands. To be sure, a model of DNA needed to be both stable and one that placed its atomic components in close proximity as biology demanded, but Pauling, by his own admission, had gone overboard. Even though they had committed their ideas to the preprinted paper, Pauling and Corey were still fiddling with their model to ease the molecular equivalent of the feeling of wearing one's clothes too tight.[12]

The second error was one Watson found so basic that "if a student made a similar mistake, he would be thought unfit to benefit from Cal Tech's chemistry faculty." Specifically, "the phosphate groups in Linus's model were not ionized, but . . . each group contained a bound hydrogen atom and so had no net charge . . . the hydrogens were part of the hydrogen bonds that held together the three intertwined chains."[13] This meant that Pauling's structure ignored the major characteristic of a nucleic acid: it is, first and foremost, "a moderately strong acid." Astonishingly, "Pauling's nucleic acid in a sense was not an acid at all."[14]

Watson knew this conjecture made no sense. Whether DNA consisted of two, three, or four intertwined chains, stable hydrogen bonds were needed to hold them together. Otherwise, the chains would simply "fly apart and the structure vanish." Based upon his reading of nucleic acid chemistry, however, Watson had determined that DNA's "phosphate groups never contained bound hydrogen atoms" and that when DNA was found in the body, or under "physiological conditions," the negatively charged phosphate groups would instead be neutralized or bonded with positively charged sodium or magnesium ions. "Yet somehow," to Watson's amazement, "Linus, unquestionably the world's most astute chemist, had come to the opposite conclusion."[15]

Crick, too, was flabbergasted by Pauling's "unorthodox chemistry." Initially, he tried to give Pauling the benefit of the doubt. Had he come up with a revolutionary new theory on the behavior of acids and bases in large biological molecules? And, if so, why was there no explanation of such a theory in his paper? Why were there not two papers, "the first describing his new theory, the second showing how it was used to solve the DNA structure?" No, this was not that sort of announcement, Watson and Crick concluded. Pauling had produced a bona fide blunder.[16] For the first time in days, they were able to breathe a bit more slowly, knowing they "were still in the game."[17]

Pauling's chemical "blooper was too unbelievable to keep secret for more than a few minutes." And so, the two men ran over to Roy Markham's virology lab "to receive further reassurance that Linus' chemistry was screwy." Markham did not disappoint. With an almost predictable sense of schadenfreude, he, too, was overjoyed to learn that the great Pauling of Pasadena "had forgotten elementary college chemistry."[18] Before Markham could regale them with the gaffes of other colleagues, including one still working at Cambridge, Watson "hopped over to the organic chemists', where [he] heard the soothing words that DNA was an acid."[19]

The third, and most serious, problem with Pauling's model was that the triple helix structure did nothing to elucidate how cells reproduced and passed on their genetic information in an orderly and predictable manner. It failed to conform with Erwin Chargaff's rule of a 1:1 ratio of purines (adenine and guanine) to pyrimidines (thymine and cytosine). As Horace Judson concluded, Pauling's "model stood mute. It explained nothing at all . . . nor did anything else spring from it to illuminate the secret of the gene."[20]

After making their rounds, having many good laughs at Pauling's expense, Watson and Crick took their bows in the Cavendish tea room. By now, Crick was at full throttle explaining the errors of Pauling's ways to Max Perutz and John Kendrew. Their revelry stopped abruptly when they realized that soon after the paper was published, everyone else in the close-knit scientific community would discover what Watson and Crick found out that January morning. This meant they "had anywhere up to six weeks" to divine the problem "before Linus again was in full-time pursuit of DNA."[21]

IT WAS THEN, like a U-boat popping up in the cold, gray mid-Atlantic Ocean, that the issue arose of informing Maurice Wilkins. Who would tell him about the Pauling paper? Should they warn him? On the other hand, if they called him right away, would the excitement in their voices alert him to their interest in joining the race? Watson and Crick were not ready to collaborate with anyone outside Cambridge. There also remained the issue of Bragg's moratorium and John Randall's proprietary attitude toward DNA research. Randall was likely to spontaneously combust if he knew what was going on a mere fifty miles away from his laboratory.

A quick ring to Wilkins simply would not do. Watson piped up that he was planning to visit London two days hence, on January 30, for a meeting with Bill Hayes. "The sensible course," he determined, "was to bring the [Pauling] manuscript with me for Maurice's and Rosy's inspection."[22] This approach bought them time to digest and capitalize on their findings. They were back on the hunt, with Perutz and Kendrew's passive approval, and were determined to win it, whether Wilkins and Franklin knew about it or not.

After enacting the 1950s' British version of a high-five, Watson and

Crick repaired to the Eagle, which they found closed until 6:30 p.m. in accordance with the Defense of the Realm Act of 1914 restricting the sales of alcoholic beverages in pubs from 2:40 p.m. to 6:30 p.m.[23] Soon after the Eagle's doors opened, however, they drank a toast to "the Pauling failure." This was not an evening for swilling beer or sipping sherry. Crick stood his young, eager colleague to a proper drink of good Scotch whisky—and drink they did. There was still a mountain of work ahead of them, perhaps the most important, concerted, creative work hours of their lives. To a large extent, Watson understood the high stakes. He knew that, if successful, he would join the modern biological pantheon begun by Charles Darwin.[24] He sipped the strong single malt Crick had bought him and from his perch at the Eagle quietly declared war: "Though the odds still appeared against us, Linus had not yet won his Nobel."[25]

Photograph No. 51

Interviewer: *What's your . . . I mean the famous story of the photographs has had several different versions. I'm not sure what your version of it is.*
Maurice Wilkins: *Oh, how I stole Rosalind Franklin's photograph and showed it to Jim.*

—INTERVIEW WITH MAURICE WILKINS, C. 1990[1]

There's a myth which is, you know, that Francis and I basically stole the structure from the people at King's. I was shown Rosalind Franklin's X-ray photograph and, Whooo! that was a helix, and a month later we had the structure, and Wilkins should never have shown me the thing. I didn't go into the drawer and steal it, it was shown to me, and I was told the dimensions, a repeat of 34 Ångstroms, so, you know, I knew roughly what it meant and, uh, but it was that the Franklin photograph that was the key event. It was, psychologically, it mobilized us . . .

—JAMES D. WATSON, C. 1999[2]

How in the world did Maurice Wilkins come into possession of Rosalind Franklin's X-ray DNA photographs? In 1990, Wilkins was still rather defensive about the whole affair: "Gosling had handed me over the negatives as part of the preparation, you know, she was clearing her stuff out, preparing to go . . . Well, I didn't ask for the photographs . . . I'm pretty sure it was [Gosling] and not she who gave it to me . . . But the awful thing was, that pattern had been taken [in]

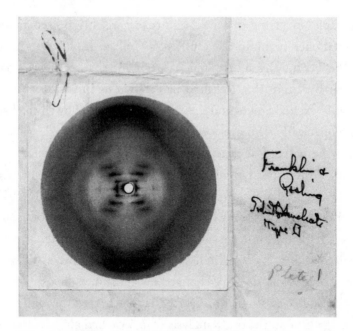

"Photograph No. 51." Franklin's X-ray crystallography photograph
of the B form of DNA, at more than 75% relative humidity,
showing the double helix.

May, nine months before . . . and so I felt, well, it was a bit naughty of
her to have had all this evidence for it . . . being helical, and not telling
us anything about [it]." Somewhat circumspectly, Wilkins added, "So I
thought, oh well . . . if you want to hop on the helical bandwagon now,
and change your tune completely, I suppose it's a free world, and one
can't make any objection . . . I was rather cynical about it. I didn't know
that she had actually done something. I did her a discredit there, but
how was I to know? So, there was a lot of people simply not knowing
what each other were doing, and sort of misunderstandings and so on."[3]

Thirteen years later, in his 2003 memoir, Wilkins mentioned that the
photograph transfer was a bitter one. "Rosalind was so negative that I did
not want to be involved in asking her for anything."[4] Yet only a few pages
later, he offered a clear recollection of making his way down the physics
department's main corridor and running into Raymond Gosling, when
"something extraordinary happened." Gosling handed him Franklin's

Photograph No. 51 and informed him that it was his to keep and use as he wished once Franklin left the King's College group. Wilkins could hardly believe his eyes at how "much clearer and sharper [it was] than the first clear B pattern that Rosalind had shown us in October 1951—the one that so excited Stokes and me. The new pattern showed the helix X-shape more clearly than ever before."[5]

With the passage of time, Raymond Gosling, too, offered varying accounts of this critical event. In 2000, he explained that "Maurice had a perfect right to that information. There was so much going on at King's before Rosalind came."[6] Yet by 2003, Gosling made the story murkier: "I cannot remember how he came by this beautiful picture. It may have been given to him by Rosalind, or it may have been me."[7] Nine years later, in 2012, Gosling's story changed once again, as he explained that since Franklin was about to leave the laboratory and would not have time to "go beyond the draft analysis which we had already drafted . . . she, therefore, decided to make a 'present' to Maurice of the original film of our best B structure." But in this telling, Gosling recalled walking down the corridor to find Wilkins "sometime in January 1953 and gave him this beautiful negative. He was very surprised and wanted reassurance that Rosalind was actually saying he could make whatever use he wished of this interesting data."[8]

Paddling water in a lake of defensive memories is no easy task. Still, a few sharp points keep emerging from the muck. Why would Franklin authorize her graduate student to freely hand over so critical a discovery to the individual at King's College she most despised? If she was going to give away her hard-won data to anyone upon her exit—a common practice for postdoctoral fellows upon leaving a laboratory because the work product is considered to be the property of the principal investigator on the grant—why did she not hand-deliver it to John Randall?

We can say with certainty that Franklin produced the "beautiful" X-ray diffraction Photograph No. 51 on May 2, 1952. She began the task by carefully bundling twenty or more snot-like DNA fibers together at the end of a tiny capillary tube (no easy task), realigned the heavy camera for many hundreds of angles and exposure shots, and endured, at least, one hundred hours of high radiation exposure. Both Wilkins and Watson have always insisted that it was Gosling who took these X-rays, which is technically true in that, as her research assistant, he performed a great deal of the grunt work. But as her sister, Jenifer Glynn, has correctly noted, "There is a big difference between designing the experiment and pressing the button of an X-ray machine."[9] Unable to discern the answers within the crystal-clear "X" she produced on the photographic plate, Franklin put this picture aside and occupied the next several months analyzing the time-consuming Patterson equations for the more regular and crystalline A form.

§

IN THE FIRST WEEK OF JANUARY 1953, she returned to interpreting the B form. Her laboratory notebooks confirm that she was working hard on both the helical nature of the wet B form and the more difficult-to-interpret dry A form, as well as trying to conform her crystallographic data with Chargaff's rules by squeezing together "the four bases of DNA into a structure with the phosphates on the outside."[10] She was scheduled to present her "exit seminar" at King's College on January 28, 1953, and this data would be central to it.

The person who should have been happiest at this occasion was Maurice Wilkins. Wilkins being Wilkins, he was not. One week before her talk, he wrote to Crick informing him about the upcoming seminar and tossing some nasty jibes Franklin's way: "Let's have some talks afterwards when the air is a little clearer. I hope the smoke of witchcraft will soon be getting out of our eyes. P.S. Tell Jim the answer to his question 'When did you last speak to her' is this morning. The entire conversation consisted of one word from me."[11]

Directly after Franklin's lecture ended on January 28, Wilkins wrote to Crick about how depressed he was over his work environment. "Rosie's colloquium made me a bit sicker. God knows what will become of all this business. [She] spoke for 1¾ hours non-stop . . . and had a unit cell big enough to sit in but nothing in it."[12] In his memoir, Wilkins refined his recollection somewhat. He still described Franklin's talk as "exceptionally long, and solely about her structure for A-DNA—she did not mention the B-form at all." All he claimed to remember was a model she constructed of "bent pieces of wire, zig-zags, and figure-of-eight shapes. Doubtless all of that was well thought out, but to me it did not add up . . . it [was] distressing to think of an able scientist like Rosalind struggling so hard in the wrong direction."[13]

This last statement, meant to expand the myth of Franklin's "sad" pursuit of "non-helical structures," is contradicted, or at least complicated, by her research notes, by Wilkins's recollections to journalists about her so-called "anti-helical talk" in November 1951, and even by his own memoir, in which, one sentence later, he describes the "question time after the talk." First to raise his hand, Wilkins asked Franklin "how the non-helical structure she had been discussing could be reconciled with the very good B-pattern she had passed on to me." Franklin, in her firm, confident, professional manner, replied that "she saw no problem: B-DNA was helical and A-DNA was not." In 2003, Wilkins recalled his surprise at Franklin's response to his question: "I was taken aback by her answer because that was the first time I heard her concede that any DNA could be helical. I was even more surprised that she thought that B-DNA was helical and A-DNA was not. I do not think it ever occurred to me that she might believe that." He was equally confused by her glib explanation that DNA would need to switch "back and forth easily between the helical to non-helical forms as its water content changed . . . [because] Stokes and I firmly believed that, if B-DNA was helical, A-DNA would be helical too. Actually, we did not think it had to be, but we felt strongly that it was very likely." Perhaps, he reminisced, he had not understood her because whenever they talked about DNA, he felt as if they "were on different wavelengths." Unable to muster a cogent response, he simply resumed his seat. To the end of his life, he maintained that the lecture had concluded thus: "No one else said anything about B patterns, but if the striking new pattern had been shown to the audience, I think there might have been some discussion. Why did she not show it?"[14]

Herbert Wilson, a postdoctoral fellow who worked under Wilkins but not with Franklin, was also at the seminar. He claimed to have taken notes of her exit lecture, but sadly, he did not preserve them for posterity. In 1988, while Wilkins, his mentor, was still alive, Wilson insisted there was "no reference to B-DNA and I was not aware of her views on its structure at the time."[15] He claimed not to know of Franklin's helical interpretation of B-form DNA until 1968, when the biophysicist Aaron Klug, Franklin's colleague and champion, wrote an article in *Nature* on Franklin's work, drawing from her laboratory notebooks.[16] Yet his professed ignorance must include mention of his extreme loyalty to Wilkins. Was Wilson telling the truth or attempting to attenuate the daily suffering of his former chief who, by the late 1980s, was being actively excoriated by many of his colleagues because of his behavior toward Franklin?

TWO DAYS LATER, on January 30, Jim Watson boarded the 10 a.m. train from Cambridge to King's Cross, London, and then took the Underground for a forty-five-minute ride and walk to the Hammersmith Hospital. There, he met with William Hayes to finalize their bacterial recombination paper. Jammed into his pocket was a copy of Pauling and Corey's prepublication paper on the triple helix model. He skipped lunch and, approaching teatime, left Hammersmith for King's College.[17]

Barging into Wilkins's office just before 4 p.m., he let loose with the news that Pauling had developed a DNA model and it "was far off base." Wilkins, who still thought of Watson as a pest and an oddball, explained he was presently occupied and, ever polite, asked him to return a bit later—probably hoping the American would disappear. Instead, Watson walked down the basement corridor to Franklin's lab.[18] Even at this late date, this seems a peculiar choice, given his feelings about her and the fact that she had little love for him. She was hardly shy about expressing her view of him as little more than a scientific dilettante who flitted and fluttered about without producing any of his own experimental data.

It was also no secret that Watson had woman problems. Despite his poor track record as a swordsman, he saw all attractive "popsies" and au pairs as potential prey. The French microbiologist André Lwoff observed his behavior askance: "Watson pursued women with a crude violence which distressed and frightened many of them."[19] As a young man anxious to explore his sexuality, the ninety-year-old Jim Watson admitted, "I was more dominated by not having a girlfriend than by science."[20]

For Watson, there were four major categories of women. The first he considered to be unapproachable and unavailable, ethereal goddesses worthy of his worship, admiration, and protection—such as his mother, Margaret Jean, his sister, Elizabeth, and the novelist and poet Naomi Mitchison (Avrion's mother). The second group included those few female scientists who were kindly, get-along types and did not threaten Watson with their brilliance, like Dorothy Hodgkin at Oxford. The third category were those he spurned as fallen trollops, an example being Herman Kalckar's young mistress, Barbara Wright. The fourth comprised the women Watson had the most difficulty interacting with, those who were contemptuous of his immaturity and doubted his intellect—and that category included Rosalind Franklin. Watson projected his insecurities by dismissing such women as hostile and unattractive shrews, not up to the excellence he believed that he himself displayed on a daily basis.[21] Resentful of how Franklin had vanquished his triple helix model in November 1951, Watson nurtured and burnished his grudge against her for decades.

Most infamously, he transmogrified her into "Rosy," the one-dimensional archenemy in the pages of *The Double Helix*. Nowhere in the book is this more the case than in the scene when he walked into her laboratory unannounced on the afternoon of January 30, 1953. Sixty-five years later, in 2018, Jenifer Glynn, Franklin's younger sister, described this passage as the unkindest cut of all: "My mother wished Rosalind had remained obscure or completely forgotten rather than have her remembered by Watson's brutal portrayal of her. She was hurt and upset. I was hurt and worried for her. *The Double Helix* is a novel. It is not history."[22] It may well be the cruelest published description written by one scientist of another that this historian has read in nearly forty years of practice.

AT APPROXIMATELY 4:05 P.M., Watson found Franklin in her lab with the door ajar. He walked in and, not surprisingly, startled her. She "quickly regained her composure and, looking straight at my face, let her eyes tell me that uninvited guests should have the courtesy to knock."[23] Indeed, such decorum was more than mere civility. By all accounts, Franklin was an intense soul; this was never a point in contention. When Watson burst in, she was in a darkened room, peering through a magnifying microscope at an X-ray diffraction pattern placed over a lightbox—the same device physicians use to look at their patients' X-rays. She was entirely focused on the task at hand, measuring the tiniest of smudges in units of an Ångstrom, which is one ten-billionth of a meter. She was understandably startled by Watson's abrupt intrusion and not apt to respond with the kindliest of manners.

According to Watson, he told her the news about Pauling and, after regaining her composure, Franklin insisted that the helical structure of DNA had yet to be scientifically proven. Relying entirely on Wilkins's assertion that Franklin was "anti-helical," as well as his misrepresentations of Franklin's ability and character, Watson proceeded to explain the Pauling paper to her. This was the same article Franklin had requested from Robert Corey a few weeks earlier and should have received promptly, considering how generous she had been in sharing her images with him the previous May. When she glanced at the second page of the preprint, she was surprised to find that Pauling and Corey's assumptions were based on Astbury's 1938 X-ray pictures, which consisted of mixed samples of the A and B forms of DNA. More egregious, Pauling and Corey added an acknowledgment that "Wilkins of King's College" had taken better X-ray pictures in 1951 but failed to reference Franklin's images—meaning that the kindness she showed Corey in May 1952 went formally unappreciated.[24] This may have further fueled her simmering anger at Watson, an emotion that soon boiled all over him.

For a brief moment, Watson toyed with Franklin, curious to see how long it would take her to figure out Pauling's mistake. He quickly sensed, however, that "Rosy was not about to play games with me." Unfortunately, he did not have the self-awareness to see how much he (and Pauling's failure to mention her work) was disturbing her. In the moment, he was incapable of taking a deep breath and slowing down his rapid-fire inquisition. He propelled his ideas at Franklin with the power of a B-52 Stratofortress, pointing "out the superficial resemblance between Pauling's three-chain helix and the model that Francis and [he] had shown her fifteen months earlier."

Completely ignorant of Franklin's new analysis of the B, or wet, form of DNA, Watson did not know that she had started to consider model building, too. Her notebooks for that week, including the entries for the very day Watson visited her and for Monday, February 2, clearly demonstrate that she was "twisting and turning from one cul-de-sac to the next in the effort to visualize what structure to build." Her frustration over being stuck is palpable in the neat lines of questions, alternatives, and objections she jotted down on the pages of her red Century notebook. Aaron Klug later described her vexation perfectly: "the stage reached by Franklin at the time is a state recognizable to many scientific workers, when there are apparently contradictory, or discordant, observations jostling for one's attention and one does not know which are the clues to select for solving the puzzle."[25]

Franklin was in no mood for Watson's exuberance, and she objected vociferously to his babbling. "Interrupting her harangue," Watson wrote, he proceeded to tell her what she already knew—that "the simplest form for any regular polymeric molecule was a helix." He continued to condescend to her until "Rosy . . . was hardly able to control her temper, and her voice rose as she told me that the stupidity of my remarks would be obvious if I would stop blubbering and look at her X-ray evidence."[26] Not incidentally, Watson had not yet seen her Photograph No. 51. Completely misjudging the woman standing before him, Watson "decided to risk a full explosion" and let fly that she was "incompetent in interpreting X-ray pictures. If only she would learn some theory, she would understand how her supposed anti-helical features arose from the minor distortions needed to pack regular helices into a crystalline lattice."[27]

"If only" . . . if only the contretemps had ended there. It did not, because Watson understood that historical skirmishes usually belong to

the victors. He had no intention of letting Franklin best him in any telling of what became *his* story. And because there were only two people in
the room at the time—Franklin did not live to read Watson's book—it
was a case of "he said/she didn't." Thus, an alleged threat of physical
attack became Watson's shot read round the world:

> Suddenly Rosy came from behind the lab bench that separated us
> and began moving toward me. Fearing that in her hot anger she
> might strike me, I grabbed up the Pauling manuscript and hast-
> ily retreated to the open door. My escape was blocked by Mau-
> rice, who, searching for me, had just then stuck his head through.
> While Maurice and Rosy looked at each other over my slouching
> figure, I lamely told Maurice that the conversation between Rosy
> and me was over and that I had been about to look for him in the
> tea room. Simultaneously, I was inching my body from between
> them, leaving Maurice face to face with Rosy. Then, when Maurice
> failed to disengage himself immediately, I feared that out of polite-
> ness he would ask Rosy to join us for tea. Rosy, however, removed
> Maurice from his uncertainty by turning around and firmly shut-
> ting the door[28]

As they retreated down the hall, Watson thanked Wilkins for preventing
Franklin from "assaulting" him. In turn, he claimed, Wilkins confided
that "this very well might have happened." A few months earlier, Franklin "had made a similar lunge toward him. They had almost come to
blows following an argument in his room." As Wilkins tried to escape,
Watson wrote, "Rosy had blocked the door and had moved out of the
way only at the last moment. But then no third person was on hand."[29]
The operative words here, so cleverly added, were, "no third person was
on hand."

The notion that Franklin would strike the six-foot-one-inch Watson
(or the over-six-foot-tall Wilkins) shreds the imagination—even one as
fertile as Watson's. In 1970, Maurice Wilkins speculated that Watson
was exaggerating: "there came a point where I feared stirring her up . . .

I don't think she would ever attack anyone physically [but] I can imagine that she might have slapped someone's face but that's not a physical
attack."[30] Only six years later, in a documentary filmed for PBS, Wilkins
revised his comments: "Oh my god! Who hit who? I don't think anyone
hit anybody. Some people may have thought they were going to be hit . . .
there certainly weren't many friendly feelings."[31] More convincing, even
though she was not present during the squabble, is the assessment of
Franklin's sister, Jenifer Glynn: "Rosalind was a little over five feet four
inches, and weighed, I would think, something under nine stone [about
125 pounds]. She would, of course, never hit anyone."[32] Regardless of
what actually happened, Watson's damage to a fellow scientist's reputation was done. The story of a raging and out-of-control Rosalind Franklin rising out of her chair and threatening to box Watson's ears sticks like
warm chewing gum on the sole of a shoe. Certainly, Watson never let
the fable of Rosalind the Termagant die as long as he drew breath. In the
summer of 2018, the ninety-year-old Watson still said with the greatest
of conviction, "I really believed she was going to strike me."[33]

THE ENCOUNTER, WROTE WATSON, "opened up Maurice to a degree I
had not seen before. Now that I need no longer merely imagine the emotional hell he had faced during the past two years. He could treat me
almost as a fellow collaborator rather than as a distant acquaintance with
whom close confidences inevitably led to painful misunderstandings."[34]
In 2001, in the opening pages of a companion memoir entitled *Genes,
Girls and Gamow: After the Double Helix*, Watson told the story even
more vividly: "Maurice—bristling with anger at having been shackled
now for almost two years by Rosalind's intransigence—let loose the,
until then, closely guarded King's secret that DNA existed in a paracrystalline (B) form as well as a crystalline (A) form."[35]

As Watson recalled in *The Double Helix*, "then, the even more important cat was let out of the bag: since the middle of the summer Rosy had
had evidence for a new three-dimensional form of DNA. It occurred
when the DNA molecules were surrounded by a large amount of water."

Now it was Watson's turn to be unable to contain himself as he practically pleaded, "what the pattern was like?" Maurice turned around and quietly walked into the next room, reached into a file drawer and retrieved a photographic "print of the new form they called the 'B' structure." The print was so freshly-made that it still reeked of the vinegar and water solution used in the dark room.[36]

What followed was the most important moment of James Watson's life. Today, it is almost impossible to find a scientist or science buff who cannot tell you the gist of the episode. In keeping with his intention of memorializing the greatest scientific discovery of the twentieth century, Watson's love of literature, motion pictures, and good storytelling kicked into high gear. The words he chose, carefully written and rewritten in many successive drafts of his memoir, are as memorable as a Mozart measure:

> The instant I saw the picture my mouth fell open and my pulse began to race. The pattern was unbelievably simpler than those obtained previously ('A' form). Moreover, the black cross of reflections which dominated the picture could arise only from a helical structure. With the A form, the argument for a helix was never straightforward, and considerable ambiguity existed as to exactly which type of helical symmetry was present. With the B form, however, mere inspection of its X-ray picture gave several of the vital helical parameters. Conceivably, after a few minutes' calculations, the number of chains in the molecule could be fixed.[37]

Seventeen years after the fact, Wilkins had a different interpretation of that afternoon. "Perhaps I should have asked Rosalind's permission, and I didn't. Things were very difficult. Some people have said I was entirely wrong to do this without her permission, without consulting her, at least, and perhaps I was—I don't know. You can say I was wrong if you want to, if that's how you see it. I don't particularly defend myself." Wilkins also tried casting the blame on Watson: "I thought there was data-pinching. It's not a nice question or a pleasant thought, but yes,

I'd have to say I think there was that. Jim was not above data-pinching. I'd say this was what he did. Now Francis—Francis doesn't do that sort of thing. But he knew where Jim was getting the stuff. He must have known that. I don't know how he justifies—we haven't discussed it . . . They couldn't have done it, <u>when</u> they did it. I'm confident of that."[38] Crick took the easy way out by simply justifying Wilkins's act. In April 2000, he wrote, "It seems to me Maurice did nothing wrong in showing the photo to Jim."[39]

Seeing Photograph No. 51 opened the floodgates of Jim Watson's mind. That first glance at Franklin's crisp, beautiful picture set him on the path to scientific glory. "If I had known that," Wilkins stated in great sorrow, "I might not have shown him the pattern." Unfortunately, Wilkins's remorse was focused on losing the race for scientific priority to "Jim the Snoop," rather than on his betrayal of Rosalind Franklin by showing her data to an outsider.[40] In 2018, Jenifer Glynn described Wilkins's multiple attempts to explain away these "lapses of decency" as "rather feeble, I think. It must have been hard being Wilkins. He spent fifty years defending himself over what happened."[41]

The participants, scientists, and historians have since pondered the ethical conundrums created by Wilkins's unauthorized gift to Watson in so many ways that is no longer possible to count them. At the risk of being simplistic, let us begin with two obvious assumptions: 1) the clear assignments of authorial or scientific priority as they existed in 1953; and 2) the standard operating procedure in modern scientific inquiry, in which one always asks the originator's permission before showing her or his research to a known competitor. The essential question to ask then is: What were Wilkins, Watson, and Crick thinking when they committed, and later struggled to justify, their behavior?

After all, Rosalind Franklin was working only a few steps down the hallway when Wilkins surreptitiously showed Watson Photograph No. 51. No matter how angry the Franklin–Watson encounter may have been a few minutes earlier, or for the past year for that matter, why didn't Wilkins shout out, "Rosalind, would it be all right if I showed your picture to Jim?" Quite simply, it was *not* all right. There exists no ethical standard whereby Franklin's permission did not need to be expressly asked, and because permission was not requested, Wilkins's showing Watson Photograph No. 51 remains one of the most egregious ripoffs in the history of science.

After salivating over the picture for a good long while, Watson and Wilkins left King's and made their way to Soho for supper. Obsessed by the possibility of Pauling hunting down the structure of DNA, Watson informed Wilkins that merely laughing at Pauling's bungled triple helix was not an option. He knew from his summer at Caltech that Pauling had a team of assistants who would soon be dispatched to take more and better X-rays, some of which, undoubtedly, would demonstrate the helical B form structure. And then the game would be over.[42]

No matter how much Watson pleaded, however, Wilkins refused to act upon the immediacy of the situation until after Rosalind Franklin left his laboratory for Birkbeck College. He dispassionately counseled Watson that following one's scientific hunches was far more important than chasing every mad dash that came his way. After the waiter served their order, Wilkins advised, "if we could all agree where science was going, everything would be solved and we would have no recourse but to be engineers or doctors."[43]

The discussion lost its momentum between Wilkins's "long drawn-out" replies and his request that they consume their meal before the food got cold. Wilkins did, at least, admit that he believed Franklin was correct in placing the bases inside the helix and the sugar–phosphate group backbone on the outside. Watson, however, "remained skeptical for her evidence was still out of the reach of Francis and me."[44]

Watson hoped that a few cups of after-dinner coffee might energize Wilkins, but this was not to be. Instead, they finished off a bottle of cheap Chablis until Watson's "desire for hard facts" diminished in the haze of alcohol. The two men left the restaurant and walked across Oxford Street. The solitary thought Wilkins articulated to Watson before retreating to his flat was his desire to move to "a less gloomy apartment in a quieter area."[45] They said their goodbyes and Watson headed back to King's Cross.

Before boarding the train, he bought a copy of the next day's *Times* to read on the journey home. As the train "jerked toward Cambridge," he pulled out a pencil from the breast pocket of his woolen jacket. On the edge of the page where the crossword puzzle was printed, he drew from memory the remarkable X-ray photograph that Rosalind Franklin had acquired with skill, sweat, too much radiation exposure, and the shrouded bruises that came from being a woman playing in an all-boys' game. By the time he finished sketching a rough version of the "B pattern," it had become *his* picture. Staring at it, he wondered whether to build a novel two-chain or another three-chain model. He may have been temporarily influenced by Wilkins's lack of enthusiasm for a two-chain molecule; during dinner, he had said that he wanted a better estimate of the molecule's water content and density before committing to one option or the other, but Wilkins was still leaning toward three chains.

Watson was so immersed in his drawing that he lost track of time. The scream of the train whistle and the conductor calling "Cambridge! Cambridge station approaching!" stirred him out of his intellectual trance. He sprang up from his seat and made a hurried exit from the carriage, then retrieved his bicycle from one of the racks at the station entrance. As he cycled the few miles to Clare College, he played with the numbers in his head, dreamed about the helical symmetry of that beautiful, heart-racing picture, and worried over the possibility of repeating the failure of his three-chain model fifteen months earlier. By the time he arrived at Clare, it was "after hours" so he "climbed over the back gate."[46]

Sometime between changing into his pajamas and brushing his teeth, Jim Watson chose to ignore Wilkins's preference for a three-chain structure and set his mind on a two-chain configuration. Lying in bed, he dreamed of telling Francis Crick what he had just seen at King's, and then convincing him of the double helix structure of DNA. "Francis would have to agree," he soothed himself into slumber. "Even though he was a physicist, he knew that important biological objects come in pairs."[47]

The Mornings After

Of course, Rosalind would have solved it. Maurice wouldn't have done, but with Rosalind it was only a question of time . . . well, the fact is that Maurice isn't terrifically bright . . . if she lacked anything, it was intuition . . . or mistrusted it . . . And she didn't know any biology. That held her up. She didn't have a feeling for the biology . . . But I don't see what Maurice would have contributed. He didn't even understand her X-ray photographs. He says now he did, but he didn't . . . Maurice says that Rosalind got the B-form by accident, but I told him it wasn't an accident he managed for himself . . . the point is to set up reasonable and intelligent experiments—you don't predict the outcome, or it's not an experiment, but you don't have a successful result unless you've set it up in a way that allows you to get that result.

—FRANCIS CRICK[1]

Rosy, of course, did not directly give us her data. For that matter, no one at King's realized they were in our hands.

—JAMES D. WATSON[2]

Jim Watson awoke early the next morning, January 31. Throwing on some clothes, he bounded across the Cam for a bowl of gray, overcooked porridge and a few cups of tea in the Clare dining hall. After swiping a starched white napkin across his face, he pushed himself away from the table and ran all the way to the Cavendish. Crashing into Max Perutz's office, he barely noticed Bragg sitting in the corner, perusing a journal. Bragg was used to Watson's silliness by now and looked forward to the day when this pesky American would finally be out of his rapidly thinning hair. Crick had not yet arrived. He never came into the laboratory early on Saturday mornings and was probably still in his flannel Marks and Spencer's pajamas, lounging in bed and reading the newspapers with his wife, Odile.

Watson could barely contain himself as he tried to convey the nuances of the intelligence he'd gathered the evening before. He kept saying the words "B form, A form" as if either Bragg or Perutz, who were not at all invested in DNA, understood the distinction. Perutz was a protein man. Although Bragg was a key creator of the field of X-ray crystallography,

Sir William Lawrence Bragg, c. 1953.

most of his work focused on inorganic lumps of metal and minerals. He had next to no formal training in biology, let alone in the nascent field of genetics. "There was no reason to believe that he gave [DNA] one hundredth the importance of the structure of metals," Watson griped, "for which he took great delight in making soap-bubble models. Nothing then gave Sir Lawrence more pleasure than showing his ingenious motion picture film of how bubbles bump each other."[3]

To better educate his chief, Watson approached the dusty blackboard, fully chalked with Perutz's scribbles. Quickly erasing the formulae scrawled on it without even inquiring as to their importance, he drew from memory the "Maltese cross" pattern of Franklin's Photograph No. 51. Bragg began peppering the gawky American with a series of questions. Watson stammered with excitement as he answered them, knowing he had the professor's full attention. Then came the one-two punch, calculated to hit Bragg right in the solar plexus: Watson broached "the problem of Linus, giving the opinion that he was far too dangerous to be allowed a second crack at DNA while the people on this side of the Atlantic sat on their hands." With the precision of a royal British salute, Watson asked permission for the Cavendish machine shop to make a slew of tin pieces representing purines and pyrimidines so that he might build a model. Knowing that he was the catalyst in the chemical equation yielding a serious break of professional protocol, he played his boss perfectly, counting the seconds it took for the old man's "thoughts to congeal."[4]

Still smarting from the defeat over Pauling's α-helix protein structure and impatient with the "internal squabbling" at King's, Bragg could not, would not "allow Linus to get the thrill of discovering the structure of another important molecule."[5] Watson and Perutz stared at their chief, knowing full well that Bragg's approval was essential if they were to move ahead and win the battle.[6] They were not disappointed. In a matter of seconds, the Cavendish Professor threw out his chest and transmogrified into Captain Corcoran from Gilbert and Sullivan's *H.M.S. Pinafore*. But instead of singing out, 'For he himself has said it, and it's greatly to his credit, that he is an Englishman," Bragg ordered his ensign to plow full speed ahead. Watson *must* find the structure of DNA for the honor of the Cavendish. He *must* succeed for the cause of British science. Most important, he *must* do it before Pauling realized his errors and corrected his model.[7]

Watson made certain to use only the personal pronoun when telling Bragg about his plans for model building. This was no mere semantic twist; he wanted Bragg to be completely in the dark about Francis Crick's inevitable involvement. Watson had been working independently for so long on tobacco mosaic virus that Bragg assumed this, too, would be a solo job. "Thus," Watson schemed, "he could fall asleep that night untroubled by the nightmare that he had given Crick *carte blanche* for another foray into frenzied inconsiderateness."[8] Before Bragg could change his mind, Watson darted out of Perutz's office, "dashed down the stairs to the machine shop," and informed the machinist he "was about to draw up plans for models [he] wanted within a week," or sooner![9]

Shortly after Watson settled at his desk in Room 103, a well-rested Crick entered with some news of his own—even though it turned out to be little more than idle gossip. The evening before, he and Odile gave a dinner party, to which they had invited Jim's pretty sister, Elizabeth, who had been living in Cambridge for the past month at Camille "Pop" Prior's boarding house. Watson had craftily sent her Pop's way so that he could dine each evening "with Pop and her foreign girls," thus saving Elizabeth "from typical English digs, while I looked forward to a lessening of my stomach pains."[10] Elizabeth brought along her latest beau, a wealthy young Frenchman named Bertrand Fourcade, whom Watson described in full homoerotic swoon as "the most beautiful male, if not person, in Cambridge." Fourcade, too, was living at Pop's "for a few months to perfect his English." His physical allure, "well-cut clothes," and continental charm "enchanted" both Elizabeth and Odile. Thus, while Watson watched "Maurice meticulously finish all the food on his plate, Odile was admiring Bertrand's perfectly proportioned face as he spoke of his problems choosing among potential social engagements during his forthcoming summer on the Riviera."[11]

Crick was still slightly hung over, but he sensed Watson's urgency.

He braced himself for another one of Watson's rants about being left behind on DNA. Instead, he was treated to Watson's vivid description of Photograph No. 51. With each successive word, the information Watson reported grew stronger and better.[12] Crick "drew the line" when it came to Watson's "assertion that the repeated finding of twoness in biological systems told us to build two-chain models" rather than a triple helical structure. As Watson later recalled, "since the experimental data known to us could not yet distinguish between two- and three-chain models, [Crick] wanted to pay attention to both alternatives. Though I was totally skeptical, I saw no reason to contest his words. I would of course start playing with two-chain models."[13]

ON FEBRUARY 2, Rosalind Franklin was seated at her desk, with her back to the door, in the corner of her little basement office. Toiling away at her Patterson calculations, she was working through an array of structural possibilities—from the position of the nucleotide bases to the form taken by the external sugar–phosphate backbones. She had already made the major (and correct) determination that the space group of the DNA crystal was of the monoclinic C2 type but could not figure out how that might unlock its molecular structure. On this day, however, her notebooks indicate a new direction. On a fresh page, she began modeling her A form and wrote, "Objection to figure-eight structure." Her reasoning was clear and her conclusion correct: "therefore, impossible."[14] She ruled out both "paired rod-shapes" and a single chain of "repeating eights," and began to reconsider helical structures.

A week later, on February 10, Franklin boldly scribbled at the top of a page in her notebook, "Structure B. Evidence for 2-chain (or 1-chain helix)?" She stared intently at her photographs, searching for helical features in the diffraction pattern. Eventually, she wrote eight lines reviewing what she knew about helical diffraction theory. Like a good scientist, she asked the data to tell her more about the possibility of a helical structure. She played some more with the mathematical formulae, but no answers were forthcoming. All she offers on the yellowing pages of her red notebooks are tantalizing but vague hints. With gusto and a large, inky "X", she scratched out the last four lines of one entry, which the historian tracking her thinking must strain to read: "and this is indistinguishable from double helix with residues on each having same z-value since 2nd chain has opp. signs in 5-7 turns (e.g. 1-2) and e.g. contains only . . ."[15] Right after this unfinished passage, we find a different set of calculations, including the water density of the wet DNA and the diffraction patterns a helical structure might form. Soon enough, she retreats from thinking about the B form and leaves a page of her notebook completely blank.

She does not return to the interpretation of Photograph No. 51 until a few weeks later, on February 23, 1953. In her notes for that day, Franklin determined the diameter of the helix and confirmed her previous finding that the backbones "were coiled at the outside of that diameter." She also wondered if there are "2 helices of different radius for simple case of whole number of residues per turn. Following Cochran, Crick and Vand (*Acta Cryst.* 5, 581, 1952) . . . 2-stranded helix with pairs of groups at opp. ends of diameter."[16] After more calculations, she again pondered whether "Structure B is single-stranded helix or 2 stranded."[17] This entry is especially remarkable because it took her only a few hours to mute the anti-helical background noise that had distracted her for nearly a year.[18]

A review of her notes of this period demonstrates how closely she was orbiting around the double helix, having analyzed both the A and B forms of DNA. Aaron Klug determined she was a mere two deductive steps away from definitively winning the race. After reviewing her scientific papers and notebooks, Klug concluded that by late February 1953, Franklin knew "there were two chains per unit cell in the A structure and she was considering a structure with eleven nucleotides per chain." When it came to the B structure, she thought it very likely consisted of two ten-nucleotide chains but "did not see the relation between the two structures, perhaps because she could not extricate herself readily from her deep commitment to solving the Patterson function without *a priori* assumptions, a course which required consideration of non-helical structures."[19] On the last page of her lab notes for February 23, 1953—

which detail her analysis of Photograph No. 51 and were written five days before Watson and Crick solved DNA—Klug annotated two words that speak volumes: "Nearly home." On the pages documenting her notes for February 24, he wrote, "R.E.F is at last making the correct connection between structures A and B."[20]

Here the paper trail abruptly stops. Franklin did not make the connection, as Watson and Crick would soon do, of how the purine and pyrimidine nucleotide bases related to one another within the helix. Nor did she recognize the biological relevance of her finding that crystalline DNA (the A form) organized itself in a C2, face-centered, monoclinic space group. This configuration, we now know, indicates the molecule's two-chain complementarity. When seated in the quiet archives, reading the last few lines of her notebook page for February 23, one almost wants to yell out "No! Go on! Please!" Alas, we cannot speak to her, now or then.

Sadly, Franklin was alone, in an existential quarantine, isolated by gender, religious, and cultural discrimination, petty office politics, patriarchal hegemony, and, yes, her fierce, self-protective, and ultimately self-defeating behavior. In 1970, Wilkins slammed her discovery of the B form: "it was a fluke, really. I don't think she quite knew what she had there."[21] Similarly, for decades Watson and Crick publicly chided her for not working with a collaborator who might have broken the log jam of her thinking and challenged the assumptions that led her astray. In 2018, Watson elaborated on why Franklin had failed to determine the double helix: "She didn't have any friends. She had no one to talk to, to share ideas with, or to force her to revise her ideas to become better ones. That's how great science progresses."[22] In real time, however, neither Watson nor Crick—let alone Wilkins—lent Rosalind Franklin anything resembling a helping hand.

IN PASADENA, ON FEBRUARY 4, Linus Pauling wrote to his son Peter about the tight and unwieldy structure of his triple-stranded model. He explained that he was busy with new adjustments because "we have found that that the atomic co-ordinates in our nucleic acid structure need to be changed a bit . . . It will be a few weeks before we have finished the job of checking over the parameters again, but I expect them to come out all right—at any rate I hope so." In the same letter, he chuckled over Peter's reports that the Cavendish crew had likened him to the big bad wolf of fairy tale lore and that, despite their childish fears, Watson and Crick were highly skeptical of his triple helix model. [23]

A few weeks later, on February 18, Pauling confessed to Peter the continued problems he and Vernon Schomaker were having: "I think that the original parameters are not exactly right." He added an uncharacteristic scrap of insecurity: "I heard a rumor that Jim Watson and Crick had formulated this structure already sometime back, but had not done anything about it. Probably the rumor is exaggerated."[24] He continued making improvements to the model and, by February 27, was ensorcelled with the results, convincing himself that the parts "just fitted together so beautifully that it had to be right."[25]

SITTING AT HIS DESK on the morning of February 4, a distracted Watson grumbled that Peter Pauling "had charmed Pop into giving him dining rights." Also ricocheting across Watson's crowded cerebrum were daydreams of "motoring in a friend's Rolls to a celebrated country house near Bedford . . . [and] moving into the fashionable world [of the Rothschilds], so that he might have a chance to escape acquiring a faculty-type wife."[26] He was startled out of his reverie by a hard knock on his office door. He looked up and saw the whisky-breathed, bent old shop machinist, who was holding a set of model phosphorus atoms in hands. They were wrapped in a dirty cloth, stained with metal shavings and lathe grease, but to Watson they represented a pot of gold. Not surprisingly, he fell on them hungrily, as if the metal pieces represented his first meal after days of fasting.

After stringing together "several short sections of the sugar-phosphate backbone," he tinkered with the C-clamps, rods, and baling wire he had pinched from a chemistry lab. Perversely ignoring Franklin's data on the location of the sugar–phosphate backbone, he wasted a full day and a half fiddling with a "suitable two-chain model with the backbone in the center." When he compared the possibilities with what he remembered from her Photograph No. 51, however, his scientific intuition told him that his latest arrangement was as wrong as the three-chain model he and Crick had built fifteen months earlier. To put it bluntly, Jim Watson was stumped.

Eventually, Watson did what many twenty-five-year-old postdoctoral fellows do when colliding with what seems to be an insurmountable intellectual impasse: he left the lab and pursued some fun. He changed into his tennis shorts and brutalized the sexy Fourcade on the courts. After a few sets, he was satisfied with his serve and Fourcade was even more satisfied with his conquest of Watson's sister. The two repaired to Clare College for tea. When Watson returned to the laboratory, Crick "put down his pencil to point out that not only was DNA very important" but also there would soon come a time when Watson would finally "discover the unsatisfactory nature of outdoor games."[27]

Thanks to the nudge from Crick, Watson extended his work days by dining with the Cricks each evening. Though he was still insisting that the sugar–phosphate backbone was in the center of the molecule, his inner voice told him to start worrying "about what was wrong." As they sipped their coffee, he admitted to his father confessor, Francis of Cambridge, that none of his reasons for keeping the backbone in the center of the model "held water." Nonetheless, he was hesitant to contend with the difficulties that would arise if he placed the bases in the center. He would have to build an "almost infinite number of models of this type," only to have the impossible task of choosing which one was right. That, to Watson, was the stumbling block. It was easier to place the nucleotide bases outside, but "if they were pushed inside, the frightful problem existed of how to pack together two or more chains" of nucleotide bases consisting of different sizes and shapes.[28]

How would they surmount this obstacle? Even Crick had to admit, as he stirred another lump of sugar into his black coffee, that he could not see "the slightest ray of light."[29] A stymied Watson excused himself from the table, walked up the stairs and out onto Portugal Place, and made his way back to his rooms at Clare. It was up to Crick, he insolently thought, to come up with "at least a semi-plausible argument before I would seriously play about with base-centered models."[30] One motivation for his stubborn refusal to accept a model with the nucleotide bases in the center and the sugar–phosphate backbones on the outside may have been that it was Franklin's idea. The placement of the backbone structure was her primary objection to his failed triple helix model more than a year previously.

The next evening at dinner, Crick expressed his vexation at Watson's "feeble" reasons for putting the sugar–phosphate backbones on the inside of the structure. In his memoir, Crick re-created the verbal interaction when he told Watson to ignore his qualms:

"Why not," Crick asked Watson, "build models with phosphates on the outside?"

To which Watson replied, "Because that would be too easy (meaning that there were too many models he could build in this way)."

"Then why not try it?'" asked Crick, "as Jim went up the steps into the night. Meaning that so far we had not been able to build even one satisfactory model, so that even one acceptable model would be an advance, even if it turned out not to be unique."[31]

Watson returned to the lab early the following morning and took apart "a particularly repulsive backbone-centered molecule," careful not to damage the fragile pieces. He spent a few more days building "backbone-out models." The shop was still grinding away at "the flat tin plates cut in the shapes of purines and pyrimidines," forcing him to shelve the seemingly impossible arrangement of the bases for at least another week.[32] Jiggling the phosphate groups, Watson saw how easy it was to twist the external backbone into a "shape compatible with the X-ray evidence." Soon enough, he realized the clarity of an external sugar–phosphate backbone and declared ownership of it. Deciding that "the most satisfactory angle of rotation between two adjacent bases was between 30 and 40 degrees," he interpreted his findings to mean that "if the backbone was on the outside, the crystallographic repeat of 34 Å had to represent the distance along the helical axis required for a complete rotation."[33]

Crick continued to grind out pages of his mind-numbing dissertation all the while he was in the same small room as Watson.[34] The proximity of Watson's generative brainstorming made him want to share his partner's joy. "At this stage," Watson later recalled, "Francis' interest began to perk up and at increasing frequencies he would look up from his calculations to glance at the model." Watson's tin tower was expanding by the hour, but on Friday afternoon, February 6, "neither of us had any hesitation in breaking off work for the weekend. There was a party at Trinity on Saturday night and, on Sunday, Maurice was coming up to the Cricks' for a social visit arranged weeks before the arrival of the Pauling manuscript."[35] Wilkins had confirmed his visit by post the day before, and recklessly wrote to Crick, "I will tell you all I can remember and scribble down from Rosie['s exit lecture]."[36] The timing of his visit was perfect—that is, for Watson and Crick.

Wilkins joined the Cricks and Watson for Sunday lunch on February 8 and "was not allowed to forget DNA. Almost as soon as he arrived from the station, Francis started to probe him for fuller details of the B pattern." Wilkins opened his mouth only to express generalities or to tuck into the meal Odile had prepared. Neither Crick nor Watson were able to extract much more information than what Watson had learned a week earlier.[37] Shrewdly, the duo had asked Peter Pauling to drop by later that afternoon. All three Cambridge men were surprised to learn that although Crick had sent Wilkins a copy of Pauling's paper, he had not read it yet. Thrusting the Pauling manuscript at Wilkins, they asked him what he thought was wrong with it.

Wilkins digested the tables and figures far more quickly than the meal he had just consumed. He, too, saw that Pauling incorrectly "had phosphates forming the core of the helix, as had Francis and Jim's ill-fated model." Searching further, he realized "there were no sodium atoms listed, and he knew quite well that crystallized "DNA does contain sodium." In response to Wilkins's observations, Crick assumed his best Professor Higgins pose and squealed with delight, "Exactly!" This heartfelt praise immediately made the insecure Wilkins feel special. As Wilkins recalled with pride: "I had spotted the flaw, with a schoolboy's pride in acing an oral exam. There was also no sign that the model fit the X-ray data."[38]

Watson and Crick beseeched Wilkins to start building models and "be the first to discover the DNA structure." Peter Pauling generously advised Wilkins that if he did not step up, his father most certainly would. Wilkins remained calm and resolute as he explained that he would begin building models as soon as Franklin left King's College, a few weeks hence. It was at this point that Crick and Watson, like lawyers about to ensnare a hostile witness, popped a question more momentous than any of their marriage proposals: "Would [Wilkins] mind if we started to play about with DNA models?"[39]

Fifty years later, Wilkins was able to summon up his response to this momentous query with great precision: "I found their question horrible. I did not like treating science like a race, and I especially did not like the idea of them racing against me . . . I do not remember thinking about the possibility that I might be The Big Shot Who Discovered the DNA Structure, but I did not enjoy making room for Francis and Jim." The obvious solution to this quandary would have been for the King's College and the Cavendish groups to collaborate, but that was not to be. Watson and Crick were confident that they had all the data and talent needed to solve the structure and, as a result, did not make such an offer. Wilkins, too, avoided the topic, fearing the tension that would invariably occur if "the London–Cambridge Rat Race was to begin again."[40]

Watson and Crick held their collective breath as they waited for Wilkins's response. Wilkins being Wilkins, he claimed later that he paused to consider the larger issue of holding back science: "DNA was not private property: it was open to all to study peacefully without any one person throwing his weight about."[41] It must have seemed like eons before he finally and slowly answered, "No," he would not mind if they started playing with models. Watson's rabbit-like pulse slowed with relief. He must have displayed a terrific poker face, because he knew

Wilkins's consent was irrelevant. Even if Wilkins had expressed reservations, Watson later admitted, they would have gone ahead with their model building.[42]

As soon as the words came out of his mouth, Wilkins regretted uttering them. His gracious permission for Watson and Crick to enter the fray marked the moment he lost the DNA race. For decades, he tortured himself with regrets. Why didn't he discuss options with John Randall first? Or, perhaps, he should have invited John Kendrew, an old friend and a superb mediator, to the Cricks' table that afternoon. Surely, Kendrew could have brokered a fair deal. Wilkins later lamented, "I was very cast down and could not conceal it. I had come to Cambridge looking forward to a carefree, jolly time, and now there was no chance of that. I just wanted to go home, and Francis had the sense not to press me to stay." Watson ran out into the street after him and "expressed his regrets; but I was not very receptive."

Wilkins gave Watson and Crick the benefit of the doubt, assuming they meant well and "were open about their general intentions, even though they gave me no details of their plans or ideas."[43] He had it wrong on both counts.

The MRC Report

Men are most virile and most attractive between the ages of 35 and 55. Under 35 a man has too much to learn, and I don't have time to teach him.

—HEDY LAMARR[1]

The weather in Cambridge in the second week of February was unusually temperate. For Jim, the springlike clime—along with more delays in fabricating the tin base model parts—inspired a cavalier attitude toward work. Although he arrived at the Cavendish each morning a full hour or more before Crick's "ten-ish entrance," his free and easy demeanor perturbed the usually freer and easier Crick. Every afternoon, after lunch at the Eagle, Watson headed for the university courts for a few sets of tennis.[2] One opponent remembered these matches well: "[Watson] had no idea of form, but he had tremendous energy and seemed to reinvent the game every time he played. And he hated to lose."[3] After tennis, he returned to Room 103 to briefly tinker with the model before dashing out the door to go to Pop Prior's for "sherry with the girls." Crick's complaints about Watson's time-wasting activities were matched by his partner's counterargument that "further refining of our latest backbone without a solution to the bases would not represent a real step forward."[4]

On some nights during this seminal period, Watson went to the movies at the Rex Cinema. Cambridge students disparaged the Rex as a flea pit, a British slang term for the dirtiest of movie houses. Formerly a roller-skating rink, by 1953 the Rex was an unkempt auditorium where the sticky floors and detritus left behind by previous filmgoers were best left unexplored.[5] For the price of one shilling and eight pence, he saw all kinds of films, from the artsy pictures of Federico Fellini, Vittorio De Sica, and Jean Cocteau to the slick products of Hollywood.

The "worst" movie he remembers seeing during this period was *Ecstasy*, starring Hedy Lamarr, an Austrian-born actress who enjoyed a successful career at Hollywood's MGM studio.[6] Like many young men of his startlingly prim era, Watson had long wanted to see this 1933 motion picture.

Even two decades later, *Ecstasy* was widely known as the first non-pornographic film to portray "romps in the nude," skinny-dipping, sexual intercourse, and the female orgasm—all simulated but clearly communicated by Lamarr's acting.[7] Watson was joined by his sister and Peter Pauling. Alas, this "wild pursuit of the celluloid backfired" when the audience discovered that "the only swimming scene left intact by the English censor was an inverted reflection from a pool of water. Before the film was half over we joined the violent booing of the disgusted undergraduates."[8]

Distractions aside, Watson "found it almost impossible to forget the bases." He finally accepted that the backbone he had constructed, with the sugar–phosphate complexes on the outside of the helix, was nicely compatible with "the experimental data" and Franklin's "precise measurements." He conveniently suppressed, however, the glaring problem that Franklin's data had been secreted out of the King's College lab without her knowing about it."[9]

THE TROUBLING ETHICAL LAPSES in the misuse of Rosalind Franklin's data only escalated after Max Perutz entered into the fray. As director of the Cambridge Medical Research Council Biophysics Unit, Perutz made and received reports on the progress of his and the other research units funded by the council. The rationale of such reports, Perutz later asserted, was "not to look into the research activities" of other labs but, instead, "to bring the different Medical Research Council units working in the field of biophysics into touch with each other."[10] In early 1953, John Randall sent Perutz a mimeographed report on the King's College MRC

Biophysics Unit. The blurry typescript included a detailed section titled "X-ray Studies of Calf Thymus DNA by R. E. Franklin and R. G. Gosling." Watson insisted, "the report was not confidential and so, sometime between February 10 and 20, Max saw no reason not to give it to Francis and me." Yet until the publication of Watson's *The Double Helix*, in 1968, the occasion of Max Perutz's "gift" remained an open secret in Cambridge that was little known outside of its city limits.

When Crick read Franklin's eleven-paragraph summation of her work in the MRC report (which did not include photographs), a chain reaction set off in his brain—each atom splitting the next, releasing more ideas, more thoughts, and more energy.[11] What started Crick's neuronal fission was Franklin's determination that the A form crystal of DNA was a monoclinic, face-centered C2 structure. Crick immediately understood that "this was the crucial fact. Furthermore, the dimensions of the unit cell, which were also in the Medical Council Report, proved that the dyad [a two-chained structure] had to be perpendicular to the length of the molecule and implied that the duplication was in fact within the single molecule and not merely between adjacent molecules in the crystal. And so, the chains must come in pairs rather than three in a molecule, and one chain must run up and the other down."[12] What Crick meant is that if one strand of nucleotide bases ran A-C-G-T from bottom to top, then the other strand would be A-C-G-T from top to bottom. The major reason this snippet of data signaled Crick so strongly was that he—and by extension his lab supervisors, Perutz and Kendrew—had long been studying another biological molecule which was also configured into a face-centered, monoclinic space group C2 crystalline structure: the oxygenated form of hemoglobin.

As Horace Judson has explained, the MRC report also afforded Crick his first opportunity to comprehend that what "ran up and down were the two backbone chains. The sequence of the bonds by which the phosphates and sugars linked up and alternated along one chain was flipped over in the chain running alongside."[13] Most important, it was "the dyadic symmetry" that afforded Crick's realization of how the backbones spiraled around the molecule's inner core, a finding essential to building an accurate three-dimensional model. At first, Watson could not accurately visualize the spatial complexities of the double helix; he predicted that each chain individually rotates halfway about the cylinder to reach the height of 34 Ångstroms and then repeats. Such a configuration was not only incorrect, it also placed the sugars too closely together. Crick corrected this error by asking Watson to "close your eyes and think about it," as he patiently explained how both backbones *had* to spiral "all the way around the cylinder, three hundred and sixty degrees, before the structure made a complete repeat." In other words, ten base pairs, each measuring 3.4 Å, represented a complete turn of the helix, for a total of 34 Å. This arrangement fit Rosalind Franklin's X-ray measurements.[14] Years later, Crick railed against Watson's dreamy insistence in *The Double Helix* that the rationale behind their double helix was the notion that biological things came in pairs. "That's just nonsense," Crick said, "we had a very *good* reason [to build a two-chain model], which Jim's forgotten!" (emphasis in original).[15] To another interviewer, Crick was even more precise in his attribution: "[We] needed a clue to get to that point, and the clue was Rosalind Franklin's data," which was found in the MRC report Perutz showed them.[16]

WHEN *THE DOUBLE HELIX* WAS PUBLISHED in 1968, Perutz's role in the transfer of information was finally made public. Soon after the book became a cause célèbre, Erwin Chargaff wrote a scathing review of it in the March 29, 1968, issue of *Science*. Even today, one can feel Chargaff bristling over Watson and Crick's "parasitic hybrid of what other people really did discover."[17] A few paragraphs into his diatribe, Chargaff expanded his attack beyond the two victors who, when he met them in 1952, did not even "know how to spell adenine." He also blindsided the mild-mannered Perutz by alleging that he had acted improperly in passing the secrets of the MRC report to Watson and Crick.[18] After seeing the galley proofs of Chargaff's review, Perutz nearly exploded with angst. Understandably, he did not want to be seen as the Julius Rosenberg of the molecular biology world. Initially, he dissembled that he had skimmed over this episode while reviewing Watson's manuscript and would have objected to it had he read the book more slowly. Soon after, Perutz took up arms against this existential threat to his career and reputation. He began by writing to John Randall of the need to "kill that ugly story of Watson's as soon as possible."[19] It was an extermination Perutz pursued with scholarly precision.

Perutz rifled through his cluttered office, only to discover he had thrown away most of the correspondence related to the MRC Biophysics Committee for that period. He then went to great lengths to find duplicate copies of these letters, which were gathering dust in the file cabinets of Landsborough Thomson, secretary to the committee, and Harold Himsworth, the MRC secretary. What Perutz sought was official confirmation that these reports were never meant to be "restricted" or "confidential" per se. This was technically true. There was no addition of the word "confidential" at the top of these reports, even though they were addressed only to MRC staff and unit directors.[20]

Once the digging through dead letters was completed, there appeared a highly orchestrated series of explanations in the June 27, 1969, issue of *Science*. Perutz offered the leaky excuse that when he showed Crick and Watson the MRC report, he "was inexperienced and casual in administrative matters and, since the report was not confidential, I saw no reason for withholding it." He quibbled further that while the report "did contain one important piece of crystallographic information useful to Crick" (an understatement at best), Watson had heard all of this data at Franklin's November 1951 colloquium. "If Watson had taken notes," Perutz wrote, he would have accessed the same information an entire year earlier.[21]

Watson, too, contributed to this set of recollections, but instead of killing the "ugly story," he amplified it by illustrating how important Perutz's sharing of the MRC report really was. "The relevant fact," he insisted, "is not that in November 1951 I *could have* copied down Rosalind's seminar data on the unit cell dimensions and symmetry, but that I *did not*" (emphasis in original). It was only after he and Crick began to grasp "the significance of the base pairs and were building a model for the 'B' structure" that Crick read Franklin's MRC report and "suddenly appreciated the dyad axis and its implication for a two-chained structure."[22]

Perutz's post hoc insistence that he acted properly became more untenable after historians began exploring the Medical Research Council papers in the British National Archives. On April 6, 1953—only days after Watson and Crick submitted their now famous DNA paper to *Nature*—Perutz wrote a misleading letter to Harold Himsworth at the MRC. In one key paragraph, he prevaricated about the timing of when Watson saw Photograph No. 51, as well as what and when Watson and Crick knew about the specific measurements Franklin described in her portion of the King's College MRC report:

> They used a certain amount of unpublished X-ray data which they had seen or heard about at King's. All these X-ray data were either poor or referred to a different form of structure, and while they indicated certain general features of the structure of DNA, they did not give a guide to its detailed character. While Watson and Crick were building their structure here, Miss Franklin and Gosling at King's obtained a new and very detailed picture of DNA. Watson and Crick only heard of this photograph when they sent the first draft of their paper to King's, but it now appears that this new photograph confirms the important features of their structure.[23]

Georgina Ferry, Max Perutz's biographer, described this letter as a "highly uncharacteristic departure from strict truthfulness," inspired by "the need to conceal Wilkins's role in revealing Franklin's photograph from the director of the Biophysics Unit at King's, John Randall." In reality, the purpose of the concealment was far more aggressive and complicit than merely keeping Randall out of the loop or protecting Wilkins. With the composition of his letter to Harold Himsworth, Perutz formally joined the conspiracy against Rosalind Franklin, one that already numbered Watson, Crick, Wilkins, and Randall as charter members. As to the propriety of Perutz's handing over the MRC report to Franklin's most formidable rivals, Wilkins described it best in an angry letter he wrote to his boss, John Randall, on January 13, 1969—all the while ignoring the fact that it was *he* who showed Watson Franklin's photograph in the first place: "if Perutz thinks that the only documents one should not show to other people are those marked 'Restricted' or 'Confidential,' he seems to me to be living in a funny world!"[24] Randall agreed and complained to Bragg that even if the report was not marked "confidential," it should have been treated as such.[25]

Nearly twenty-five years later, in 1987, Perutz was still wringing his guilty hands. As an unguent for his self-inflicted wounds, he wrote a long essay, "How the Secret of Life was Discovered," for the *Daily Telegraph*. Instead of admitting to his part in the process, he aimed his anger at Watson for "maligning that gifted girl who could not defend herself. [But] I could not get him to change it." The weirdest part of this confession was his remark on Rosalind Franklin's appearance: "Not that she was unattractive or did not care about her looks. She dressed much more tastefully than the average Cambridge undergraduate."[26]

BY LATE FEBRUARY 1953, the Cavendish duo had the sugar–phosphate backbone configuration well in hand. It was now time for them "to puzzle out the mystery of the bases," which turned out to be their most brilliant leap toward solving DNA's structure.[27] Historians of science often find their fun in exploring the books and articles their subjects were studying at the time of an important discovery or experiment. This situation is no exception. Even at this late date, it is difficult not to be impressed by the depth of Watson's reading of the literature on nucleic acids. In an era when everything had to be looked up in a library card catalogue or published indexes, Watson managed to ferret out every extant morsel of information.

One source he consulted was *Biochemistry of the Nucleic Acids*, published in 1950.[28] Its author was James N. Davidson, a Scottish biochemist at the University of Glasgow who built up "one of the most active centers for research into the biochemistry of nucleic acids on this [the British] side of the Atlantic."[29] Davidson wrote many textbooks, including two on nucleic acids with Erwin Chargaff.[30] Watson owned a dog-eared copy of Davidson's slim volume. Looking up the appropriate pages to make sure he had the correct schematics, he carefully drew out the chemical structures of DNA's purines and pyrimidine bases in his tiny script on a sheet of the Cavendish Laboratory's ecru-colored stationery.[31]

Watson's approach was sound. His task was to figure out how to position the nucleotide bases in the center of the helix "in such a way that the backbones on the outside were completely regular;" the sugar–phosphate groups bonded to the nucleotide bases needed to be arranged in "identical three-dimensional configurations." What made this so difficult was that the purine and pyrimidine bases had entirely different shapes. It was impossible to fit them together perfectly, in the manner of a jigsaw puzzle. Twisting one base a few degrees made it difficult to fit the next one into place. In "some places the bigger bases" touched each other, while "in other regions, where the smaller bases would lie opposite each other," there were inexplicable gaps that caused the backbones to buckle inward toward the center. In short, every configuration he tried resulted "in a mess."[32]

Watson also struggled to properly arrange the hydrogen bonds holding the two strands of the double helix together. The nature of intra-molecular hydrogen bonds versus inter-molecular bonds was not well understood at this time. Such chemical fine points were way beyond the imagination and knowledge base of both Watson and Crick, who only a year earlier had "dismissed the possibility that bases formed regular hydrogen bonds." Instead, they initially posited that "one or more hydrogen atoms on each of the bases could move from one location to another."[33]

Watson then consulted the DNA acid–base chemistry studies of John Mason Gulland and Denis Oswald Jordan of Nottingham. Their work persuaded him to "finally appreciate the conclusion that a large fraction, if not all, of the bases formed hydrogen bonds to other bases." He now realized that "these hydrogen bonds were present at very low DNA concentrations, strongly hinting that the bonds linked together bases in the same molecule."[34]

Isolating himself in his rooms at Clare, Watson spent the next day or two entirely within the fourteen or fifteen centimeters of his oblong skull. His "doodling of the bases on paper" went "nowhere . . . even the necessity to expunge [the motion picture] *Ecstasy* from [his] mind did not lead to passable hydrogen bonds." At one point, he sought out some social interaction at an undergraduate party in Downing College, hoping it "would be full of pretty girls." His libidinous hopes were dashed as soon as he arrived to find "a group of healthy hockey players and several pallid debutantes." Feeling out of place, he stayed "a polite interval before scooting out." Spying Peter Pauling as he left, he made certain to inform his happy-go-lucky pal that he was "racing [his] father for the Nobel Prize."[35]

WATSON'S "DOODLING" CONTINUED for almost a week, when he hit upon a configuration he erroneously thought was the solution. He recalled reading that pure adenine crystals and the molecules within them were held together by hydrogen bonds. What if, he asked, each adenine residue in a DNA molecule contained the same bonds? There would be two hydrogen bonds between two adenine residues, with a 180-degree rotational axis. This like-with-like, symmetrical hydrogen bonding also worked well "to hold together pairs of guanine, cytosine, or thymine." The only problem was that the same-to-same pairing jammed up the backbone structure he had carefully built the week before. There was too much buckling, inward and outward, "depending upon whether pairs of purines or pyrimidines were in the center."[36]

By evening, Watson resolved to ignore the shambolic backbone in favor of the larger implications of his model. There was no way, he reasoned, that Mother Nature would intertwine two chains "with identical base sequences" by chance. It was a compelling hypothesis. The grander meaning was that at some point in cell replication one chain "served as the template for the synthesis of the other chain. Under this scheme, gene replication starts with the separation of its two identical chains. Then two new daughter strands are made on the two parental templates, thereby forming two DNA molecules identical to the original molecule." He continued to play with his "bombshell" of a structure well into the night, even though he still could not arrange everything quite right: "I could not see why the common tautomeric form of guanine would not hydrogen-bond to adenine. Likewise, several other pairing mistakes should also occur. But since there was no reason to rule out the participation of specific enzymes, I saw no need to be unduly disturbed."[37]

He again felt his "pulse began to race" and, "as the clock went past midnight," he fantasized about the acclaim that would come on the morrow. For Watson, this did not involve hitting a grand slam in the ninth inning of the final game of the World Series or catching a touchdown pass in the fourth quarter of the Super Bowl. He chose to imagine himself delivering grand lectures and gracefully accepting the applause of his scientific elders. After all, the structure he had dreamed up was so darned interesting. Eventually, he wanted some rest. He had a big day ahead of him explaining his latest idea to Crick. Undoubtedly, they would argue over its good and bad points as they rearranged the model pieces, and then argue again. By the day's end, Crick would undoubtedly agree and a celebration at the Eagle would ensue, followed by more heated days and nights of writing up their findings before sending them off to a premier journal. Neither Watson's excitement nor his anxiety afforded him much rest that night. He spent the next two hours awake "with pairs of adenine residues whirling in front of [his] closed eyes." Even after he finally fell asleep, stinging moments of fear shot through his thin frame telling him "that an idea this good could be wrong."[38]

Base Pairs[1]

Let's face it, if the fates hadn't ordained that I share an office with Watson and Crick in the Cavendish in 1952–1953, they'd still be puttering around trying to pair "like-with-like" enol form of the bases.

—JERRY DONOHUE[2]

Watson began his Friday, on February 20, with a series of mundane chores. After his ablutions, which may or may not have included running a comb through his unruly, coiled hair, he ventured out to the Whim for some bacon and eggs. He returned to his rooms to respond to Max Delbrück's letter about the bacterial genetics paper he'd asked Delbrück to sponsor for publication in the *Proceedings of the National Academy of Sciences* (and to do so promptly, so that it would appear in print before the summer symposium at Cold Spring Harbor that July).[3] Delbrück told Watson that his paper was full of holes and more than likely wrong in its conclusions but he'd sent it along to the editors anyway, warning the always-in-a-hurry Watson, "it will do you good to learn what it means to publish prematurely."[4] Despite Delbrück's warning, he wanted that extra notch on his curriculum vitae, come what may.[5] He later claimed that Delbrück's admonition had a purposefully "unsettling effect," but it was soon superseded by the joy of potentially solving DNA. Ignoring the gastric gurgles Delbrück may have inspired, he demonstrated one of his worst demons: the voice whispering in his ear to jump a few rungs up the academic ladder and ignore the potential costs of mishap. His cavalier excuse for the bacterial genetics paper—which, once in print, would be shown to be wrong—was, "I would still be young when I committed the folly of publishing a silly idea. Then I could sober up before my career was permanently fixed on a reckless course."[6]

In the three-page reply he wrote to Delbrück, Watson reprised the confidence in his paper about how "bacteria mated." With a student's insistence on topping his professor, Watson later wrote in *The Double Helix*, he "could not refrain from adding a sentence saying that I had just devised a beautiful DNA structure which was completely different from Pauling's." In the actual letter of February 20, however, he wrote far more than that—including a long paragraph declaring how "extremely busy" he was working on DNA's structure and that "I believe we are close to the solution." Watson went on to remark that when reading Pauling and Corey's *PNAS* paper he had discovered it "contains several very bad mistakes." He even threw a verbal punch Rosalind Franklin's way. Specifically, Pauling may have picked the "wrong type of model," but, at least, it was offered "in the proper mood and the type of approach which the people at King's College London should be taking instead of being pure crystallographers." He also moaned about being taken off the DNA scent for over a year because the "King's group did not like competition or cooperation." Now that Pauling had announced that hunting season was officially open, Bragg had finally allowed him to join the search party, and "I thus intend to work on it until the solution is out."[7]

For two, maybe three seconds, Watson considered expounding a tad further by adding a few more tidbits of information, but just as quickly decided against it. He knew Delbrück worked closely by Pauling at Caltech and would inevitably hand-carry such news to him at the speed of middle-aged footsteps. Instead, he simply said he was optimistic over his "very pretty model, which is so pretty I am surprised no one had thought of it before." He finished his letter by saying that he was working out the atomic coordinates to make sure they corresponded with "the X-ray data" (but not confessing that it was Franklin's data he was using). Even if his new model was wrong, Watson concluded, it still represented a huge step forward and he would soon be sending Delbrück all the details.[8]

Watson dropped the letter into a bright-red, oval mailbox newly embossed with "ER II," marking the new reign of Queen Elizabeth II.[9] His caution in not spilling all of his ideas out to Delbrück served him well. Long before the postman unlocked the mailbox with his big iron turnkey and retrieved the pile of letters at the bottom, Watson's model would be "torn to shreds" and categorized as "nonsense."[10]

The person doing the shredding was the fourth office mate in Room 103 and, perhaps, the most important minor character in this tale. His name was Jerry Donohue; he was a sarcastic, lumbering, square-jawed man who hailed from Sheboygan, Wisconsin. Donohue had received his AB, magna cum laude, from Dartmouth College (1941) and earned both an MA in chemistry (1943) and a PhD in theoretical chemistry and physics (1947) under Linus Pauling at Caltech. This irascible genius spent the winter 1953 term at the Cavendish on a Guggenheim Fellowship endorsed by Pauling. Soon after his arrival, John Kendrew assigned Donohue—just as he had Peter Pauling a few months earlier—to sit alongside Watson and Crick. The four desks were arranged around the perimeter of the small room: Crick's was on the right side, near a window overlooking a dingy courtyard below; Watson was directly opposite, on the left side; and the newcomers were situated in the corners closest to the door, facing plastered walls with peeling paint. A narrow lab bench was placed in the center.[11] Watson's senior by eight years and Peter Pauling's by eleven, Donohue was often annoyed by their flibbertigibbety behavior, especially when they gossiped about the parties they planned to attend and the "popsies" they hoped to pop.

Years later, as a professor of chemistry at the University of Pennsylvania, Donohue recalled, "When I went to Cambridge I didn't even know what a nucleic acid was." What he did know, however, was structural chemistry, "inside and out," with a special expertise in the nature of hydrogen bonds.[12] While in England, he studied purines and the bonds they formed, advancing the work of June Broomhead, a Canadian–British crystallographer who earned her doctorate in the subdepartment of crystallography at the Cavendish Laboratory in 1948. After completing her degree at Cambridge, Broomhead worked in Dorothy Hodgkin's laboratory at Oxford, where she solved the structure of adenine and guanine crystals. Watson and Crick, too, had read her doctoral thesis and papers.[13]

In 1952, Donohue closely reviewed Broomhead's crystallographic data on how pure guanine formed "hydrogen bonds from molecule to molecule in a regular, repeated configuration."[14] During his sabbatical, Donohue determined that the "positions of the hydrogen atoms were fixed [and] did not, after all, jump from place to place." This was the opposite of the shifting hydrogen bond theory for nucleic acids that Watson and Crick had incorrectly elaborated a year earlier. They had predicted that purines and pyrimidines within the DNA molecule exhibited *tautomerism*—wherein one or more of the atoms of a molecule move freely, and interchangeably, from one isomer (a variant of a molecule, with the same constituent atoms but in a different arrangement) to another—thus giving the isomers different chemical, biological, and physical properties. Adenine, guanine, cytosine, and thymine, the nucleotide bases of DNA,

Tautomeric forms of guanine and thymine.

have the chemical potential to exist in two forms: an *enol* isomer and a *keto* isomer. Donohue's essential conclusion, however, was that *none* of the bases in the nucleic acids went through tautomeric shifts, meaning that they probably remained in their *keto* form, the more stable of the two.[15]

Immediately after Watson told Donohue about his brilliant like-with-like bonding scheme, the chemist informed him that "the idea would not work." Watson was crestfallen as Donohue explained how the tautomeric forms Watson had so carefully copied out of Davidson's *Biochemistry of the Nucleic Acids* were "incorrectly assigned" and that the diagram in the book depicted the nucleotide bases as *enol* isomers rather than the more chemically stable *keto* isomers. As soon as Donohue completed what he had to say, Watson fired back that "several other texts also pictured guanine and thymine in the *enol* form." This sharp retort, Watson admitted, "cut no ice with Jerry."[16]

Donohue did not have demonstrable data or a "foolproof reason" to back up his inklings. He could give only one example of a far simpler crystal structure, "diketopiperazine, whose three-dimensional configuration had been carefully worked out in Pauling's lab several years before. Here there was no doubt that the *keto* form, not the *enol,* was present." Applying "quantum-mechanical arguments," Donohue insisted that the same "should also hold for guanine and thymine." He told Watson that "for years organic chemists had been arbitrarily favoring particular tautomeric forms over their alternatives on only the flimsiest of grounds. In fact, organic-chemistry text books were littered with pictures of highly improbable tautomeric forms."[17] He "firmly urged" Watson "not to waste more time with [his] harebrained scheme" of pairing like-with-like bases, before turning back to the papers cluttering his desk.[18]

Jim Watson always had a high opinion of his own ideas, but one of his best attributes as a scientist was the ability to distinguish a good idea from a "bogus" one or, to use his favorite derogatory word for faulty science, "crap."[19] He loved his like-with-like configuration so much that he briefly clung to the hope that "Jerry was blowing hot air," but he knew full well that he could not play this game for very long. A day or so later, in a moment of humility and clarity, he was forced to conclude that, other than Pauling, "Jerry knew more about hydrogen bonds than anyone else in the world." Donohue had already had great success at Caltech in determining the structures of a number of organic molecules, so, like it or not, Watson had to admit that "I couldn't kid myself that he did not grasp our problem. During the six months that he occupied a desk in our office, I had never heard him shooting off his mouth on subjects about which he knew nothing."[20] Yet even after accepting Donohue's pronouncement, Watson struggled with the fact that the adenine and guanine bases are physically larger molecules than the thymine and cytosine bases. The former two each contain two carbon rings; the latter two contain only one. Watson shifted the hydrogen atoms to comply with the *keto* configuration of the bases, but he remained unable to fit them together in a like-with-like fashion without some imaginative bends and bows of the model's backbone.

ON MONDAY AFTERNOON, February 23, Rosalind Franklin walked up two flights of stairs to the King's College library. She went straight to the journal rack—a complex wooden array of slots, each labeled with a journal's title and containing its latest copy—and pulled out the February 21 issue of *Nature*, to which Pauling had, in an unusual step, sent a letter announcing the triple helix DNA model he would soon be publishing in the February issue of the *Proceedings of the National Academy of Sciences.*[21]

Franklin's disciplined reading habits, developed while a student at Cambridge, served her in good stead. In this pre-photocopying or scanning-machine era, one marked by long waits between the time a reader mailed a postcard to an author requesting a reprint of their paper and that author sending it out with his or her compliments, she adopted a habit of regularly reading the latest issues of the crystallography, chemistry, and physics journals and recording any pertinent citations on individual sheets of loose notebook paper, along with her notes and comments. Then she filed the pages for future reference and, when she did receive a reprint copy, she clipped her notes to it.

Franklin had detected problems in Pauling's model when Watson showed her the preprint three weeks earlier, on January 30. In the sheet recording her comments on Pauling's *Nature* letter, she noted an error that, when corrected, dictated a two-chain model rather than a three-chain one. In her spiky handwriting, she asked, "Not clear why structure which is so empty in its outer parts wd give by X-rays the outside diameter." She knew from her encounter with Watson that Pauling was using old data—the fuzzy X-rays William Astbury had taken in 1938 and revisited in 1947. She also understood that because Astbury had not identified the two different forms of DNA, his pictures contained elements of dry form A mixed with elements of wet form B—an artifact that was reflected in Pauling's calculations as well. In a manifestation of her characteristic allegiance to truth-seeking, often at the cost of amiable relations with her colleagues, she wrote to Pauling that day to tell him that his model had misplaced the phosphate groups. That she, as a thirty-two-year-old postdoctoral fellow actively being pushed out of one laboratory for another position, was contradicting the world's greatest chemist did not intimidate her. She had hard data on her side and she explained it, calmly and collegially, to Pauling.[22] What she could not do, without the express approval of the chief of her lab, John Randall, was send him the unpublished manuscripts of what ultimately became her three most important publications on DNA. Two were for the journal *Acta Crystallographica*; one demonstrated the existence of DNA in the two forms, dry or crystalline (A) and wet paracrystalline (B), and the second detailed the method of experimentally transitioning one form to the other. The third paper was the summation of her work on the A and B forms, which ultimately appeared in *Nature* alongside the famous Watson and Crick paper and the Wilkins paper.[23] Pauling politely wrote back that while he stood by the veracity of his model, he hoped to meet with her when he next visited England. His next overseas letter was to his son Peter, telling him about the three DNA papers Franklin was completing.[24]

WHEN DESCRIBING THE WEEK OF February 20–28 in his memoir, *The Double Helix*, Watson collapses the timeline of events like a toy telescope, making it seem like he discovered the implications of nucleotide base pairing in a matter of twenty-four hours. In this telling, soon after Donohue's impromptu seminar on the *keto* configuration of the bases, Crick arrived at the lab and shot down Watson's "like-with-like" nucleotide base structure on crystallographic grounds, demonstrating that his proposed structure did not correspond to Rosalind Franklin's X-ray data, which clearly showed a 34 Å crystallographic repeat. Watson's like-with-like model required a complete three-dimensional rotation every 68 Å, with a rotation angle of only 18 degrees, making it physically impossible. More problematic, Watson's model did not explain Chargaff's rules (the 1:1 ratio of adenine to thymine, and guanine to cytosine). Even Watson had to accept the fact that his tentative like-with-like structure, which had seemed so wonderful a few hours earlier, did not work.

In Crick's memory, it was not until a week later, on Friday, February 27, that Watson finally gave up on his like-with-like model. Some time that afternoon, he and Donohue were squinting at whatever they had last chalked up on the blackboard while Crick sat hunched over his desk. In a mutual flash of clarity, they all came to the same conclusion: "Well, perhaps we could explain 1:1 ratios by pairing the bases," Crick recalled. This, too, "seemed too good to be true," but "all three of us" decided to accept the premise that "we should put the bases together and do the hydrogen bonding and it was the next day that Jim came in and did it."[25]

Watson later claimed that he did not listen closely to Crick's contributions that afternoon (regardless of its precise date) because "he regularly talked so much, throwing out so many ideas. . . . always trying to organize the others in the office . . . thus, Crick's advice for the next step fell on deaf ears."[26] He also remained miffed over Donohue's sound destruction of his like-with-like theory and, instead, shifted his "thoughts to why undergraduates could not satisfy *au pair* girls." He had little enthusiasm for fiddling with Donohue's *keto* forms, fearing he would again "run into a stone wall and have to face the fact that no regular hydrogen-bonding scheme was compatible with the X-ray evidence." So, he dawdled and gazed out the window at the blooming crocuses, hoping that a better idea would soon strike.

After lunch, Watson and Crick learned, once again, from the shop machinist that the model pieces were taking longer than expected to fabricate. They would not be ready for several more days. These pieces were essential if they were to solve the irksome issue of hydrogen bonding between the purine and pyrimidine bases at the core of the helix. The realization that there would be more wasted time on his hands led Watson to while away the afternoon by "cutting accurate representations of the bases out of stiff cardboard."[27]

By the time he was done with his game of organic origami, dusk was falling. The dulcet sounds of boy sopranos at Evensong resonated in the air, reminding Watson that he had a date to join Pop Prior and her girls for an evening of theater. The play was Richard Sheridan's classic 1775 comedy *The Rivals*, which introduced the world to a character named Mrs. Malaprop, whose habit of replacing an incorrect word with a similar-sounding one, to comic effect, enjoys a permanent place in our vernacular.[28]

THE FOLLOWING MORNING, Saturday, February 28, Watson arrived to find his office empty. He cleared away the papers, pencil stubs, and dirty teacups cluttering his desk to create a large flat surface for the newly-cut cardboard base pairs. Once satisfied with his version of a tabula rasa, he began arranging and rearranging the puzzle pieces, wondering where hydrogen bonds might exist. Once more, he tried his beloved "like-with-like prejudices" before accepting that they "led nowhere."[29]

More than two millennia earlier, the Greek mathematician Archimedes (287–212 BCE) is said to have stepped into a bathtub and noticed how the water level rose. Struck by how the volume of the water he displaced was equal to the volume of the submerged portion of his body, he supposedly exclaimed, *"Eureka, Eureka!"* (Εύρηκα, "I have found it!").[30] Whether this lovely tale is true or not, the word has since become the gold standard for scientific discovery. Soon after Donohue's chemistry lecture in Room 103, Watson's eureka moment arrived with the resounding force of a deus ex machina.[31] After pairing the *keto* forms of adenine and thymine with two hydrogen bonds, Watson could not believe his eyes. The shape this molecular combination made was "identical in shape to a guanine–cytosine pair held together by at least two hydrogen bonds." Better still, "the hydrogen bonds seemed to form naturally; no fudging was required to make the two types of base pairs identical in shape." Watson called out to Donohue, asking for his blessing over what he now possessively referred to as "my new base pairs."[32]

Donohue carefully reviewed the cardboard pieces and declared he had no problem whatsoever with the arrangement. His pronouncement

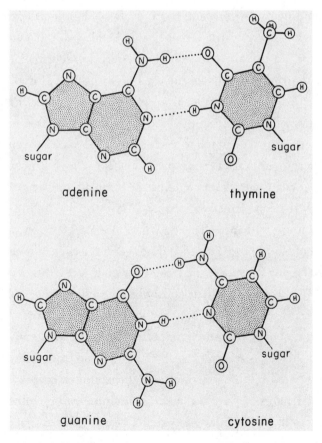

"Eureka!" The correct adenine–thymine and guanine–cytosine (*keto* form) base pairs used to construct the double helix.

sent Watson's pulse rocketing into the stratosphere. He had solved the riddle of the Chargaff rules, an enigma so difficult that not even its crusty namesake could unravel it. At this point, like a spear-carrier in a Shakespeare history play, Donohue makes his exit from the story of DNA, never to reappear. In 2018, Watson regretted how "we certainly mistreated Jerry Donohue. His work was so important he might have been a co-author [on the 1953 Watson and Crick DNA paper]. That was the breakthrough. It was very early in quantum chemistry for anyone to know about the *keto* and *enol* forms."[33]

Donohue had served his purpose. Watson's mind was now clearly focused on how the number of purines equaled the number of pyrimidines, by way of the hydrogen bonding. This translated into adenine always binding to thymine and guanine always binding to cytosine in a regular arrangement in the center of the double helix. It was so simple and elegant. Best of all, it suggested "a replication scheme far more satisfactory than [the] briefly considered like-with-like pairing." The solution screamed out to Watson: if adenine always paired with thymine and guanine with cytosine, "the base sequences of the two intertwined chains were complementary to each other. The base sequence of one chain determined that of its partner. Conceptually, it was now easy to visualize how a single chain could be the template for the synthesis of a chain with the complementary sequence."[34]

As if waiting for Santa Claus on Christmas Eve, Watson stood at the threshold of Room 103: "Francis did not get more than halfway through the door before I let loose that the answer to everything was in our hands."[35] In keeping with their collaborative method, where each maintained a healthy skepticism about the ideas of the other and the right to tear them down in an entirely impersonal, grudge-free manner, Crick examined Watson's arrangement up and down and from side to side. He, too, was floored by the now obvious, but heretofore cryptic, insight of how similarly shaped were the adenine–thymine and guanine–cytosine pairs. It was like a Saturday *New York Times* crossword that had previously stumped the entire family, only to be solved by its youngest member while the others were sleeping.[36]

While Crick compared the crystallographic data with the configurations on Watson's desk, he was able to confirm his insight that the glycosidic bonds (i.e., the bonds joining a nucleotide base to a sugar) "were systematically related by a dyad axis perpendicular to the helical axis. Thus, both pairs could be flip-flopped over and still have their glycosidic bonds facing in the same direction."[37] Thanks to Donohue's mighty morsel of stereochemistry, Watson's cardboard puzzle pieces now perfectly displayed the C2 symmetry of the unit cell described by Rosalind Franklin in her section of the 1952 King's College MRC report and which Perutz had shown to Crick and Watson. When Crick saw what Watson had modeled, he slapped him on the back and squealed with excitement, "Look, it's got the right symmetry."[38] Finally, Watson understood what Crick had been nattering on about over the past week, specifically, that "a given chain could contain both purines and pyrimidines. At the same time, it strongly suggested that the backbones of the two chains must run in opposite directions."[39]

There was more work to be done. Watson and Crick still needed to determine how these adenine–thymine and guanine–cytosine pairs would fit into the backbone configuration they had designed over the past two weeks. Thankfully, there was a large space in the center of the helix for the base pairs to reside. It was now Watson's turn to be cautious; perhaps, as historian Robert Olby contends, because "he lacked Crick's intimate grasp of crystallography, he did not share the confidence Crick possessed from the support afforded to their structure by the C2 symmetry."[40] In reality, both men understood that their work would not be completed until they had built a new Tinker Toy model "in which all the stereochemical contacts were satisfactory. There was also the obvious fact that the implications of its existence were far too important to risk crying wolf."[41]

It is at this point in the carefully woven tapestry of *The Double Helix* that Watson describes Crick winging "into the Eagle to tell everyone within hearing distance that we had found the secret of life."[42] Although he long denied making this grandiose statement, Crick's confident announcement of their work—no matter how it was actually articulated—ran the risk of being wrong and, as a result, made Watson feel "slightly queasy."[43] Fifty years later, on the anniversary of the greatest day of his life, Watson was interviewed by the BBC. After decades of accolades, he recalled that moment as if it were last week. "When we saw the answer we had to pinch ourselves. Could it really be this pretty? When we went to lunch, we realized it probably was true because it was so pretty. The discovery was made on that day, not slowly over the course of the week. It was simple; instantly you could explain this idea to anyone. You did not have to be a high-powered scientist to see how the genetic material was copied."[44]

It's So Beautiful

There was in the early fifties, a small, somewhat exclusive biophysics club at Cambridge, called the Hardy Club, named after a Cambridge zoologist of a previous generation who had turned physical chemist . . . Jim was asked to give an evening talk to this select gathering [on May 1, 1953] . . . The food there was always good but the speaker was also plied with sherry before dinner, wine with it, and if he was so rash to accept them, drinks after dinner as well. I have seen more than one speaker struggling to find his way into his topic through a haze of alcohol. Jim was no exception. In spite of it all he managed to give a fairly adequate description of the main points of the structure [of DNA] and the evidence supporting it, but when he came to sum up he was quite overcome and at a loss for words. He gazed at the model, slightly bleary-eyed. All he could manage to say was "It's so beautiful, you see, so beautiful!" But then, of course, it was.

—FRANCIS CRICK[1]

After declaring, in one form or another, that they had discovered the secret of life, Crick and Watson quickly consumed their meal at the Eagle. Within the hour, they returned to the Cavendish and committed themselves full-time to completing their DNA model. How could they think about anything else? For the rest of the day, Crick gushed about the implications of their discovery, and often to just himself. Occasionally he paused his palaver to leap out of his chair and jiggle the model. And then, like a new father, he stood back and beamed at it with satisfaction. Watson usually luxuriated in his partner's unrelenting bombast, but now he shook his head in disapproval at how Crick's words "lacked the casual sense of understatement known to be the correct way to behave in Cambridge." Apparently, Watson fooled no one. He, too, was thrilled after realizing that "the DNA structure was solved, that the answer was incredibly exciting, and that our names would be associated with the double helix as Pauling's was with the alpha helix."[2] In 2018, a ninety-year-old Watson recalled with crystal (pun intended) clarity, "I had a sense, yes, that—you know, I'm now with Darwin."[3]

They kept strictly to themselves for the remainder of the afternoon and did not emerge until 6 p.m., when the Eagle opened for dinner on Saturday evenings. At their usual table, they discussed their work agenda for the next few days. Crick declared that speed was of the essence. So, too, was accuracy in the construction of a three-dimensional model that met all of the stereochemistry requirements; the lengths and angles of the bonds between the atoms, as well as the spaces between the atoms themselves, *had* to conform with existing knowledge. Despite his exhilaration, Watson could not stop worrying not only about building an accurate model but, even more so, about the possibility that Linus Pauling would figure out his mistake and "stumble upon the base pairs before we told him the answer."[4]

Little further progress could be made that evening because the machine shop had still not finished making the metal nucleotide model pieces. This was the rate-limiting step, because there was no way they would convince Perutz, Kendrew, and Bragg, let alone the King's College group, of the validity of their work using pieces of crudely-cut cardboard held together by wires. As a result, Watson and Crick accommodated the civilized British custom of taking time off on Saturday evening and all day Sunday.

Later that evening, Watson bicycled over to Pop Prior's boarding house for supper. Incapable of maintaining Crick's order of silence, he told his sister and her beau, the beautiful Bertrand Fourcade, that he and Crick "had probably beaten Pauling to the gate and that the answer would revolutionize biology." "Genuinely pleased," Elizabeth beamed with pride.

Fourcade, who would one day become publicity director of *Vogue* magazine, loved the idea of telling his cadre of monied playboys "that he had
a friend who would win a Nobel Prize." Seated next to Watson was Peter
Pauling, who "was equally enthusiastic and gave no indication that he
minded the possibility of his father's first real scientific defeat."[5]

Crick's recollection of how he spent the evening of that wonderful
Saturday was far tamer: "[The model building] started about Wednesday
and finished on a Saturday morning, by which time I was so tired, I just
went straight home and to bed."[6]

INSTEAD OF "THE FOLLOWING MORNING" (which was Sunday), as Watson reports in his memoir, it was probably on Monday morning, March
2, that he returned to the laboratory.[7] Whether it was on March 1 or 2,
Watson awoke feeling "marvelously alive." Again adopting the imag-
ery of the Hollywood films he adored, he stared up at "the gothic pin-
nacles of the King's College [Cambridge] Chapel" as they reached for
"the spring sky," certain in the belief that there was no limit to the illuminating power of knowledge. Stopping to gaze at "the perfect Georgian features of the recently cleaned Gibbs Building," he reflected on
his long walks with Crick through the Cambridge colleges and their fre-
quent trips to Heffer's bookstore to "unobtrusively read the new books."[8]
Entering Room 103, he found Crick already working on adjustments to
their model.

By morning's end, Watson and Crick were satisfied that "both sets
of base pairs neatly fitted into the backbone configuration." At various
points, Max Perutz and John Kendrew "popped their heads into the
room to see if we still thought we had it." In reply, Crick presented
a rapid-fire, high-pitched DNA lecture, which he would deliver many
more times in the days to come. As he spoke, Watson made his way
down to the machine shop in the hope that the nucleotide pieces would
finally be completed that afternoon. After "only a little encouragement,"
the machinist told him that "the brightly shining metal plates" would be
ready in "the next couple of hours."[9] When they arrived later that after-

noon, Watson and Crick unwrapped the newspaper-covered pieces as excitedly as little boys ripping open a birthday gift.

In an hour, maybe less, the first formalized Watson and Crick model of DNA was fully erect. It was nearly six feet tall and more than three feet wide. Composed of brass rods and thin sheet-metal pieces cut precisely to specification, it was held together by brass sleeves that were fitted over the rods and screws—"a fiddly business with a spidery, skeletal result."[10] The towering structure was so unwieldy that it could only be manipulated by one person at a time. Hence, as Watson made his spatial corrections, Crick impatiently bounced around the room muttering suggestions. When Crick took the wheel, his job was to make certain that "everything fitted," and to detect slight discrepancies from the published bond angles. There were also "brief intervals" when Crick frowned to himself and Watson's stomach again "felt uneasy." But "in each case he became satisfied and moved on to verify that another interatomic con-

Watson and Crick's double helix DNA model, 1953.

tact was reasonable." As they fiddled with the atomic contacts, careful not to twist the backbone too far, there was the risk that, like a house of cards, the entire structure might fall to the ground.

The atoms were all placed "in positions which satisfied both the X-ray data and the laws of stereochemistry. The resulting helix was right-handed with the two chains running in opposite directions." Because all the technology-dependent measurements had already been made by Rosalind Franklin, Watson and Crick's laboratory armamentarium was no more advanced than a grade-schooler's pencil box—pencils, a ruler and a compass—plus a carpenter's plumb line "to obtain the relative positions of all atoms in a single nucleotide."[11] These simple utensils, however, cannot diminish their accomplishment since they had no road maps or diagrams beyond the King's College X-ray diffraction data. It was Watson and Crick's brilliance, curiosity, and intuition that built the three-dimensional model now so recognizable around the world.

AT DUSK, CRICK AND WATSON broke off their intricate labors and walked the short distance to Portugal Place for supper.[12] The topic of conversation was entirely DNA. Odile Crick later recalled not believing a word of the "big discovery." Years later, she told her husband, "you were always coming home and saying things like that, so naturally I thought nothing of it."[13] Instead of trying to convince his wife of the veracity of his revolutionary findings, Crick shifted the discussion to the delicate issue of how to "let the big news out." They knew that Maurice Wilkins should be told immediately—which would require a great deal of diplomacy. There was more than discretion at stake: Watson and Crick did not want a repeat of the triple helix flop sixteen months earlier. Neither relished another round of Rosalind Franklin's sharp words, let alone the opportunity to again make (in Crick's words) "asses of ourselves."[14] To avoid another fiasco, Watson insisted, it made sense to keep the King's group "in the dark." They still needed to obtain "exact coordinates . . . for all the atoms." What they wished to avoid at all costs was fudging a "successful series of atomic contacts so that, while each looked almost acceptable, the whole collection was energetically impossible. We suspected we had not made this error but our judgment conceivably might be biased by the biological advantages of complementary DNA molecules."[15]

By the time they finished their after-dinner coffee, Odile Crick had accepted the importance of the day's events. She asked her husband if "they would still have to go into exile in Brooklyn," given how "sensational" DNA was; she even encouraged him to ask Professor Bragg if he might allow Crick and Watson to "stay on in Cambridge to solve other problems of equal importance." Watson tried to reassure this cultured Frenchwoman that American mores were not as horrible as advertised. She might even enjoy visiting the "wide-open spaces" of the United States, he said, "where people never went."[16] Such pleasantries, however, fell on deaf ears. Odile Crick clearly wanted to stay put in Cambridge.

On Tuesday morning, March 3, Crick beat Watson to the laboratory for the second day in a row and was already at work on the model, moving the atoms "back and forth." By now they had determined that "Maurice and Rosy were right in insisting they model the Na+ salt of DNA"—meaning, that in order for DNA to form a stable, salt-crystal, it needed to shed a positive-hydrogen bond from the acidic portion of the molecule and join bonds with a positively charged cation commonly found in the body, such as sodium.[17] Watson could barely stay seated at his desk, thinking of the triumphant letters he would soon write to Max Delbrück, Salvador Luria, and, best of all, Linus Pauling. Absorbed by Technicolor dreams of scientific glory, Watson ignored the disapproving glances Crick threw his way for not directing his full attention to their model.

DURING THE FALL AND WINTER of 1952–53, influenza stalked the planet with a vengeance. According to epidemiologists at the World Health Organization, the virus began its path, independently, in the United States and Japan before spreading to western Europe.[18] Although the 1952 influenza virus was not a particularly deadly strain, it did make millions of people very ill. One of them was Sir William Lawrence Bragg;

another was Rosalind Franklin, who was so sick she lost a crucial month of research in the last weeks of 1952.[19]

No mere cold, influenza is a rip-roaring attack on a person's respiratory, neurological, and immune systems. The sixty-three-year-old, sedentary Bragg, who contracted the virus in early March, battled a raging fever of 104° Fahrenheit and coughed up a chestful of thick, tenacious mucus, the perfect medium for a secondary infection of bacterial pneumonia. His flabby, middle-aged body ached as if someone had swung a cricket bat at every inch of it. Thus, he had good reason to be completely absent from the Cavendish while Watson and Crick toiled away on their model. Confined to his bedchamber, Bragg had "completely forgotten all about nucleic acids."[20]

On Saturday, March 7, Crick declared that the model was ready for inspection. After hearing this news, Max Perutz promptly picked up the telephone and called Bragg at his home. Between the coughs and wheezes blowing his way from the other end of the receiver, Perutz asked his chief to come in for a quick peek at the model. Kendrew chimed in that Watson and Crick had "thought up an ingenious DNA structure which might be important to biology."[21]

ON MONDAY, MARCH 9, a wobbly and visibly ill Bragg made his way to the Cavendish. "During his first free moment" there, he "slipped away from his office for a direct view" of the double helix model. He had not entered Room 103 since assigning it a year earlier to the talkative Crick and the weird Watson. Slowly, carefully, he inspected the structure, admiring the handiwork of his machinist. "Immediately," Watson recalled, "he caught on to the complementary relation between the two chains and saw how an equivalence of adenine with thymine and guanine with cytosine was a logical consequence of the regular repeating shape of the sugar–phosphate backbone."[22] Watson did not let a beat of time elapse before pointing out the importance of Chargaff's rules—the 1:1 ratio between the purine and pyrimidine bases—and "the experimental evidence on the relative proportions of the various bases." His emotive antennae were perfectly pitched and, with each word, he sensed that Bragg "was becoming increasingly excited by its potential implications for gene replication."[23]

The conspiracy to deprive Rosalind Franklin of scientific priority shifted into overdrive—only it was now Bragg holding the throttle. The professor asked Watson where the X-ray evidence had come from, and Watson answered truthfully. Bragg quietly nodded his head, indicating that "he saw why we had not yet called up the King's group."[24] Kendrew and Perutz witnessed this brief moment, too, and failed to object to the data heist.

In fact, what worried Bragg at that particular moment was not the ethical dilemma of purloined data but why Watson and Crick "had not yet asked Todd's opinion." Alexander Todd was the professor of organic chemistry at Cambridge and one of the world's experts on nucleotide chemistry. Crick's assurances that he and Watson had "got the organic chemistry straight did not put [Bragg] completely at ease." Bragg knew from experience that Crick's nonstop yammering was often not redacted until it was too late. There was always the possibility that his graduate student had used "the wrong chemical formula" and bollixed the basic facts of the model.[25] With a snap of fingers, a student was dispatched to run to Todd's laboratory on Pembroke Street and bring the chemist to the Cavendish post-haste.

In his rambling autobiography, Todd gave a more detailed explanation for the urgent consultation beyond his brilliant (and Nobel Prize–winning) chemical determination of nucleotides. He emphasized "the almost total lack of contact between physics and chemistry in Cambridge—a lack of contact which is all too common in universities."[26] Surprisingly only to those who work outside of academia, the distance between buildings, or even between the floors of one building, can be insurmountable in facilitating much-needed communication among researchers; the streets separating these buildings are, figuratively, the widest avenues in the world.

This separation of scientific powers factored into Bragg's 1951 humiliation over Linus Pauling's delineation of α-helical structure of proteins.

"I well remember Bragg coming over to see me in the chemical laboratory (for the first time since my arrival in Cambridge)," Todd wrote,
"and asking me how Pauling could have chosen the α-helix from among
three structures all equally possible on the basis of X-ray evidence, and
all of which he (Bragg) had indicated in a paper with Perutz and Kendrew." Todd "shattered" Bragg's self-confidence when he "pointed out
that any competent organic chemist, given the X-ray evidence, would
unhesitatingly have chosen the α-helix."[27] A direct result of this encounter was Bragg's decree that "no nucleic acid structure based on X-ray evidence would go out from his laboratory without it first being approved
by [Todd]."[28]

HAPPILY FOR ALL INVOLVED, Professor Todd approved the structure.
From his first glance at the model, he recognized Watson and Crick's
"brilliant imaginative jump."[29] By the following evening, they had completed the "final refinements of the coordinates," though they still did
not have access to "the exact X-ray evidence" to verify that their atomic
"configuration was precisely correct." But they were less concerned about
those details than about establishing "that at least one specific two-chain
complementary helix was stereochemically possible. Until this was clear,
the objection could be raised that, although elegant, the shape of the
sugar–phosphate backbone might not permit its existence."[30] In 1968,
Crick explained how difficult it was to predict bond distances and angles
in the pre-computer era: "Being a bit lazy and not having at my finger-
tips the formula for an angle between three points, I never checked the
angles. So, you will find that the distances are pretty good but some of
the angles are really a bit off."[31] Nonetheless, Crick and Watson felt in
their guts that their model was true. Almost trance-like, they kept telling
themselves "a structure this pretty just had to exist," and off again they
went for lunch at the Eagle.[32]

   Watson told Crick that "later in the afternoon I would write Luria
and Delbrück about the double helix," but first he headed to the tennis
courts for a few sets with Bertrand Fourcade.[33] In what threatened to

Crick and Watson having a cup of tea in Room 103 of the Austin Wing, Cavendish Laboratory.

become a pattern of loafing on Watson's part, Crick insisted to deaf ears that there was still more hard thinking to be done. Crick was also worried that the clumsy Watson might be "killed by a tennis ball" before their work was completed.[34]

EVEN AFTER THE TENSION OF deducing the model's structure had eased, Watson and Crick studiously avoided telling Maurice Wilkins about their discovery. In his 1968 memoir, Watson casually recalled their slimy dodging and how they "arranged" John Kendrew to call Wilkins and invite him to "see what Francis and I had just devised. Neither Francis nor I wanted that task."[35]

In a story chock-full of coincidences (most of them quite good for Watson and Crick, more of them quite awful for Rosalind Franklin and Maurice Wilkins), another weird twist of fate occurred that Monday morning. The postman delivered a letter to Crick that Wilkins had written on Saturday, March 7—the day Watson and Crick completed their model. As Watson recalled in his memoir, Wilkins told Crick he "was now about to go full steam ahead on DNA and intended to place emphasis on model building."[36] Alas for Wilkins, it was too late.

[28]

Defeat

My dear Francis,
Thank you for your letter on the polypeptides.
 I think you will be interested to know that our dark lady
leaves us <u>next week</u> and much of the 3-dimensional data is
already in our hands. I am now reasonably clear of other
commitments and have started up a general offensive on
Nature's secret stronghold on all fronts: models, theoretical
chemistry, and interpretation of data, crystalline and
comparative. At last the decks are clear and we can put all
hands to the pumps!
 It won't be long now.
Yours ever,
M[aurice Wilkins]

P.S. May be in Cambridge next week.[1]

After months of enduring Randall's edicts, "Rosy's tyranny" and all the mishigas framing their relationship, Wilkins still worried that Franklin might try to take credit for *his* DNA. He was more than ready to bid Franklin farewell and "put all hands to the pumps." The words he scrawled to Crick on Saturday, March 7, 1953, on a flimsy, five-by-seven-inch piece of King's College's notepaper, reveal a palpable excitement: "It won't be long now." Wilkins had no idea how prescient those words were—even if they in no way applied to him.

In the same letter, Wilkins creates a new nickname for Rosalind Franklin: "our dark lady." To many modern-day readers, this might seem to be yet another bitchy epithet in the faux morality play the male actors wrote up after her death. But for those reading these three words in 1953—especially on "this blessed plot, this earth, this realm, this England"[2]—the phrase was an easily recognized allusion to the luminous "Dark Lady" of Shakespeare's sonnets.[3] As the Shakespearean scholar Michael Schoenfeldt has noted, although the "Dark Lady" does not possess the traditional British attributes of beauty (fair skin, blond hair, blue eyes), she is a symbol of forbidden love, darkness, sexual appetite, and lust.[4] She can also represent a malady, as Sonnet 147 declares: "My love is as a fever, longing still; For that which linger nurseth the disease."[5] At this late date, one can only speculate about what Wilkins meant when he labeled Franklin "our dark lady." Ray Gosling wrote off the term as simply a reference to Franklin's raven hair, dark brown eyes, and "dun" or olive-colored skin—attributes common in people of Ashkenazi Jewish descent.[6] Given how well known these sonnets were to men of Wilkins's vintage, however, it is difficult not to suspect a subtext of deep feelings—perhaps of love, lust, or utter sexual confusion—rumbling through his cluttered mind.

THERE WAS ONE OTHER loser in the scientific sweepstakes. In California, Linus Pauling was still fussing with his unwieldy triple helix model, which contained "several unacceptable [interatomic] contacts that could not be overcome with minor jiggling."[7] On March 4, Pauling gave a research seminar for the Caltech faculty. Unlike on previous occasions, when he made majestic announcements of other molecular structures, the reception was cool at best. No one was more critical than Max Delbrück, who had already received news from Cambridge about the Watson–Crick model and Watson's assessment that Pauling's structure contained "some very bad mistakes."[8]

Pauling was unwilling to listen to Delbrück's objections. In contrast to the siloed departments of Cambridge, the Caltech physicists and chemists worked closely together, and the professional relationship between Pauling and Delbrück was an excellent example. The predicament now was more specifically a Linus problem. He was, at this point, so famous, so confident, and so often grandiose that he rarely endured stinging critiques from his colleagues, such as the ones Watson and Crick subjected themselves to daily. Watson best described the power dynamics at Caltech during the 1950s: "Linus's fame had gotten himself into a position where everyone was afraid to disagree with him. The only person he could freely talk to was his wife, who reinforced his ego, which isn't what you need in this life."[9]

More than a week later, Peter Pauling broke the news gently to his father, in a letter describing the excitement Watson and Crick's model was generating. He offered few structural details to the one man who needed them most, appropriately explaining that "they (W.C.) have some ideas & shall write you immediately. It is really up to them and not to me to tell you about it."[10] Criticizing the King's College team's lack of effort with youthful scorn, Peter told his father, "Morris [sic] Wilkins is supposed to be doing this work; Miss Franklin evidently is a fool. Relations are now slightly strained due to the Watson–Crick entering the field." He concluded by reporting that he had given Watson a copy of the Pauling–Corey paper and that they had had difficulty building Pauling's version, which "was pretty tight. Perhaps we should try the new one. They are getting pretty involved with their new efforts, and losing objectivity."[11] Crick, too, wrote to Pauling and thanked him for the advance copy of his paper. Yet he could not help adding a dig he knew would be felt in Pasadena: "We were struck by the ingenuity of the structure. The only doubt I have is that I do not see what holds it together."[12]

MARCH 12, 1953, may have been the worst day of Maurice Wilkins's life. That morning, John Kendrew, "helpful as usual," called to invite him to "see the new model Jim and Francis had built and he briefly told [Wilkins] what it was like." Wilkins was on "a train to Cambridge straight away."[13] A few hours later, when he walked into Room 103, he sensed that "this was not like the relatively carefree time 16 months before when Francis called me to see their first model. Now there was tension in the air."[14] Given pride of place, Watson and Crick's new model

The third man: Maurice Wilkins.

was "standing high on a lab bench." Wilkins carefully inspected what he referred to as "the W-C model," a Freudian joke incorporating the abbreviation for "water closet," the European euphemism for a tiny room with a flush toilet. He also chose to see connections to Bruce Fraser's flawed triple helix model, with the "phosphates on the outside and bases stacked in the middle and joined by hydrogen bonds."[15]

Wilkins was confused by Crick's flood of words and constant references to the dyad, or dual, axis, peppered by Watson's laughs and giggles. He needed a moment to process what he was seeing and how to respond. What came through clearly—just as it had for Watson and Crick, and a few days later for Bragg, Perutz, and Kendrew—was "the extraordinary way in which the two kinds of base pairs had exactly the same overall dimensions." Although Wilkins had been in steady contact with Erwin Chargaff for more than a year, neither man made the critical connection of base pairing that Watson and Crick made in a few weeks. Now, Wilkins was forced to admit the obvious implications of these complementary strands with respect to heredity. In 2003, still confounded by the spectacle of that long-ago afternoon, Wilkins recalled, "it seemed like an incredible new-born baby that spoke for itself, saying, 'I don't care what you think—I know I am right' . . . It seemed that non-living atoms and chemical bonds had come together to form life itself."[16] Wilkins was "rather stunned by it all" and could not yet know that he would spend the next seven years confirming and amending the W-C model until almost all of its spatial details had been thoroughly vetted by ever sharper X-ray studies.[17]

Looking up at the tower of tin, brass, and wire, Wilkins did not question "the decision to put guanine and thymine in the *keto* form. Doing otherwise would destroy the base pairs. He accepted Jerry Donohue's spoken argument as if it were a commonplace." Unfortunately for

Wilkins, there was no Jerry Donohue at King's College to alert him "that all the textbook pictures were wrong." Donohue's abilities were a rare commodity and, as Watson recounted, the only other person in the world who "would have been likely to make the right choice and stick by its consequences" was Linus Pauling. "The unforeseen dividend of having Jerry share an office with Francis, Peter, and [Jim] though obvious to all, was not spoken about."[18]

Bragg later claimed, "Wilkins, of course, nearly committed suicide because he had been working on it for so long."[19] Wilkins resented this accusation and angrily denied it in a 1976 letter to Max Perutz: "The most unpleasant thing of all [is] where Bragg is quoted as saying I 'almost committed suicide' because I lost priority on the double-helix structure. Although I have been very keen on my scientific research, priority questions have never loomed so large in my mind. If Bragg did in fact say that, I am sorry to think he thought me so small-minded."[20] Suicidal ideation aside, from the first moment he laid eyes on the W-C model, Wilkins knew he had missed the "final great step." British decorum demanded he act like a gentleman and say, "what mattered was scientific progress." Yet when facing the Watson–Crick DNA structure for the first time, he was lost, unable to "think clearly in all the excitement."[21]

PRESSING THEIR CASE FURTHER, Watson and Crick enlisted Wilkins's help to confirm their double helix by checking it against Rosalind Franklin's X-ray diffraction patterns. Trapped in a haze of sorrow, Wilkins nodded numbly and agreed to measure the "critical reflections." He must have done an excellent job of hiding his feelings, as Watson later complimented him for not showing a "hint of bitterness," but this encomium may represent a sense of relief rather than good fellowship. As do so many people who stab, or simply poke, others in the back, Watson wanted absolution for his sins. "There was no trace of resentment on his face," Watson wrote—ignoring Wilkins's palpable sadness—"and in his subdued way he was thoroughly excited that the structure would prove of great benefit to biology."[22]

To alter the chalky taste of the moment, Crick offered Wilkins (but not Franklin) the opportunity to collaborate on their paper, which would be submitted to *Nature* above all of their names. Wilkins recalled his great confusion over this proposal: "Now, having been completely absorbed in examining the model, I needed a rest. I had little energy and was not prepared for discussing authorship questions." He eventually told Crick he could not be the co-author as he "had not taken part directly in building the model." Crick readily agreed, explaining that the co-authorship had been Watson's idea.

Just before the visit ended, the usually buttoned-down Wilkins indignantly asked, "how much did Francis and Jim's model-building depend on work done at King's?" Crick offered the stunning retort that Wilkins was being "unfair" and, astoundingly, Wilkins privileged Crick's objection over his own. True to character, Wilkins never stopped castigating himself for his indignation that day. In his 2003 memoir, the physicist formally expressed regret over his behavior and disappointment that "he had not been more involved in the final great step." On those pages, he openly thanked Watson for not mentioning his angry outburst in *The Double Helix*.[23]

A throng of King's physicists awaited Wilkins's return to London, eager to learn about Watson and Crick's latest folly. They were unprepared for what came next. Wilkins "told everyone at King's what the main features of the [Watson–Crick] structure were" and asked Gosling to pass "the news on to Rosalind, who was by now working at Birkbeck, a mile or so to the north of King's in Bloomsbury."[24] As evidence of how completely frozen out of the King's College circle she now was, the news was not communicated to her until a week later. When it came to DNA, in Wilkins's view, Franklin no longer mattered.

The morale at King's College dropped a few floors lower than the basement level of the physics department. After John Randall heard the awful news, Willy Seeds claimed, he was as angry as a "scalded rat."

Geoffrey and Angela Brown described Wilkins as "devastated." Gosling, too, felt "quite upset, quite scooped."²⁵ Watson and Crick's great good fortune translated into an inescapable wave of loss in the laboratory just off the Strand. With the distance of both observer and partial facilitator, Jerry Donohue described the King's College defeat best: "Had it been the other way around, if someone anywhere had done the same with the data collected by the MRC group at the Cavendish, the resulting eruption would have paled that of Krakatoa to a grain popping."²⁶

MARCH 12 TURNED OUT TO BE a bad day for Pauling, too, even if he did not realize it for a few more days, because that same afternoon Watson wrote Max Delbrück a detailed letter describing his and Crick's DNA model. Clear and concise, the elegant facts contained on these yellowed pages signify biology's Magna Carta and the Declaration of Independence rolled into one. Watson's letter exuded wonder at how living organisms pass on their genetic information to the next generation. He drew by hand both the pyrimidine and purine base structures and discussed the reasoning for choosing the *keto* over the *enol* forms, the model's stereochemical considerations, and the need to "obtain collaboration from the group at King's College London [again, not specifically mentioning Rosalind Franklin] who possess very excellent photographs of a crystalline phase in addition to rather good photographs of a paracrystalline phase." The postscript made a polite request: "we would prefer your not mentioning this letter to Pauling. When our letter to *Nature* is completed we shall send him a copy."²⁷

Watson's request for confidentiality may have been a play of reverse psychology. It certainly turned out that way. Delbrück was so impressed by the letter's graceful truth that, soon after reading its final words, he showed it to Pauling. Delbrück later explained that "Pauling had made him promise to let him know the minute he heard" from Watson. Equally important, Delbrück detested "any form of secrecy in scientific matters and did not want to keep Pauling in suspense any longer."²⁸ Watson was by now confident enough in the correctness of his model to craftily use Delbrück as a back channel to gain Pauling's endorsement—an act of validation that vested Watson and Crick's victory over both King's and Caltech.

ON MONDAY MORNING, MARCH 15, Wilkins called Crick to tell him that he had spent the weekend comparing the model to the King's X-ray data and confirmed that the data "strongly supported the double helix."[29] Later that afternoon, a telephone parley was arranged between an aggrieved Randall and the triumphant Bragg. Randall had to play the situation just right, however, and had little room for self-righteous anger. Thanks to the Medical Research Council, he had built Britain's largest biophysics laboratory with the goal of ascertaining DNA's structure. Yet here were these two knuckleheads from Cambridge—one a perpetually smart-aleck graduate student and the other an infuriating American—who had beaten his team to the punch. Randall could not risk the added embarrassment of his lab being left out of the publication process. In search of a gallant resolution, the two chiefs agreed that Watson and Crick would hold off sending their paper to *Nature* while Wilkins readied a report of his own. If there was little chance of Wilkins co-authoring a paper with Watson and Crick, the likelihood that he and Rosalind Franklin would both affix their names to the same manuscript was less than zero. Thus, when the deal was struck, there was no accommodation nor mention of how to give due credit for Franklin's work.

WHEN FRANKLIN HEARD FROM GOSLING about the "W-C model," nearly a week after Wilkins saw it, she was busy arranging her cramped lab at Birkbeck College. Anne Sayre claimed that "the news that the Cavendish had cracked DNA was an irrelevant parting gift."[30] Not exactly. Soon after she found out about the plans to publish a King's and a Cavendish paper, Franklin contacted Randall to demand that "she and Gosling publish their material on the B form simultaneously."[31] They were already hard at work on a paper (illustrated by Photograph No. 51), which was made ready within a week.

On March 19, Franklin and Gosling took the train to Cambridge to inspect the Watson and Crick model for themselves. In his memoir, Watson recalls being "amazed" by her "instant acceptance of our model." Initially, he "had feared that her sharp, stubborn mind, caught in her self-made anti-helical trap, might dig up irrelevant results that would foster uncertainty about the correctness of the double helix." This unsympathetic prediction reveals how poorly he understood Franklin's commitment to finding scientific truth through cold, hard, reproducible facts. She did not display "fierce annoyance" because there was nothing to be fiercely annoyed about, scientifically speaking.[32] The model looked to be correct. It was interesting. It answered questions about the data she had gathered but was heretofore unable to fully interpret. In a 2013 interview for *Nature* on the sixtieth anniversary of the double helix, Gosling recalled Franklin's response as "gracious and sanguine: "She didn't use the word 'scooped'. What she actually said was, 'We all stand on each other's shoulders.'"[33]

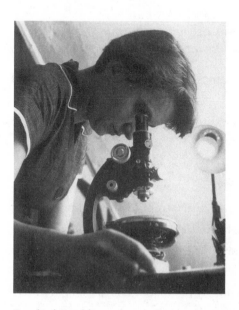

Rosalind Franklin at the microscope.

Rosalind's "pleasure" in having her "first rate crystallographic ability" recognized by the Cavendish group baffled Watson. Flummoxed for years by what he misperceived to be a behavioral conversion, he offered up a posthumous, left-handed, and somewhat self-serving compliment at the tail end of his 1968 memoir. Her anti-helical views, he admitted, "reflected first-rate science, not the outpourings of a misguided feminist." He now told the story of how "Rosy's transformation" was a newfound "appreciation that our past hooting about model building represented a serious approach to science, not the easy resort of slackers who wanted to avoid the hard work necessitated by an honest scientific career."[34]

The New Yorker journalist Horace Judson coldly summed up Franklin's DNA story in two short sentences: "It is easy to feel sympathy with Franklin. The fact remains that she never made the inductive leap."[35] In 2018, Jim Watson was far blunter: "I'd call her a loser . . . I use the word loser not in the sense she was a lowlife or bad person. She blew it. She blew it! And that sounds like an awful thing to say but she threw it away, that it was—she had no reason to do what she did except she hated the idea that the A form was helical."[36]

Francis Crick, too, derided Franklin for not being able to climb the same wall of scientific deduction he and Watson scaled so nimbly, writing that her "difficulties and her failures were mainly of her own making." She may have seemed brisk and confident but, at root, she was "oversensitive, and ironically, too determined to be scientifically sound and to avoid short cuts. She was rather too set on succeeding all by herself and rather too stubborn to accept advice easily from others when it ran counter to her own ideas. She was proffered help but she would not take it."[37]

Such summations are as unfair as they are callous. As Brenda Maddox observed, Rosalind Franklin was carefully trained, from childhood on, as a student at St. Paul's School for Girls, at Cambridge, and especially as a scientist, "never to overstate the case, never to go beyond hard evidence. An outrageous leap of the imagination would have been as out of character as running up an overdraft or wearing a red strapless dress."[38]

Perhaps it was one of the most minor characters in this tangled yarn who best assessed Franklin's complex nature. In 1990, a Welsh physicist and X-ray crystallographer named Mansel Davies described an exchange he had with her in 1952, while visiting the King's College physics department. He was eager to meet with her, having worked on DNA with William Astbury at the University of Leeds from 1946 to 1947. As Franklin graciously showed him her X-ray images, his "pulse raced." (There are those words again, the same ones Watson used a year later when Wilkins showed Franklin's pictures to him.) Davies quickly "realized she was showing me a key to the solution of the DNA problem." What Watson, Crick, or Wilkins could never solve, Davies observed, was *their* "Rosie problem." To begin, Franklin's and Watson's approaches to scientific inquiry were vastly different: "one soberly conscientious, with an unbending professional attitude to her work; the other a bright spark with a devil-may-care attitude." Davies admitted that "Rosie almost certainly made a mistake; Watson, for all his rudeness, could well have given her clues to the solution of the DNA structure." Yet he also noted that "only if Rosie had been an angel could they have hit it off to indulge in useful exchanges." But Davies insisted, it was unjustified to "write Rosie off as 'difficult.' The epithet has arisen because she was an individual orientated by her own scientific interests, and happiest when pursuing them without unnecessary interference." All that was required to smooth these relations, Davies explained, was "some element of understanding."[39] Alas, none was forthcoming from any of the men she circled at King's or the Cavendish.

THE IDES OF MARCH blossomed into a perfect English spring and, with each passing day, more members of the Cambridge scientific community came to Room 103 to gawk at the W-C model. At each showing, Crick presented a jubilant tour through the structure and "its implications lost none of its zest for having been given several times each day for the past week. The pitch of his excitement was rising each day."[40] Crick's voice got louder with every iteration until the "physicists upstairs commented [that] 'steam' was rising from the floor below."[41] One of the visitors was the eighty-eight-year-old experimental physicist G. F. S. Searle, who had worked at the Cavendish with J. J. Thomson in the 1890s. After Crick explained that DNA was "the basis of human heredity," Searle is said to have remarked, "No wonder we're such a queer lot!"[42] Soon enough, it was not only Bragg who wanted to stay out of earshot of Crick's braying laugh and high-pitched squeals. Whenever Donohue or Watson heard "the voice of Francis shepherding in some new faces, [they] left [the] office until the new converts were let out and some traces of orderly work could resume."[43]

To Crick's consternation, on March 13, Watson flew off to Paris for a week of fine dining and relaxation with the Institut Pasteur geneticists Boris and Harriet Taylor Ephrussi, a trip which "had been arranged some weeks earlier." Watson was exhausted and saw no reason to cancel his much-anticipated visit to the City of Light. He had already purchased a plane ticket, still a novel form of travel from London to Paris, and looked forward to telling the Ephrussis and their friends about "his" double helix.[44] Crick, unhappy over Watson's literal flightiness, told him that "a week was far too long to abandon work of such extreme significance." Watson later explained his youthful, oppositional behavior: "A call for seriousness, however, was not to my liking—especially when John [Kendrew] had just shown Francis and me a letter from Chargaff in which we were mentioned. A postscript asked for information on what his scientific clowns were up to."[45] On reflection, it seems impossible to imagine anyone on the brink of announcing such a prodigious discovery taking a vacation. Perhaps Watson meant what he said when he wrote to Delbrück about his need for one last week before the spotlight of deoxyribonucleic acid permanently shined on his every scientific move and made it impossible "to concentrate on other aspects of life."[46]

When Watson returned from Paris, Crick was waiting impatiently for him. Now it was Crick's turn to commandeer his partner's full attention, in the task of writing up their model for publication.[47] Crick did, however, find the time to compose a seven-page letter, complete with diagrams, to his twelve-year-old son, Michael, on March 19. It may not be the first handwritten description of DNA, but it is easily the loveliest:

Dear Michael,
Jim Watson and I have probably made a most important discovery.
We have built a model for the structure of de-oxy-ribose-nucleic-acid

(read it carefully) called D.N.A. You may remember that the genes of the chromosomes—which carry the hereditary factors—are made up of protein and D.N.A. Our structure is very beautiful. D.N.A. can be thought of roughly as a very long chain with flat bits sticking out. The flat bits are called bases . . . Now we have two of these chains winding round each other—each one is a helix—and the chain, made up of sugar and phosphorus, is on the <u>outside</u>, and the bases are all on the <u>inside</u> . . .

Now we believe that the D.N.A. is a code. That is, the order of the bases (the letters) makes one gene different from another gene (just as one page of print is different from another). You can now see how Nature makes copies of the genes. Because if the two chains unwind into two separate chains, and if each chain then makes another chain come together on it, then because A always goes with T, and G with C, we shall get two copies where we had one before . . .

In other words, we think we have found the basic copying mechanism by which life comes from life. The beauty of our model is that the shape of it is such that only these pairs can go together, though they could pair up in other ways if they were floating about freely. You can understand that we are very excited. Read this carefully so that you understand it. When you come home we will show you the model.

Lots of love,

Daddy[48]

It Has Not Escaped Our Notice

We wish to suggest a structure for the salt of deoxyribose nucleic acid (D.N.A.). This structure has novel features which are of considerable biological interest . . . It has not escaped our notice that the specific pairing we have postulated immediately suggests a possible copying mechanism for the genetic material.

—JAMES D. WATSON AND FRANCIS CRICK[1]

Cambridge custom dictated that Crick and Watson personally inform Pauling about their double helix, even though neither was foolish enough to think that Pauling did not already know of their success. On March 21, they finally wrote a long letter telling the world's most powerful chemist that they had bested him. Discretion—a personality trait rarely displayed by Watson and only occasionally by Crick—was essential to this task.

First, they had to invent a reason for the delay in communicating directly with Pauling. At Crick's insistence, Watson added a few excuses: "One of us (J.W.) has been away in Paris and we have also been delayed because Professor Bragg had been down with the flu."[2] Both pretexts were feeble. Watson's Parisian excursion occupied only six days, March 13–18. Blaming Bragg's influenza seemed the safer course, because decorum demanded they inform their laboratory chief about their work before telling outsiders. Although Bragg was still feeling the remnants of his viral infection when they wrote to Pauling, he had visited the lab directly from his sickbed on March 9 to review the W-C model. The real reason for the delay was that Watson and Crick were not ready to share the specifics of their result with the man they considered to be their most daunting competitor.

At the end of March, when Watson and Crick did send a draft of their paper to Pauling, they politely requested permission to mention his structure, even if they could not "suppress [their] doubts about" it. If changes had been made to the Pauling–Corey model, they added, "we can always qualify our remarks in proof." They further informed Pauling that the "Kings College workers will publish some of their experimental data at the same time as our [paper]" and that Wilkins (again, without mention of Franklin) would soon send him a copy of his paper in final draft. The letter's concluding sentence was an outright lie: "We are looking forward very much to your visit and the opportunity for a full discussion about DNA. Would you mind treating this as confidential for a few days as Professor Bragg has still not been able to hear about it."[3] A few days later, on March 24, Watson wrote to his parents that he was so nervous about Pauling's response and, more broadly, the scrutiny of the scientific community that he was unable to look at his "very great discovery . . . objectively. I thus tend to try to forget and instead play tennis."[4]

PAULING WAS SCHEDULED TO VISIT London and Cambridge before going to Brussels for the ninth conference of the Institut International de Chimie Solvay on April 6–14. The topic for that meeting was proteins. Bragg, too, was invited and he planned to present Perutz and Kendrew's work on hemoglobin. Given recent events at the Cavendish, he asked for and received permission to present a complementary note on the Watson and Crick double helix model.[5]

Once again, Pauling wrangled with the U.S. State Department when applying for a passport. The dreaded Ruth Shipley erected a blockade after discovering the transcript of his November 1951 testimony before the Industrial Employment Review Board, in which he stated under oath, "I recognize that my political activities and associations are such to indicate unreliability as a repository of classified information." The two-year-old comment set off another round of letter-writing, with Pauling again declaring that he was not a Communist and that top-secret security clearance was not required for his current work or travel. Mrs. Shipley had less to stand on than she had during their battle of 1952, because the IERB subpoena had been made in error and the entire case was dismissed. Still, she harassed Pauling for a week before quietly approving his application.[6]

Pauling wrote to Bragg that he wanted to inspect both the Watson and Crick model and the King's X-ray data for himself. He had by now read Watson's "DNA letter" to Max Delbrück of March 12, and sent signals that he was about to concede defeat—a retreat that proved to be a "genuine thrill" for Watson and Crick.[7] The public face Pauling exhibited was remarkable given that he was "burning" inside from being beaten by such an unlikely pair of colleagues.[8] After all, as he once testified under oath before the U.S. federal government, "I have, I think a broader grasp of science as a whole—mathematics, physics, chemistry, biology, and geology (minerology)—than any other man in the United States."[9]

Nature, the prestigious British weekly "magazine" of science, was the obvious venue for rapid publication of Watson and Crick's work, especially because both Bragg and Randall vouched for the DNA paper and were chummy with the editors.

The co-editor of *Nature*, Lionel J. F. ("Jack") Brimble, was an active member of the Athenaeum, the most prestigious gentlemen's club in London. So, too, was John Randall. Over sturdy crystal glasses of good Scotch whisky, Brimble listened to Randall's woes and became "the first to feel sorry for King's having been pipped at the post."[10] Randall sensed his opportunity and talked Brimble into publishing "Wilkins' paper alongside Watson and Crick's; Franklin's paper was added only after she petitioned for its inclusion." When Randall returned to King's College from his lunch date, he ordered his troops, "Get writing!"[11] Remarkably, peer review was skipped entirely, paving the way for the paper to be edited, typeset, proofed, and published within a month of its receipt. A. J. V. Gale, the other co-editor, later recalled the "outstanding importance" of the Watson–Crick article, but there the paper trail ends. Unfortunately, *Nature*'s extensive editorial records, including all communications with its eminent contributors from 1869 to 1963, were thrown out during an office move in 1963 and so the editorial correspondence concerning the April 25, 1953, issue is lost.[12]

A few weeks earlier, on March 17, Crick sent Wilkins a draft of his and Watson's paper, which had not yet been approved by Bragg. In an accompanying letter, Crick asked for permission to reference some of Kings College's unpublished work, and broached the subject of how to approach the gummy issue of acknowledgments—a problem that would only become stickier in the days to come. The letter ended with the news that "Jim has gone to Paris, lucky dog."[13] The following morning, Wilkins wrote to Crick, with a clear eye on history:

> I think you're a couple of old rogues but you may well have something. Thanks for the MSS. I was a bit peeved because I was convinced that 1:1 purine pyrimidine ratio was significant and had a 4 planar group sketch and was going to look into it and as I was back on helical schemes I might, given a little time, have got it. But there is no good grousing—I think it's a very exciting notion and who the hell got [the DNA structure] isn't what matters . . . We should like to publish a brief note with a picture showing the general helical case alongside your model publication . . . I can have the whole thing ready in a few days. I think the two publications would look nice side by side . . . [I] hope you won't mind the slight delay that may result in your own publication. Just heard this moment of a new entrant in the helical rat-race. R.F. [Franklin] & G. [Gosling] have served up a rehash of our ideas of 12 months ago. It seems they should publish something too (they have it all written). So at least 3 short articles in *Nature*. As one rat to another, good racing.[14]

Wilkins's use of the phrases "old rogues," "rehash of our ideas," and "one rat to another" give some sense his disappointment over being beaten so roundly. These resentments would not end any time soon. More than a month after Watson and Crick's paper was published, Wilkins wrote to an equally disgruntled Erwin Chargaff, "I will admit I am not the only person who rather hopes the model will be wrong but so far we have no good evidence against [it]."[15]

ORCHESTRATING THE "BUNDLING" OF all three DNA papers for *Nature* required a great deal of negotiation, given the bad feelings between Wilkins and the Cavendish duo, and the worse ones between Wilkins and Franklin. But Bragg and Randall set a strict deadline and all three papers arrived in the *Nature* editorial offices on April 2.[16]

The DNA symphony, published in the April 25, 1953, issue of *Nature*, had three movements: the Watson and Crick model appeared first, loudest, and most memorable—*molto allegro*. The order of the authors, with Watson first and Crick second, was settled "by the flip of a coin."[17] It was an entirely theoretical paper, without a scintilla of original research data. They had yet to handle, prepare, or actually look at a single fiber of DNA.[18] Consisting of 842 words, the article's terse but clear tone belied the fact that it would soon set off "a string of depth charges in a calm sea."[19]

The original manuscript was typed in a pica font, neither by Watson or Crick nor by the Cavendish secretary—who, for some unrecorded reason, was not available—but by Elizabeth Watson. She agreed to perform the chore over the last weekend of March both out of love for her brother and because he informed her that "she was participating in perhaps the most famous event in biology since Darwin's book."[20] To complete this tableau of stereotypical gender roles, Watson and Crick stood over her shoulders as she pecked out each word, shouting with glee at the sentences they particularly liked and correcting her whenever she made an error.

Added to the text were six references (to the work of Pauling and Corey, Furberg, Chargaff, Wyatt, Astbury, and Wilkins and Randall), a legend for the picture of the double helix (drawn by Odile Crick), and acknowledgments to "Dr. Jerry Donohue for constant advice and criticism, especially on interatomic distances. We have also been stimulated by a knowledge of the general nature of the unpublished experimental results and ideas of Dr. M. H. F. Wilkins, Dr. R. E. Franklin and their co-workers at King's College, London. One of us (J. D. W.) was aided by a fellowship from the National Foundation for Infantile Paralysis."[21]

The next movement, in *pianissimo*, was by Wilkins, Alec Stokes, and Herbert Wilson. Their paper was, essentially, a recapitulation of Stokes's helical theory and Wilkins's X-ray diffraction studies, which he had presented at the Naples Zoological Station in May 1951 and again in Cambridge in the summer of 1952.[22] Convoluted and jargon-filled, the paper almost dared the reader to finish, let alone comprehend it.

In the final, *a cappella* slot—a position reader-theorists would argue all but guaranteed it to be the least read—was the Franklin and Gosling data-driven paper on the A and B forms of DNA.[23] The order of the three papers was arranged to Franklin's disservice because she was no longer attached to King's College and had no one to advocate for her. Reading about X-ray crystallography is always a highly technical, challenging, and arduous endeavor, and Franklin didn't help matters with her turgid prose and lengthy sentences. Most of the paper, which was twice as long as Watson and Crick's contribution, had been drafted *before* she traveled to Cambridge on March 19 to inspect the model. We know from her archived papers that she completed an earlier, and fairly complete, draft on March 17.[24]

Franklin did add one sentence to the final manuscript which could only have been written *after* seeing the W-C model: at the close of the penultimate paragraph, in her tiny, spidery handwriting, is the simple, careful, but declarative sentence: "Thus, our general ideas are not inconsistent with the model proposed by Watson and Crick in the preceding communication." In the same paragraph, she noted that the B structure "is probably helical."[25] In 2002, her biographer Brenda Maddox reported with outrage (and post facto irony, given how the phrase "me too" has evolved in our contemporary vernacular): "the alteration transformed her own fundamental findings into a 'me-too' effort." Should there be any surprise that Franklin's results were consistent with those presented by Watson and Crick? They used her X-ray diffraction measurements to build their model. More ironic, she used her Photograph No. 51 of the B form of DNA to illustrate the *Nature* paper, but nowhere in the Watson and Crick communication was the admission that Watson had not only seen it but was "inspired by it."[26]

UPON ARRIVAL IN CAMBRIDGE on April 4, Linus Pauling took his son Peter's ill-considered recommendation and booked a room in Pop Prior's boarding house. Watson smugly recalled how Pauling chided his son for the less than deluxe accommodations and ordered him to make reservations at a superior hotel because "the presence of foreign girls at breakfast did not compensate for the lack of hot water in his room."[27]

The next day Watson and Crick invited Pauling to view the DNA model dominating Room 103: "All the right cards were in our hands and so, gracefully, he gave his opinion that we had the answer."[28] Bragg beamed with pride. His beloved Cavendish Laboratory had finally beaten the Wizard of Caltech. Equally rewarding for the British physicist was that, despite the fact that the hard data establishing the double helix originated in another laboratory, forty years earlier he and his father had developed the methodology—X-ray crystallography—that "was at the heart of a profound insight into the nature of life itself."[29]

That evening, Peter and Linus Pauling and Elizabeth and Jim Watson dined at the Cricks' home, where Odile served up a sumptuous celebratory feast and all drank "a fair amount of Burgundy." Crick was unusually quiet in Pauling's presence. To keep things lively, Watson encouraged the chemist to flirt with Odile and Betty. Given the time difference and the ravages of jet travel, Pauling had trouble deploying his usual charm. Later in the evening, Watson sensed that Pauling preferred to speak directly to him, because he was still " an unfinished member of the younger generation" while Crick was less impressionable. At any rate, the long journey to England soon took a firm hold on Pauling's mind and body and the party dispersed by midnight.[30] Pauling left for Brussels the following morning.

On April 6, after the first day of the Solvay Conference, Pauling retired to his hotel room and wrote to his "dearest little love," Ava Helen

Neuvième Conseil de Chimie Solvay, April 1953. (In the front row, two from the left is R. Signer, five from the left is W. L. Bragg, five from the right is L. Pauling.)

Pauling. He reported, "I have seen the King's College nucleic acid pictures, and talked with Watson and Crick, and I think that our structure is probably wrong, and theirs right."[31] During Bragg's presentation on April 8, Pauling recorded his notes with a blunt pencil: "Bragg then discussed Watson & Crick's N.A. [nucleic acid] I said that I am pretty sure W & C are right. I explained why we went wrong."[32] Later that summer, as he crisscrossed the continent visiting eminent scientists in Germany, Sweden, and Denmark, Pauling became even more convinced. In his July–August diary is a brief entry: "Watson and Crick's structure explains everything."[33]

Pauling often told his students, "Don't be afraid to make mistakes, too many scientists are so cautious, if you're never wrong then you're working in a field that is too easy for you . . . there are thousands of scientists who have nothing better to do than show that you've made a mistake. If you've got something important, publish it."[34] When it came to DNA, Pauling's mistake was a whopper. One of the biggest factors foiling his model was a miscalculation of the molecule's water density. This was the same issue that so vexed Franklin, and later, Wilkins, Watson, and Crick. As Pauling later explained, he did not realize that the DNA preparations used for Astbury's old X-ray photographs were 33 percent water: "So the calculation that I made ignoring the water gave three strands. And if you correct for the water—I just hadn't realized there was so much hydration—then it turns out to be two strands."[35] Another glaring problem had to do with Pauling not having access to the pristine photographs taken by Franklin and Gosling and, instead, relying on the older, blurry photos made by Astbury, which superimposed the A and B forms. Later still, he blamed his error on not knowing enough about the chemistry of purines and pyrimidines.

On and on it went for years, until Ava Helen Pauling grew weary of her husband's excuses and bluntly asked, "If [solving DNA] was such an important problem, why didn't you work harder on it?"[36] His answer to his wife, and eventually to the world, was a modest one: "I don't know, I guess that I always thought that the DNA structure was mine to solve, and therefore I didn't pursue it aggressively enough."[37] Pauling believed he was brilliant enough to get away with a speed round of discovery and capture the flag for what became one of the biggest prizes in modern science. His biographer Thomas Hager reduced his historic flop to a binary equation: "There were two reasons Pauling failed with DNA: hurry and hubris."[38]

FOR THOSE CONDEMNED TO the legal profession, a conspiracy represents a covert criminal act involving more than one person. With their eyes and ears peeled for a word's coining and usage over time, the editors of the *Oxford English Dictionary* define conspiracy more broadly as "a combination of persons for an evil or unlawful purpose; an agreement between two or more persons to do something criminal, illegal, or reprehensible (especially in relation to treason, sedition, or murder); a plot."[39] The double helix collusion scheme was nothing short of a plot among men of mutual interests, cultural beliefs, and entitlements. A long trail of conspiratorial dominoes was carefully put in place by the participants long before the Watson and Crick paper was published in *Nature*. How those dominoes toppled one after another with such precision, and the machinations by Watson, Crick, Wilkins, Randall, Perutz, Kendrew, and Bragg to conceal the fact that the W-C model was predicated on Rosalind Franklin's data, fits the definition of conspiracy all too well.

Wilkins widened this net by lobbying to include Bruce Fraser's unpublished work on the triple helix theory in the *Nature* bundle of papers—the very study Wilkins said was not ready for publication two years earlier.[40] He must have spent a small fortune, given the cost of long-distance communications in the 1950s, cabling and phoning Fraser, who was by then working in his native Australia. Fraser labored all night typing up his results and drawing by hand a diagram of his proposed model, and eagerly cabled the results back to London shortly after the sun rose the following morning, hoping to become part of scientific history.[41] Wilkins made certain to share the paper with Crick, who insisted that it could not possibly be included in the *Nature* bundle. He thought Fraser's model too "feeble" to be paired with his and Watson's beautiful findings. A compromise was eventually reached, with Watson and Crick agreeing to mention the Fraser model in their paper as one "in the press," accompanied by the dismissive comment, "This structure as described is rather ill-defined, and for this reason we shall not comment on it."[42] The original sentence was, in Wilkins's memory, even harsher until he got the men from Cambridge to tone it down by admonishing Crick, "Why be bitter about it?" As it turned out, Fraser's paper was so "feeble" that it was never published. Insisting that Watson and Crick include a mention of it in their paper was Wilkins's none-too-subtle attempt to show that King's was working on DNA's helical structure for at least two years before Watson and Crick arrived on the scene, as well as a means of diminishing Franklin's subsequent work.[43]

In a "suggested modification to your MSS," Wilkins asked Watson and Crick to obscure the fact that Franklin's superb X-ray data confirmed their theory:

Could you delete the sentence "It is known that there is much unpublished experimental material?" (This reads a bit ironical.) Simply say "The structure must of course be regarded as unproved until it has been checked with fuller experimental material . . ." Delete *very beautiful* and say, "We have been stimulated by the work at King's or something."[44]

By March 23, Wilkins was despondent over the whole affair. Franklin had not only demanded to be included in the publication "rat race" but also to meet with Pauling when he came to Britain. Wilkins feared the double humiliation of being shown up by her both on the pages of *Nature* and directly to Pauling. Squeezed on one side by Watson and Crick's success, and on the other by Franklin's demands for fair play and access, Wilkins vented to Crick:

> It looks as though the only thing is to send Rosy's and my letters as they are and hope the editor does not spot the duplication. I am so browned off with the whole madhouse I don't really care what happens. If Rosy wants to see Pauling, what the hell can I do about it. If we suggested it would be nicer if she didn't that would only encourage her to so do. Why is everybody so terribly interested in seeing Pauling? . . . Now Raymond [Gosling] wants to see Pauling too! The hell with it all, M.
>
> P.S. Raymond and Rosie have your thing so everybody will have seen everybody else's.[45]

IN CAMBRIDGE, far more sinister manipulations were being penciled into the historical record. Throughout Watson and Crick's *Nature* paper, there is ample evidence of an academic crime known as citation amnesia, wherein the authors do not cite all or some of the published or unpublished work they most certainly used to build their model. The lack of a formal citation of Rosalind Franklin's contribution to their work is the most egregious example of their negligence.[46] The deletions and omissions they made in successive drafts of the paper—which are documented in the letters Wilkins, Watson, and Crick wrote to one another at the time—are disturbing at best, and at worst worthy of notes of retraction, explanation, and sanction.[47] There has even been some post hoc speculation that had Brimble and Gale, the editors of *Nature*, known the precise details of how Watson and Crick arrived at their theoretical model, they would have insisted that Rosalind Franklin be listed as one of the principal authors of their paper. Sadly, because both editors died before being queried, and because of the destruction of their papers by the good people at the Macmillan publishing house, we shall never know.[48] But if an editor knew of an author purposefully failing to cite the appropriate sources or to give full authorial credit to all of the paper's contributors, and the editor failed to correct such an omission, he, along with the authors of the paper, would be guilty of academic malpractice.

Watson and Crick's misdirection is best seen in the two sentences describing the role the King's College research played in their work, as compared with the role of the previously published literature. They did cite the 1947 Astbury paper and a 1953 paper by Wilkins and Randall.[49] But in a preceding paragraph they disingenuously told the world (partly in language composed by Wilkins) that "the previously published X-ray data on deoxyribose nucleic acid are insufficient for a rigorous test of our model. So far as we can tell, it is roughly compatible with the experimental data but must be regarded as unproved until it has been checked against more exact results." And here's the most incriminating line: "Some of these are given in the following communications [i.e., in Wilkins's and Franklin's *Nature* papers]. We were not aware of the details of the results presented there when we devised our structure, which rests mainly though not entirely on published experimental data and stereochemical arguments."[50]

An informed reader of these last thirty-nine words can only utter an astonished, "What?" Some have argued, in Watson and Crick's defense, that they were *technically* telling the truth in that they had not yet read the papers Franklin (or Watkins) submitted to *Nature* when they devised the W-C model. Such a defense feels lawyerly, if not downright Talmudic. Opposing counsel would have a far better case by noting that Watson and Crick discussed the helical structure of DNA ad infinitum with Wilkins for nearly two years, and frequently used (their word was "pumped") him to find out what Rosalind Franklin was doing; that Wilkins showed Watson Franklin's Photograph No. 51—the very picture that set his heart racing; and that Max Perutz gave Crick and Watson a copy of the MRC report containing Franklin's most important measurements and results.

In their pursuit of scientific priority, Watson and Crick could not openly confess how critical Rosalind Franklin's data was in making their great discovery. Nor, at this point, would they admit in print that they had never asked permission to use her data. They would not come close to such a confession until a year later, in a 1954 issue of the *Proceedings of the Royal Society*. At the bottom of the first page of that article, Crick and Watson added a footnote that almost came clean: after thanking Wilkins and Franklin, they appeared to own the fact that without *their* data "the formulation of our structure would have been most unlikely, if not impossible." But the following sentence skitters away from the truth and whizzes back to their claim of primacy, declaring, "We should at the same time mention that the details of their X-ray photographs were not known to us, and that the formulation of the structure was largely the result of extensive model building in which the main effort was to find any structure which was stereochemically feasible."[51]

What cannot be debated is that the powerful and brilliant final sentence of Watson and Crick's *Nature* paper has captivated and inspired biologists ever since. Their "coy" published closing statement actually represents a heavily toned-down version of what the ever-confident Crick was "keen" to expand upon—palliated by Watson, lest they be proven wrong and make asses of themselves. Crick "yielded to [Watson's] point of view but insisted that something be put in the paper, otherwise someone would certainly write to make the suggestion, assuming we had been too blind to see it."[52] Watson's qualms aside, their claim to scientific immortality was clearly made: "It has not escaped our notice that the specific pairing we have postulated immediately suggests a possible copying mechanism for the genetic material."[53]

Once Watson and Crick figured out the two last, crucial steps in defining the double helix—complementarity and base pairing—the story ceased to be about Rosalind Franklin or her data. It was now about their beautiful mechanism, by which genes make copies of the information they carry. The names Watson and Crick were etched into history as deeply as Newton, Darwin, Mendel, and Einstein. If life were fair—and it is not—we would be calling it the Watson–Crick–Franklin model of DNA, instead of "Watson and Crick."[54]

ROSALIND FRANKLIN'S LAST DAYS at King's College were anything but momentous. There was no goodbye party, no celebratory cakes and ale, not even a valedictory speech. She packed up what few personal effects she had there and, before her final exit, thanked the lab photographer, Freda Ticehurst, for her help and friendship. In a matter-of-fact tone, Franklin told Ticehurst, "I'm not wanted here. We [i.e., she and Wilkins] could never work together. It's impossible for me to stay."[55]

Several weeks later, on April 17, John Randall wrote to warn her in oblique fashion:

> You will no doubt remember that when we discussed the question of your leaving my laboratory you agreed that it would be better for you to cease to work on the nucleic acid problem and take up something else. I appreciate that it is difficult to stop thinking immediately about a subject on which you have been so deeply engaged, but I should be grateful if you could now clear up, or write up, the work to the appropriate stage.[56]

Franklin plaintively asked Anne Sayre, "But how could I stop thinking?"[57] The request was almost comical: "Just the sort of thing they do there."[58] Wilkins later attempted to lay the blame for her banishment from King's at his boss's feet: "Randall, of course, was quite capable of doing dreadful things."[59] That said, Wilkins was eventually honest enough to confess, "I wasn't very easy myself."[60] Both assessments were accurate.

On a scientific level, Randall's request was especially silly, since Professor J. D. Bernal had asked Franklin to work on tobacco mosaic virus, and RNA, a nucleic acid, is an essential element of that microbe. A week later, on April 23, Franklin tactfully wrote to Randall:

> I am only too anxious to get the DNA work written up as soon as possible, but it is not the sort of thing that can be done in a hurry. As I told you before I left your laboratory, there is a lot to write up, and a good chance that new ideas will emerge during the writing of it . Gosling and I have both already started preparation for the writing. I hope I shall have the opportunity of discussing these things with you some time . . . Professor Bernal tells me that he is inviting you to visit this laboratory some time, and perhaps that would provide a suitable opportunity for talking things over.[61]

Rosalind Franklin could not "stop thinking about DNA" but she probably could not yet imagine the role DNA would soon play in almost every aspect of genetics, biology, evolution, and medicine. Nor did she likely dream that, decades later, King's College would memorialize her DNA work—and marry her surname with that of the man she battled so bitterly—on a new structure known as the Franklin–Wilkins Building.[62] According to her sister, Jenifer Glynn, Rosalind Franklin had no idea of how her data had been inappropriately shared by Wilkins with Watson and pilfered, without her permission, when Perutz showed Watson and Crick her MRC report. In a 2012 memoir of her sister, Glynn concluded that Franklin died without knowing the truth about the appropriation of her work. "As it was," Glynn wrote, "she was impressed and not at all angry, when she saw the final model—though she must have been sorry that she herself had not quite got there first. Not surprisingly, it fitted her results."[63] In May 2018, Glynn made the point even more clearly: "If she was aware, there would have been a great row over it. I don't doubt that. Her fury would have been understandable and alarming."[64]

A few months later, in July 2018, Jim Watson offered a slightly different account. "[Franklin] was very generous for never saying that we stole something from her . . . I think she realized we hadn't taken advantage of her. She had lowered herself by not looking at the B photo."[65] Alas, absence of evidence does not always translate into evidence of absence. In the end, we shall never know precisely what ran through Rosalind Franklin's mind in the years after the discovery of the DNA double helix.

§

ON THE DAY THE "Molecular Structure of Nucleic Acids" issue of *Nature* was sent out into in the world, April 25, 1953, John Randall threw a celebratory party in honor of his DNA researchers. In the basement of King's, wine, sherry, and beer flowed, corks popped out of bottles of cheap Champagne, and laughter swept the room. At one point during the revelry, Freda Ticehurst, the lab photographer, scanned the crowded room and asked, "Where's Rosalind?" The only response, she recalled, was "some looks."[66] Franklin was by now in exile at Birkbeck College and under strict orders from Randall to no longer contemplate the topic of nucleic acids.[67]

The harsh reality was that women scientists of Rosalind Franklin's era experienced subtle, and not so subtle, forms of denigration and oppression on a daily basis. Given her personality, she was, understandably, extremely sensitive to such criticism in the laboratory. On too many occasions, this trait led her to becoming her own worst enemy. Try to imagine how these white, entitled, English academic lords responded or ignored her complaints about the workplace and her frank but brutal descriptions of Wilkins. Add to that her demands for privacy during her work, immune to Wilkins's constant interference. Few of her quirks and insistence on ever more data would likely have been commented upon so critically if she had been male, or if her chosen field had not valued theoretical conceptualization over the long slog of obtaining the scientific proof underpinning the grand theories. Franklin's rivals, both at King's and at Cambridge, were a decidedly odd group of young and immature men. Still, none held Maurice Wilkins's multiple neuroses against him. Francis Crick may have always insisted on being the smartest man in the room but few resented him for it—save his boss, Sir William Lawrence Bragg. And when it came to Jim Watson, his flouting of the rules of British research conduct, taking any shortcut available in order to arrive at a preferred result, and his peculiar personal habits did little more than cause his colleagues to snicker and grumble behind his back.

For Rosalind Franklin, a Jewish woman in the masculine domains of high-stakes physics and British academia, there was no room for error. The slightest mistake in her calculations would represent a gaping rent in the experimental fabric she wove, through which her male competitors would happily climb. This onerous dynamic eventually crippled her progress in ways Watson and Crick could never understand and, instead, spent decades belittling. If either man had a milligram of Samuel Johnson's blistering wit, or at least the inclination to quote him, they might have applied his 1763 comment to James Boswell about women preachers to female physicists: "Sir, a woman's preaching is like a dog's walking on his hind legs. It is not done well; but you are surprised to find it done at all."[68] Instead, the King's and Cavendish boys' club relied on juvenile nicknames, nasty office pranks, condescension, and corrosive taunting. To them, Franklin was incapable of making scientific flights of inductive brilliance and would never soar in their skies. For her, the risk that she might be wrong was intolerable.

WATSON WROTE HIS 1968 MEMOIR, *The Double Helix*, with the conceit of telling the tale "the way I saw things then, in 1951-1953: the ideas, the people, and myself." In actuality, Watson carefully crafted the book more than a decade after the fact. He depicted himself as an odd, impossibly young, and brilliant interloper. Or, as Wilkins interpreted it, "Jim plays himself as the holy fool." No fan of the book, Wilkins made sure to add, "most of the time Francis is smart, too. The rest are bloody clots."[69] Some of those "bloody clots" got off much easier than others. Watson's malice toward Rosalind Franklin, and his cartoonish portrayal of her as an overemotional, angry, incompetent woman, was set into the literary equivalent of concrete by the time he composed his memoir's final line.

In 2018, Watson recalled that in late 1967 one of his editors at Harvard University Press, Joyce Leibowitz, insisted that "you have to say something nicer about Rosalind." To accommodate Leibowitz, whom he referred to as "this smart Jewish woman," Watson pasted on a short epilogue.[70] In its two pages, he praised Rosalind Franklin for her "superb" work and admitted, "my initial impressions of her, both scientific and personal (as recorded in the early pages of this book), were often wrong." He credited her work in distinguishing the A from the B form of DNA which, "by itself, would have made her reputation; even better was her 1952 demonstration, using Patterson superposition methods, that the phosphate groups must be on the outside of the DNA molecule." He went on to compliment her work on tobacco mosaic virus. She "quickly extended our qualitative ideas about helical construction into a precise quantitative picture, establishing the essential helical parameters and locating the ribonucleic chain halfway out from the central axis."[71] In the few years of life she had left, Franklin became good friends with Crick, whom she admired as possessing the brilliance, knowledge, and creative thought she demanded of great scientists.[72] She even built cordial relations with Watson and consulted him on some of her grant proposals.[73] "By then," Watson gently reminisces in his epilogue, "all traces of our early bickering were forgotten and we both came to appreciate greatly her personal honesty and generosity, realizing years too late the struggles that the intelligent woman faces to be accepted by a scientific world which often regards women as mere diversions from serious thinking. Rosalind's exemplary courage and integrity were apparent to all when, knowing she was mortally ill, she did not complain but continued working on a high level until a few weeks before her death."[74]

Yet if the epilogue is true—and it is—why publish a tome so injurious to Franklin's reputation? Was winning the Nobel Prize not enough for him? How much fame and acclaim must one receive in a lifetime to be satiated? Why shroud "her exemplary courage and integrity" in so many best-selling pages inked from the wells of misogyny, callousness, competition, discrimination, anti-Semitism, patriarchy, cultural and class differences, immaturity, harassment, and nonsense? In short, despite all the success and enjoyment derived from the superbly crafted *The Double Helix*, why would a decent man write it at all? The best answer to these queries can be found in a confidential letter the novelist and physical chemist C. P. Snow wrote to Bragg, in 1968, after reading *The Double Helix*: "The interest of Jim Watson's book lies very largely in the fact that he is not at all a nice man."[75]

In real time, Watson credited Franklin's essential work in establishing DNA's structure only in a letter he wrote to Max Delbrück on the morning of April 25, 1953—the very day his paper and hers appeared in *Nature*: "For your convenience I believe it would be best to quote the following paragraphs from Miss R. Franklin's (of King's College–London) note to *Nature* which shall appear in the same issue by Crick and I." Watson then went on to quote the four key paragraphs from Franklin and Gosling's *Nature* paper, which detailed the X-ray diffraction data she developed, the double helical nature of the B form molecule, the placement of the phosphate groups on the exterior of the helical backbone, the transformation of the A to the B form of DNA with hydration, and key atomic measurements, which served as proof of his and Crick's theoretical model. He tentatively concluded, "Thus I am inclined to believe our structure has a good probability to be correct. However, I'm not as yet ready to commit myself that it is right. Thus, at present, I'm more concerned with seeing whether it is correct than in following up its implications."[76]

The last comment is baffling. If Watson was so uncertain of the veracity of his "beautiful" model, why would he publish it in one of the most prestigious scientific journals in the world? If he was unconcerned about its "implications," why were he and Crick already hard at work on a follow-up paper, to be published in the May 30 issue of *Nature* under the title "Genetical Implications of the Structure of Deoxyribonucleic Acid," a report which most definitely hypothesized and explained the profound inferences of their April 25 paper?[77] And why did Watson only honestly acknowledge the importance of Franklin's work to his model in a private letter? Publicly, he assumed a far less modest pose by declaring, "I solved it, I guess, because I was the only one paying full-time attention to the problem . . . I had nothing else to do. I was totally underemployed. And I was the one who got it . . . The only other one who might have, would have been Francis, [had he] gone back that night and . . . done it."[78]

The day after the *Nature* papers were published, April 26, Watson flew off again to Paris. This time, he was accompanied by his sister, who would soon return to the United States and marry "an American she had known in college." Watson wrote wistfully of his beloved sibling, "These were to be our last days together, at least in the carefree spirit that had marked our escape from the Middle West and the American culture it was so easy to be ambivalent about." While strolling down the elegant Faubourg Saint-Honoré, they stopped "at a shop full of sleek umbrellas" and he bought her a wedding present for what he presumed would be her rainy days of matrimony.[79]

The effervescent Peter Pauling joined them the next day, April 27. Although Watson's twenty-fifth birthday was on April 6, they celebrated it belatedly that evening. Pauling left to explore his own romantic interests and Watson walked along the Seine back to his hotel." A decade and a half later, when recalling his rather solitary birthday celebration, he composed the far too precious final line of his memoir: "But now I was alone, looking at the longhaired girls near St. Germain des Prés and knowing they were not for me. I was twenty-five and too old to be unusual."[80]

BY APRIL'S END, the thirty-three-year-old Rosalind Franklin was well along with her work in the Birkbeck crystallography laboratory, situated within two old "slum-like," conjoined, and war-damaged townhouses at 21 and 22 Torrington Square in Bloomsbury. Franklin's lab was in 21. J. D. Bernal, the head of the department, lived on the top floor of number 22, where he often entertained female students and a stream of leftist celebrities, including Pablo Picasso and Paul Robeson, in his messy bedroom. Franklin admired Bernal's scientific prowess and even some of his politics but was unlikely to have approved of his co-educational activities.[81]

There, as at King's, she had little tolerance for the ignorance of others. In January 1955, she wrote to her boss, Bernal, in her usual forceful manner about how the pharmacists working on the floor directly above her lab "have been running grave risks in the way of setting the place on fire." A few months later she blasted the pharmacists again "about the serious inundation [they caused and how] a rapid stream of water poured directly from the ceiling on to the main carbonization apparatus and a fragile and expensive evacuated glass apparatus."[82] In July of the same year, she complained about unequal pay compared to her male colleagues and her lack of a permanent academic position: "In view of the fact that I have no security of employment, and nevertheless hold a position of considerable responsibility, this seems to me to be entirely unjust."[83]

Still, Franklin forged ahead with her research. She was happiest when conducting experiments. By all accounts, she loved working at Birkbeck—a mere mile and a half in physical distance from King's and a universe away in conduct. The smaller, less renowned college turned out to be the perfect spot for her to move forward in her work. There, she was as she always was: biting, sharp, impatient with fools, brilliant, bold, and full of life. By the mid-1950s, she was far more confident in her abilities and intuition as a scientist. Indeed, she regularly admonished the great Crick when he made some of his flightier assumptions. Her reply to one such episode of Crickian flimflammery was short, incisive, and entirely in character: "Facts are facts, Francis!"[84]

During her five years at Birkbeck College, 1953–58, Rosalind Franklin aimed her pathbreaking X-ray diffraction studies at the structure of tobacco mosaic virus, poliovirus, and RNA. One admirer of her virus research was Sir William Lawrence Bragg. In 1956, Bragg was developing the British Science Exhibit for the 1958 World's Fair, or "Expo '58," to be held in Brussels. Knowing full well the subtext of Watson and Crick's now famous paper, he was careful not to revisit the battles of 1953. Ever the diplomat, he first wrote to Crick asking how to invite Franklin and Wilkins separately to do something for a planned exhibit on "the living cell." Crick wrote back on December 8, 1956: "With regard to the Brussels exhibition Miss Franklin will cover viruses and Wilkins will, I think, be responsible for DNA. I should be quite happy to look after collagen, but I think that to collaborate with King's on this would create unnecessary friction."[85] Six months later, Bragg sent Franklin a formal invitation, asking her to build a five-foot-tall model of the tobacco mosaic virus for display in the International Science Hall, which she did to great acclaim.

Over the summer of 1957, while touring the United States, she may (or may not) have had a romantic encounter with an American molecular biologist named Donald Caspar. The two were undoubtedly close friends, having worked together on tobacco mosaic virus research, but there remains dispute between Franklin's biographers and her sister, Jenifer Glynn, as to how far the relationship progressed.[86]

That August, Franklin experienced abdominal pain on two separate occasions and consulted a physician in California, who prescribed painkillers. She ignored his advice that she should be admitted to hospital and carried on with her trip. When she returned to London in the fall of 1957, she had developed abdominal fullness to the point that her clothes no longer fit her once trim figure. Her friend and general practitioner, Dr. Mair Livingstone, asked if she was pregnant and Franklin replied, "I wish I were." Hoping it was merely an ovarian cyst, Dr. Livingstone referred her patient to University College Hospital for a full examination. The news was not good.[87]

Franklin had ovarian cancer, an aggressive malignancy which may have been the result of the massive radiation exposure she experienced during her laboratory work. Raymond Gosling often worried about her recklessness in getting "in the beam" of the machine in order to obtain the best X-ray images. Louise Heller, the volunteer assistant from Syracuse University who was working at King's during Franklin's tenure, was also concerned but remained silent because Franklin "had the sort of drive that the work was more important than anything else."[88] Other factors contributing to Franklin's cancer may have been an inherited mutation in the genes BRCA 1 and 2 which is often found in Ashkenazi Jewish women, bad luck, or all of the above. The surgeons removed tumors from both ovaries; the right one was the size of a croquet ball (about 3.5 inches in diameter), the left one the size of a tennis ball (about 2.6 inches in diameter).[89]

Cancer treatment in the late 1950s was more akin to medieval medicine than what oncologists offer today. Instead of bleeding and dosing her with poisonous herbs, Rosalind Franklin's doctors offered several months of debilitating cobalt radiation treatments, which used gamma rays to kill tumor tissue but also caused severe burns on the skin, vomiting, diarrhea, and internal bleeding; and a course of industrial-strength chemotherapy, which made her even more nauseated and ill. She also underwent tumor-debulking surgical operations, including the removal of her uterus and both ovaries. Between hospital stays, increasing weakness, and painful bouts of ascites (the collection of large amounts of fluid in the abdominal space), she valiantly returned to the laboratory and continued her work. At her most dire times, she convalesced at her brother Roland's home in London or at Francis and Odile Crick's house in Cambridge. Some have tried to use these recuperative stays "as evidence of bad feelings between her and [her] parents," but as Jenifer Glynn insists, "it was not that at all; she had found our mother's obvious concern and distress hard to bear."[90] The person who gossiped the loudest about such "bad feelings," incidentally, was Jim Watson. In a 1984 speech he delivered at Franklin's school, St. Paul's School for Girls in London, he told the lie that "she had terrible relations with her family. Indeed, she went to stay with the Cricks after her hospital treatment."[91]

Still, the harsh treatments brought about ten months of remission. Franklin's scientific output during this period was remarkable. Between 1957 and 1959, she published eleven new, peer-reviewed papers and a book chapter—several of which appeared in print after she died. Jenifer Glynn reported that while her sister was working at Birkbeck, she "showed a total lack of resentment against the Cavendish team." Aaron Klug, her daily lab mate there, corroborated this claim and recalled "that he never heard her complain about Watson or Crick, but she had great admiration for both of them."[92] Free of both resentments and harassment, Franklin's work evolved so splendidly that, in early 1958, Max Perutz "came personally to Birkbeck and invited her and Klug to move their work to Cambridge." He wanted them to join his team at a splendid new research facility being built to house the Cavendish Medical Research Council Molecular Biology Unit. Still putting in long days of work, Franklin expected to be able to make the move. After years of trudging from grant to grant, she was finally about to advance to a secure, permanent faculty position at her beloved Cambridge University.[93]

Everything came to a halt on March 28, 1958. During a family dinner, Muriel Franklin was in the garden when her son Ellis came out and told her, "Rosalind is in trouble again." She was experiencing a profound attack of pain, one of the hallmarks of severe cancer of the abdomen,[94] and was taken by ambulance to the Royal Marsden Cancer Hospital in Chelsea, on Fulham Road. The surgeons there conducted an emergency exploratory operation but closed the incision soon after discovering that the cancer had spread throughout her liver, colon, peritoneum, and small intestine. Heavily sedated on morphine and heroin, Franklin became cachectic and sallow. Her once thick black hair began to fall out in clumps, and what was left lost its luster. Her olive complexion turned a metallic greenish yellow, the result of too much chemotherapy and acute liver failure. Her abdomen was so distended with fluid and malignant tumors that had she not been so ill, she might have appeared to be nine months pregnant. At one point, she could not move her arm and was worried that she might have contracted poliomyelitis—the virus she was studying in her laboratory.[95] While she drifted in and out of consciousness, nearing the abyss, her mother tried to comfort her by cooing over and over again, "It's all right. It's all right." Ever true to her code of total honesty, Franklin "roused from her semi-conscious state to retort indignantly, 'It's not all right, and you know very perfectly well it isn't.' "[96]

As many comatose patients do, Rosalind Franklin developed bronchopneumonia. She breathed her last on April 16, 1958. On the death certificate, the attending physician listed her cause of death as "carcinomatosis" (disseminated, or metastatic, cancer) and "carcinoma of the ovary." The doctor added a single, stark, incomplete sentence of description of her life: "A Research Scientist, Spinster, Daughter of Ellis Arthur Franklin, a Banker."[97] J. D. Bernal, in his obituary of her in the July 19, 1958, issue of *Nature*, was far more eloquent:

> As a scientist Miss Franklin was distinguished by extreme clarity and perfection in everything she undertook. Her photographs are among the most beautiful X-ray photographs of any substance ever taken. Their excellence was the fruit of extreme care in preparation and mounting of the specimens as well as in the taking of the photographs. She did nearly all this work with her own hands . . . Her early death is a great loss to science.[98]

Visitors to Franklin's grave, in the Willesden Jewish Cemetery in the London borough of Brent, are often surprised to see no mention of DNA on her tombstone:

In memory of
Rosalind Elsie Franklin
מ' רחל בת ר' יהודה
Dearly Loved Elder Daughter of
Ellis and Muriel Franklin
25th July 1920–16th April 1958.
Scientist
Her Research and Discoveries on
Viruses Remain of Lasting Benefit
to Mankind.
ת' נ' צ' ב' ה'[99]

In early May 2018, Jenifer Glynn explained that whenever people ask her to revise Rosalind's headstone, "I have always responded with a resolute

NO! It's a matter of history, you see, and the historical context of the time when the epitaph was written. And so, I believe, it should remain."[100]

An accomplished historian in her own right, Glynn wrote a lovely book about Franklin that afforded her the opportunity to eulogize her sister in a manner so sound and unsentimental that it is unlikely to be supplanted by anything better:

> So, Rosalind became a symbol, first of an argumentative swot [a grind or a wonk], then of a downtrodden woman scientist, and finally of a triumphant heroine in a man's world. She was none of these things and would have hated all of them. She was simply a very good scientist with an ambition, as she told [her brother] Colin from her hospital bed, to be a Fellow of the Royal Society before she was 40. But she died at thirty-seven.[101]

PART VI

THE NOBEL PRIZE

The ancient commission of the writer has not changed. He is charged with exposing our many grievous faults and failures, with dredging up to the light our dark and dangerous dreams for the purpose of improvement.

—JOHN STEINBECK, NOBEL PRIZE
BANQUET SPEECH, 1962[1]

Stockholm

The last thing I would like to say is that good science as a way of life is sometimes difficult. It often is hard to have confidence that you really know where the future lies. We must thus believe strongly in our ideas, often to [the] point where they may seem tiresome and bothersome and even arrogant to our colleagues. I knew many people, at least when I was young, who thought I was quite unbearable. Some also thought Maurice was very strange, and others, including myself, thought that Francis was at times difficult. Fortunately, we were working among wise and tolerant people who understood the spirit of scientific discovery and the conditions necessary for its generation.

—James D. Watson, 1962 Nobel Prize Banquet toast[1]

On the afternoon of December 10, 1962, Nobel men and women were dressed in the manner in which noble men and women were supposed to dress: elegant gowns and impossible hairdos for the women, white tie and tails for the men—despite the fact that the ceremony began nearly two hours before the prescribed time when a gentleman properly wears full evening dress. All were there to watch the King of Sweden, Gustaf VI Adolf, formally award the 1962 Nobel Prizes in Physiology or Medicine, Chemistry, Physics, and Literature. Watson described this day as the "glittering, grand finale of his fairy tale."[2]

For nearly a week before the ceremony, Watson, Crick, Wilkins, and their families enjoyed the magical city of Stockholm. Well lodged in the city's Grand Hotel, they were feted at party after party. In between events, they walked into a winter wonderland, along streets lined with colorful flags snapping at the whim of the Arctic wind. Christmas lights on the buildings and festooning the public spaces flickered and swayed in the darkness. Surrounding all was the shimmering, icy-cold Lake Mälaren as it lapped onto the fourteen islands composing the city's shoreline.

During his working life, Alfred Nobel earned a fortune manufacturing dynamite and other powerful explosives. His last will and testament bequeathed the accrued interest of his wealth to be "annually distributed in the form of prizes to those who, during the preceding year, shall have conferred the greatest benefit on mankind."[3] Although Nobel did not leave a formal explanation of why he created the awards, many have speculated that his bequest was due to his remorse at having invented a slew of agents used to such lethal effect during the wars of his era.[4] Each year, the awards are presented on the anniversary of Nobel's death on December 10, 1896.

Since 1936, these annual exercises in pomp and circumstance have taken place in the Stockholm Concert House. The hall is an imposing, rectangular neoclassical structure anchoring the bustling Hötorget, or Haymarket. Home of the Royal Stockholm Symphony Orchestra, the Concert House's exterior is faced with glazed blue brick, punctuated by ten stately Corinthian columns. After climbing a series of steps and crossing the cavernous marble lobby, the audience members filed into the auditorium and took their assigned seats, which were lushly upholstered in red velvet with gold brocade trim. In Stockholm, a ticket to the Nobel Prize award ceremony is as treasured as a box seat for the World Series in Watson's hometown of Chicago.

Watson brought his father, James, Sr., and his sister, Elizabeth, as his guests. His mother, Margaret Jean, had died in 1957 of congestive heart failure, after a lifetime battle with rheumatic fever. Now a thirty-four-year-old biology professor at Harvard, Watson had hoped to "tactfully" bring along his twenty-year-old former research assistant, a Radcliffe College junior named Pat Collinge whose "fey urchin manners, together with her intense, catlike blue eyes, were likely to have no equal in Stockholm."[5] After proffering his creepy invitation, one that today would surely ring alarms and raise objections, Watson was disappointed to learn that Collinge "now had a Harvard boyfriend of literary

James Watson with his sister, Elizabeth, and his father, James, Sr., arriving in Copenhagen for the Nobel Prize ceremony, 1962.

aspirations, whom I was unlikely to supplant." He did, however, enlist her help to "master the waltz steps that I would need for the customary first dance."[6] During his time on the Harvard faculty, Watson pursued or proposed several romances with female undergraduates. At the age of forty, in 1968, he hung up his spurs and married his former laboratory secretary, a nineteen-year-old Radcliffe coed named Elizabeth Lewis. They remained happily together for the rest of his life.

The DNA party was filled out by Crick with his wife, Odile, their two daughters, twelve-year-old Gabrielle and eight-year-old Jacqueline, and his twenty-two-year-old son from his first marriage, Michael. Maurice Wilkins travelled to Stockholm with his second wife, Patricia, their toddler, Sarah, and their infant son, George.[7]

The award recipients participated in a dress rehearsal that morning, then were instructed to return to the Concert House by 3:45 p.m. At

4:15 p.m., the proud winners and their excited introducers were ushered backstage and lined up by seasoned event planners according to the
order of the presentations.

Promptly at 4:30 p.m., a spotlight followed the symphony's principal conductor, Hans Schmidt-Isserstedt, as he took his position at the
head of the Royal Stockholm Orchestra. He raised his white baton and
led the musicians in playing the rousing Royal Hymn, "Kungssången."
The cue for King Gustav VI Adolf's entrance came when the audience
sang the hymn's first line, "*Ur svenska hjärtans djup en gång*" ("Once
from the depths of Swedish hearts"). When the hymn ended, Schmidt-
Isserstedt cued the trumpeters to blast out a rousing fanfare, which
prompted the 1962 Nobel Prize winners to emerge onto the stage to
take their seats.

The Physiology or Medicine prize went to James Watson, Francis
Crick, and Maurice Wilkins "for their discoveries concerning the molecular structure of nucleic acids and its significance for information transfer in living material."[8] The prize in Chemistry was awarded to Max
Perutz and John Kendrew of the Cavendish Laboratory "for their studies of globular proteins," specifically the structures of hemoglobin and
myoglobin. The winner of the Literature prize was the American novelist
John Steinbeck. His 1939 masterpiece, *The Grapes of Wrath*, still speaks to
the social iniquities that result from unbridled capitalism.[9] Both Watson
and Wilkins were great admirers of Steinbeck's work and were thrilled to
meet him.[10] The Physics laureate, Lev Davidovich Landau of the Soviet
Union, was honored for his "pioneering theories on condensed matter,
especially liquid helium,"[11] but was unable to attend, having been seriously injured in an automobile accident nearly a year before.

After the introductory speeches by distinguished scholars representing the Nobel Prize committees, each laureate walked to a prescribed
spot to receive his award. When reviewing the motion picture footage
taken during the ceremony, one thrills at the formality exhibited by the
DNA troika. As they bowed before him, King Gustav handed Watson,
Crick, and Wilkins their medals and diplomas. Although the film is

The Nobel laureates, 1962. Left to right: Maurice Wilkins, Max Perutz, Francis Crick, John Steinbeck, James Watson, John Kendrew.

silent, one can see the king's mouth moving, no doubt uttering a few well-rehearsed words of congratulation.

The medals featured a bas-relief of Alfred Nobel's profile on one side and, on the other, a representation of "the Genius of Medicine holding an open book in her lap, collecting the water pouring out from a rock in order to quench a sick girl's thirst." Below these figures, the recipient's name is engraved, along with the year of the award. Circling both is a motto adapted from Virgil's *Aeneid*: "*Inventas vitam iuvat excoluisse per artes*" ("It is beneficial to have improved human life through discovered arts"), and the phrase "Reg. Universitas Med. Chir. Carol," referring to the academic body granting the medal, the Karolinska Institutet Medical University.[12] Each 200-gram, 66mm-diameter, 23-carat gold medal (worth about $10,000 in today's market), struck by the Swedish Royal Mint with dies that are then locked away to guard against counterfeiters, came nestled in a red leather box handcrafted by the Anders Erikkson atelier. The medals are so valuable that Nobel Prize winners are given bronze replicas to display without fear that the original—which most winners keep in a bank vault—might be stolen. A handful of winners (including Watson and Crick's heirs) have sold their medals for extraordinary sums at auction.[13]

Calligraphed in bright blue, black, and gold ink, the 1962 Nobel Prize diplomas listed the recipients' names in the order of Watson, Crick, and Wilkins. The borders of their certificates are adorned with moons and stars. On the opposite leaf of the diploma is a drawing of a toga-wearing chap carrying a stalk that looks a lot like a double helix. The sun's light falls on him where he stands framed by cypress trees, olive branches, and a plump bunch of purple grapes.

And then there was the money. The fortune Alfred Nobel left behind in 1896 was 31 million Swedish krona; with the power of compound interest, that sum is today worth well over 1.7 billion krona, or nearly 200 million U.S. dollars. Jim Watson's one-third share of the prize amounted to 85,739 krona, or $16,500, a sum now worth about $107,000.[14] He used the cash as a down payment on "an early nineteenth-century wooden house within walking distance of Harvard Square."[15]

DIRECTLY AFTER THE LAST AWARD was presented, an orderly exodus of honored guests and audience members left the auditorium. Outside waited a fleet of limousines, engines running, to whisk away the honorees to the Stockholm City Hall for the sumptuous Nobel banquet. Built between 1911 and 1923, the municipal building required more than eight million dark red bricks, known as *munktegel*, or "monks' bricks"—the same type used in the construction of many of Sweden's churches and monasteries. Anchored by a 106-meter-high bell tower and topped by the nation's emblem of three crowns, the City Hall stands impressive guard over the eastern tip of Kungsholmen Island.

The 1962 banquet began in the Blue Hall, a massive space designed to represent an open courtyard but was roofed and windowed because of the long Swedish winters. At one end of the room is a grand staircase that reliably imparts a sense of grandeur when the monarchs of Sweden and the Nobel Prize winners walk down its steps. The staircase was constructed to be gradual and easily negotiated without tripping—an especially important safety feature for women wearing gowns with long trains. Awaiting them on the evening of December 10 was an applauding throng of 822 invited guests and 250 University of Stockholm students, the latter having won their tickets in a lottery.

In the 1960s, the banquet was still held in the magnificent Gold Hall on the second floor.[16] Its walls are adorned with 18 million gold mosaic pieces, requiring four kilograms of gold foil fused between two thin layers of hand-blown Venetian Murano glass. Sixty-five long tables were draped in the finest white and gold linen cloths and decorated with sprays of yellow mimosa and red carnations. The table of honor ran the entire length of the Gold Hall and was set for 124 royals, the Nobelists, and their families.

The feast began with a toast by the chairman of the Nobel Foundation, Arne Tiselius, who offered the traditional words of praise for the king and queen. King Gustav then rose to offer his own toast and ask for a minute of silence in memory of Alfred Nobel. The dinner lasted three and a half hours, but the white table candles, which could only be of a specific height before posing the risk of toppling over and setting the tables on fire, needed to be discreetly replaced after two hours. The kitchen, directly below the Gold Room, was connected by two elevators transporting 210 white-gloved waiters wearing cerulean blue tailcoats adorned by gold epaulettes and buttons. So well trained was this corps of waiters that the time that elapsed between the king and the farthest-seated guest being served was a mere three minutes.

The menu, prepared by seventy chefs, cooks, and assistants, featured *Truite de rivière fumée à la Parisienne* (smoked river trout), *Poularde rôtie, sauce madère au foie gras* (roast chicken in madeira sauce with duck liver), *Pommes rissoles* (fried apples), *Salade de saison* (seasonal salad), and a dessert of *Pêche au Grand Marnier* with *Crème Chantilly* (peaches in Grand Marnier liqueur with whipped cream). The sommeliers uncorked nearly a thousand bottles of Château Bellevue, 1955, St. Emilion, Pommery &

Greno Brut Champagne during the meal. With the dessert course, the guests sipped coffee, Liqueur Marie Brizard & Strega, and Courvoisier Cognac. The university students, who ate downstairs in the Blue Hall, were served a simpler menu of open salted salmon sandwiches and moose steak with blackberry jelly. They appeared to enjoy their meal as much as the honored guests, and later serenaded all with a medley of Swedish folk songs.[17]

SOMETIME BETWEEN FINISHING THEIR stewed, spiked peaches and long before audience members reached into their pockets to find their coat-check tickets, the Literature Prize winner made his toast. Steinbeck approached the podium with an unsteady gait, no doubt the result of consuming too many shots of aquavit. He looked positively Mephis-tophelean, with his widow's peak of slicked-back hair dyed jet-black, along with an equally black, manicured mustache and pointy goatee. He stumbled through a few paragraphs before finding his voice, and closed his speech by paraphrasing St. John: "In the end is the Word, and the Word is Man—and the Word is with Men."[18] Forty-five years after the fact, Watson recalled enjoying Steinbeck's speech as "a cry for sanity and reason in a time of great stress and irrationality . . . I think it was better than [William] Faulkner's [who won the 1950 prize], but [Steinbeck] was nervous about how people would react."[19]

Next came Arne Engström, a cell biologist and chairman of the Nobel Prize committee for Physiology or Medicine, who was tasked with intro-ducing Jim Watson's toast. Engström described DNA as "two interwo-ven spiral staircases . . . in which no one can climb . . . [but one that contains] the genetic code . . . translating the A–T–G–C hieroglyphics to the language of protein structures."[20]

Wilkins and Crick had asked Watson to deliver the banquet toast on their behalf, a fitting tribute given that his unquenchable ambition pow-ered the race they won. Watson wore a finely sewn "suit of tails," newly purchased from the Cambridge, Massachusetts, branch of the venerable New Haven haberdasher J. Press.[21] The tailors did their best to contain and flatter his tall, thin physique, marred by an inner tube of fat around his middle. In photographs from the evening, he looks like a pop-eyed, awkward version of Fred Astaire. At thirty-four, there was far less hair emerging willy-nilly from Watson's scalp than when he was at Cambridge a decade earlier. His hairline foreshadowed worse days to come—less receding than marching in rapid retreat.

Watson struggled to speak because of a sore throat that, only a few hours earlier, had required a visit to the Karolinska Institutet. The ear, nose, and throat man saw nothing worrisome—and informed Watson that he was on the Nobel Prize committee and had voted for him. During his toast, poor Watson's nerves were so jangled that he neglected to speak directly into the microphone. Jim Watson was always audible, if not loud, in casual conversation, but at the lecture podium he took on a soft tone that was at times difficult to hear. December 10, 1962 was no different, and the audience had trouble following exactly what he said.[22]

Fortunately, we have the text of Watson's speech, which he hoped to match "the cadence of one of J.F.K.'s better speeches."[23] Holding several sheets of linen paper embossed "Grand Hotel Stockholm" and filled with his tiny, neat script, he began by saying that his was a difficult task, especially in explaining the feelings of Crick and Wilkins. For him, the evening represented "the second most wonderful moment in my life. The first was our discovery of the structure of DNA. At that time we knew that a new world had been opened and that an old world which seemed rather mystical was gone."

He proceeded to describe how they used "the methods of physics and chemistry to understand biology." After reminding the audience that he was the youngest member of the trio, Watson insisted that the discovery "could have only happened with the help of Maurice and Francis." He next thanked two other men who understood that "the techniques of physics and chemistry" could make "a real contribution to biology": his old chief, Sir William Lawrence Bragg, and—oddly, given his briefer-than-cameo appearance in these proceedings—Niels Bohr. "The fact that these great men believed in this approach," he stammered, "made it much easier for us to go forward." He did not acknowledge Max

Delbrück, Salvador Luria, his other colleagues from the Phage Group, Jerry Donohue, John Randall, or Linus Pauling. Most egregiously, he neglected to mention Rosalind Franklin.[24] After he sat down, Crick passed him a note hastily written on the back of his place card: "Much better that I could have done,—F."[25]

The following morning, neither Watson nor Crick acknowledged Rosalind Franklin during their thirty-minute Nobel Prize lectures. Crick had insisted to Watson that they focus on their ongoing and future work rather than delivering a précis of past efforts. Wilkins made no such concession and presented some historical perspective on his current DNA research. He mentioned Franklin only briefly during his lecture: "Rosalind Franklin (who died some years later at the peak of her career) made valuable contributions to the X-ray analysis." In the acknowledgment section of the published version of his lecture, he noted, "my late colleague Rosalind Franklin, who with great ability and experience of X-ray diffraction, so much helped the initial investigations on DNA."[26] For the Franklin family, Wilkins's mealy-mouthed attempt to distance Franklin from the Nobel Prize was upsetting, to say the least. A few years later, John Randall wrote to Raymond Gosling, "I have always felt that Maurice's Nobel lecture did rather less than justice to this setting [the King's College biophysics laboratory] and particularly to the contribution of yourself and Rosalind."[27] When receiving his 1982 Nobel Prize in Chemistry, Franklin's friend and collaborator at Birkbeck College, Aaron Klug, valiantly corrected the historical record by insisting, "Had not her life been cut tragically short, she might well have stood in this place on an earlier occasion."[28]

The festivities ended on Saint Lucia's Day, December 13, 1962. Watson, Crick, Wilkins, Kendrew, Perutz, and John Steinbeck were each awakened that morning "by a girl in a white robe and a crown of flaming candles singing the eponymous Neapolitan hymn that long ago became virtually synonymous with this Swedish winter festival."[29] At the age of seventy-nine, Jim Watson lasciviously warned future prizewinners not to expect "a flirtatious Santa Lucia girl." These lovely young women were accompanied by a bevy of photographers and "the moment her singing stops, she will be off to another laureate's room. Leaving you several hours more of darkness to endure before the winter sun peeks above the horizon."[30]

At nearly every event that week, Watson trolled for young countesses and the daughters of Swedish dignitaries. On the evening of Saint Lucia Day, he and Elizabeth attended the Lucia Ball at the Stockholm

Francis Crick with his wife, Odile, and daughters Gabrielle, age twelve, and Jacqueline, age eight, at the Nobel Prize dinner, 1962.

Medical Association. After a formal dinner of roast reindeer, there was much dancing and flirting. Watson recalled having even more success at a "much smaller private affair" held directly after the ball, where he took the opportunity to "banter with Ellen Huldt, a pretty-dark haired medical student with whom [he] then arranged to have dinner the next night."[31] Horace Judson best described Watson's skirt-chasing: "The pictures of the ball that were seen around the world were those of Crick doing the twist with one of his young daughters, and of Watson in the arms of one of the pretty Swedish princesses in a low-cut gown."[32]

Closing Credits

When the legend becomes fact, print the legend.

JAMES WARNER BELLAH AND WILLIS GOLDBECK,
THE MAN WHO SHOT LIBERTY VALANCE[1]

During more than forty years of research and scholarship, I have visited hundreds of libraries and historical repositories. None of them has proven more difficult to enter than the archives of the Nobel Prizes. The documents they contain are actually kept in three separate archives, situated in different corners of Stockholm: the Physiology or Medicine Prize Committee's papers are filed away at the Nobel Forum of the Karolinska Institutet; the Physics and the Chemistry Prize Committee's materials are deposited at the Center for the History of Science of the Royal Swedish Academy of Sciences; and the Literature Prize records are administered by the Swedish Academy.

These collections are not meant to be places of historical study; rather, they are working archives for the members of the various committees, so they can look up past nominations and reports whenever a nominee's name reappears on their docket. Only nominations older than fifty years are available to be viewed, and access required a more than year-long application process, including the presentation of academic credentials, five letters of reference, and a detailed research plan. As one archivist warned me early in the process, "we receive a large number of applications every year and the Nobel Assembly [can] only approve very few."[2]

Gaining admission to the archives of the Physiology or Medicine Prize Committee, the one I initially most wanted to review, was the most arduous. One roadblock is the legendary secrecy surrounding the discussions leading to an award. A more cogent reason is the small staff working at the vast Nobel Forum. Most of the work that goes into awarding the Nobel Prizes is conducted by unpaid volunteers from each academy. The Physiology or Medicine Prize's secretary general, Thomas Perlmann, is a busy molecular biologist on the Karolinska Institutet staff. He works "on duty" as a volunteer, as do his fifty review committee members. There is but one full time employee, the administrator–archivist, Ann-Mari Dumanski.[3] When I was at the Nobel Forum, taking notes in the same room where the prize winners are debated every year, Ms. Dumanski and I were the only two people in the entire building. Such frugality is by design. According to Alfred Nobel's will, the bulk of the monies accrued from his estate must go to the winners rather than to those charged with administering the prizes.

On April 24, 2019, I received an email from Ms. Dumanski inviting me to visit the archives in June and giving three potential dates. After wrangling permission from my medical school to make a sudden overseas journey (no easy task), I winged my way, not to the Eagle pub, but to Stockholm. This trip, without doubt, represents the closest I will ever get to the Nobel Prizes. Unfortunately, it was not until the afternoon before I left that I remembered that the Physics and Chemistry papers were housed in a different location. So, I stayed up until 3 a.m. to contact Professor Karl Grandin, the chief archivist at the Center for the History of Science of the Royal Swedish Academy of Sciences, for an appointment to search those files, too. Thanks to a lucky break, my email caught him between airplanes, returning from Düsseldorf to Stockholm. Dr. Grandin graciously allowed me to look at the archives later that week.

This "accidental" addition to my research represents the essence of the historian's craft and art. One never knows what one is going to find when looking through an archive, no matter how precise the finding aids may appear. More times than not, I have read one file only to be pointed in the direction of another that contains the materials I need to review the most. On-site conversations with the archivists are also essential, because they know far more about the holdings than their occasional visitors do, and typically give excellent suggestions for further inquiry. As it turns out, Watson, Crick, and Wilkins were twice nominated and reviewed for the Chemistry Prize, in 1960 and 1961, before winning the prize in Physiology or Medicine in 1962. Consequently, there was much gold to be mined at the Royal Swedish Academy of Sciences.

Almost everyone is aware of the Nobel Prize's importance, but few are familiar with its arcane rules and regulations. Each year there are boxes of self-nominations and recommendations from self-appointed nominators that never even cross the desks of the committee members. Typically, nominations from previous Nobel Prize winners and those from men and women the committee has formally requested to submit nominations carry the weight needed to advance a candidate to the next step of the process. Each September, experts in academia, government, literature, and other fields are invited to write nominations.[4]

Another rule is that "in no case may a prize be divided between more than three persons." When prizes are shared in a given year, the money is divided equally among the two or three winners. The most diversity of numbers can be found among the Physiology or Medicine prizes. As of 2020, 111 prizes have gone to 222 laureates since 1901; 39 of them were presented to one laureate, 33 were shared by two individuals, and 39 shared by three.[5]

Perhaps the most misunderstood rule surrounding the Nobel Prize is that it may not be given posthumously. In other words, you must be alive to come and collect the medal in Stockholm the year it is presented. This rule has evolved a bit over the past century or more, but only glacially. Before 1974, the Nobel Prize was awarded posthumously twice, but both winners were nominated while they were still alive: Dag Hammarskjöld (Nobel Peace Prize, 1961) and Erik Axel Karlfeldt (Nobel Prize in Literature, 1931). According to the Nobel Prize by-laws, "from 1974, the Statutes of the Nobel Foundation stipulate that a Prize cannot be awarded posthumously, unless death has occurred after the announcement of the Nobel Prize."[6]

Rosalind Franklin died in 1958 and was never nominated for a Nobel Prize. None of the other DNA contenders were nominated until 1960. Her premature death, in essence, precluded any possibility of her winning the prize, even before the change in the statute in 1974.[7] This sad fact has always been the legitimate and final answer to the query: "Why didn't she share in the 1962 prize?" Yet there was another harsh reality to her exclusion that has haunted me ever since I left the archives of the Royal Swedish Academy of Sciences on a sunny midsummer day in Stockholm.

When reviewing the papers for the Chemistry Prize nominations for 1960, I discovered the letters of many prominent men advocating in Watson and Crick's favor. To keep things on a courteous basis between the Cavendish Laboratory and King's College, Sir William Lawrence Bragg, the grand old man of British science and, by then, director of the august Royal Institution of Great Britain, "put every ounce of [his] weight" to ensure that Maurice Wilkins shared in the award.[8] Several other nominators, however, felt that Wilkins was not Nobel-worthy and were against the prize being split three ways with Watson and Crick.

One of those objectors was Linus Pauling. In March 1960, he offered his opinion of Bragg's nomination of Watson, Crick, and Wilkins for the Chemistry prize. Pauling argued that his colleague Robert Corey should share the prize with Kendrew and Perutz for their work on "the structure of polypeptide chains of proteins." Although, he recognized the importance of Watson and Crick's work, he insisted that "the detailed nature of the structure of DNA is, I think, still uncertain to some extent, however, whereas that of polypeptide chains in proteins is now certain." He was far less kind to Maurice Wilkins. He acknowledged that Wilkins demonstrated "virtuosity in having grown better fibers of DNA than any that had been grown before and in having obtained better X-ray photographs than were available before." Nonetheless, Pauling had serious doubts that this work alone "represents a sufficient contribution to chemistry to permit him to be included among recipients of a Nobel Prize."[9]

The rest of the letters of recommendation pertaining to the discovery of the double helix represent a mélange of opinion. Between 1960 and 1962, there were seven additional Chemistry prize nominations and ten Physiology or Medicine nominations. In total, eleven of them nominated only Watson and Crick; five nominated Watson, Crick, and Wilkins, including the nomination by Bragg; and one other person nominated Crick and Perutz, but neither Watson nor Wilkins.[10]

I experienced great joy at being allowed to thumb through the over-sized, black, leather-bound volumes containing these Nobel Prize nominations written by some of the most accomplished and well-informed scientists of the day. Ending that happiness, like the sudden smack of a car collision, was my realization that not a single letter commented on the work of Rosalind Franklin. To be sure, she was already dead and no longer eligible for an award. Still, there exists a sense of decency one acquires before entering the first grade, which is reinforced again and again through life. If any of these distinguished scientists merely mentioned Franklin's contributions in their nominations, it would have taken nothing away from the accomplishments of Watson, Crick, and Wilkins. It would have been the appropriate, scholarly, and honorable thing to do. Alas, none aimed their pens in her direction.

IN SIR ARTHUR CONAN DOYLE'S 1893 short story "The Gloria Scott," he has his archetypal detective, Sherlock Holmes, introduce the phrase "smoking pistol" into the English language. Now known as "the smoking gun," it refers to an unshakable proof of a crime.[11] While not exactly Holmesian evidence, there is a sheaf of papers in the Chemistry Prize archives that, unexamined, leaves the history of the double helix incomplete. The file in question is the internal report for the Chemistry Prize of 1960, written by Arne Westgren, the Secretary for the Nobel Committees in Physics and Chemistry of the Royal Swedish Academy of Sciences, a professor of chemistry and a pioneer in "the application of X-ray diffraction methods in physical metallurgy." He was also well-versed in X-ray crystallographic studies of biological macromolecules, such as DNA and proteins.[12] His fourteen-page report offers a crisp analysis of who deserved the most credit for finding DNA's structure and who among them was most worthy of the Nobel Prize.

Westgren's scientific brief acknowledges that "there is a certain diffi-

culty in evaluating the question of a prize for elucidating DNA's structure because of the many contributors. It is problematic to decide who
among the many scientists involved have participated in such a decisive
way in the developments that they, in particular, deserve to be recognized." Westgren conceded that Watson and Crick put forth an "ingenious" hypothesis of great importance, but he worried that "they have
hardly performed any experimental investigations in the field of their
own—and the testing [of their hypothesis] has been completely allocated
to others." Westgren went on to insist that experimental data trumped
theory and modeling (the italics below are added):

> Those that deserve most credit in this context are, on the one hand,
> Wilkins and his large research group, within which he without
> doubt has a leading role, and on the other hand, Franklin and
> Gosling. A reward for Watson and Crick passing by the researchers
> who experimentally have confirmed their proposal for a structure
> would not be worthy of consideration. Among the latter, Wilkins
> without doubt is in a class by himself. *Those who come next are
> Rosalind Franklin and Gosling, among whom the first mentioned is
> deceased. If she had survived she could well have had claims to receive
> her part of the prize. Bragg, who closely followed the research in the
> field, did not include Gosling in his proposal and it is therefore likely
> that his contribution has not been of decisive importance in the work
> by the research duo of Franklin and Gosling. There are no reasons to
> question the well-founded opinion of Bragg, that if a prize should be
> considered to award the research discussed, it should be split between
> Watson, Crick and Wilkins . . .* the important identifications of
> structures are without doubt of importance in chemistry. However,
> the major importance of the achievements made is within the field
> of genetics and a prize in physiology or medicine therefore seems
> to be most appealing.[13]

Although Bragg and all the other nominators neglected to note Franklin's work in their proposals, her work was admitted into the Nobel Prize

record—but only because Arne Westgren included her.[14] "If she had sur-
vived she could well have had claims to receive her part of the prize."
Thus spoke the Nobel Prize Committee of 1960. If only their successors'
legendary silence would be broken more definitively.

DURING MY FOUR-YEAR QUEST to understand the nuances of this convoluted tale, it was a series of interviews with James Watson in July 2018 that
took me the farthest distance. Like Haydn's Farewell Symphony, which
ends with each player snuffing out the candle on his music stand and leaving the stage except for two muted violins, all that was left of the triumphant DNA team were hundreds of boxes of archival remains and the still
very much alive Watson.[15] A week before we met, he had completed filming
an episode of *American Masters*, the Public Broadcasting System series on
major contributors to American culture and society. He did not disappoint the film's producers. The documentary, purportedly about his scientific career, quickly devolved into an exposition of his racist views. In the
film, Watson was asked if he had repudiated his infamous 2007 statements
claiming the gene-based intellectual superiority of whites over people of
color. He replied, "No. Not at all. I would like for them to have changed,
that there be new knowledge that says that your nurture is much more
important than nature. But I haven't seen any knowledge. And there's a
difference on the average between blacks and whites on I.Q. tests. I would
say the difference is, it's genetic."[16] After the show aired on January 2, 2019,
his beloved Cold Spring Harbor Laboratory, which he had built up from
a ragtag summer camp for obscure geneticists into a world-class research
facility, rescinded his academic titles and formally cut all ties with him.[17]

In person, Watson was a charming, brilliant, and highly likable man.
During my week with him, he did not refrain from articulating his
repugnant views on Africans, African Americans, Asians, and other ethnic groups, including mine, Eastern European Jews. Having met him
years before, I was prepared for such comments but was far more interested in discussing the three-year period, 1950 to 1953, when he so avidly
chased after DNA.

When we first sat down in this office, he began the conversation by observing that the youthful thoughts and feelings he recorded about Rosalind Franklin—and other women—in *The Double Helix* were similar to those that too many young men had expressed during the 1950s. He simply had the misfortune, he said, of writing down his misogynistic observations in a best-selling book that is still widely read by the scientific community but during a very different era of conduct: "Guilty as charged. So, I was never trying to maintain any gender thing. I didn't think about it. Now, I realize that women are certainly as intelligent as men, but they lacked testosterone . . . That's all."[18]

For a week, we shared several meals together, he invited me to his home for dinner and drinks, and we spent hours discussing molecular biology and the history of science in his oak paneled, cathedral-ceiling office where he sat behind a huge desk, the pope of DNA. Above him hung his impressive, gilt-edged Nobel Prize diploma. Pear-shaped, with layered bulges of fat rippling out along his torso, he appeared to be a geriatric version of the Michelin Man. Each day, he appeared in weird outfits of brightly-colored shorts and expensive dress shirts, unbuttoned at the collar and with open French cuffs. Even at age ninety, Jim Watson still managed to make a memorable impression. I was at times charmed and at other times repelled by the fountain of ideas and views he spouted forth. Yet, unlike so many other subjects I have interviewed, I could not help but come back for more discussions and would gladly do so again.

Toward the end of our first interview I asked, "Given a perfect world and if Rosalind Franklin was alive in 1962, would it not be better for her and Wilkins to share the Chemistry or Physics Prize and you and Crick the prize in Physiology or Medicine?" I was proud of myself for having the courage to pose such an on-target question to James Watson. He had not yet made a formal judgment on this dilemma even though he hinted around its edges in his 2002 memoir, *Genes, Girls and Gamow: After the Double Helix*, when he wrote "that Maurice was included pleased both Francis and me, but we had to wonder how the prize would have been divided if Rosalind Franklin had not died so tragically young."[19] The solution I suggested gave him the cover to bless everyone's work and, hopefully, correct the damaged historical record.

I was completely unprepared for what came next. His eyes bulged, staring directly at me, his mottled skin turned red, and the veins in his sun-spotted pate looked as if they would burst. He slowly rose from his chair and, with one finger pointing directly at me, pronounced from on high, "You don't usually win the Nobel Prize for data you can't interpret."[20] His riposte, harsh as it was, was difficult to argue. Franklin could not and did not make the last two intuitive steps to definitively solve the puzzle: the C2 face centered monoclinic crystal suggesting anti-parallel complementarity and the hydrogen bonding of adenine to thymine and cytosine to guanine to fulfill Chargaff's rules. Watson and Crick did that, even if they were only able to do so using data they appropriated from her.

Later that evening, after dinner with Jim and his charming wife, Elizabeth, I returned to my motel room and rifled through the stack of his books I had brought with me for him to inscribe. Picking up a volume of his collected essays, *A Passion for DNA*, I turned to a short 1981 piece that he had mentioned during our meal together. The essay "Striving for Excellence" described how, when writing, he always sought to create "an idea or book that people I respect will want to read or talk about." A subsequent paragraph struck me as a weird confession that he, consciously or subconsciously, wanted me to read before completing our interviews:

> In the spring of 1962 I had given a public lecture in New York on how the DNA structure was actually solved. It provoked much laughter and I knew I had to put it into writing. Initially, I daydreamed that *The New Yorker* might print it under the rubric "Annals of Crime," because there were those who thought Francis and I had no right to think about other people's data and had in fact stolen the double helix from Maurice Wilkins and Rosalind Franklin.[21]

On the end of the second day, we were both relaxed after lunch and a long conversation about science. In the middle of our discussion, he took a call from a colleague on how best to collaborate with a wily scientist from a nearby, competing institution. "Go over to his office," Watson urged his associate, as he winked at me. "Make him feel important, instead of asking for a command performance here. And then get back to me and I will give him a call." I thought that this was excellent advice and told him so after he hung up.

Sensing his elated mood, I improvised another approach to the Nobel Prize question: "I have been thinking about what you said the other day regarding Rosalind's being unable to interpret her data and it struck me that Maurice Wilkins wasn't able to interpret her data either. He had all of her X-rays duplicated and read them long before showing you Photograph No. 51 but he came up with nothing. After you saw it, you quickly realized DNA was a double helix and you and Crick completed the structure within a few weeks. So, why did Wilkins receive a Nobel Prize?" Watson chuckled at my naiveté and replied, "We *wanted* Maurice to get the Nobel, too, because we all *liked* him and we wanted to be friendly with the King's group."[22] As he smiled in my direction, I sat uncomfortably. I had run straight into the misogynistic rampart of the old boys' network—a set of practices that became far clearer after I reviewed the Nobel Prize nominations in Stockholm a year later. Honoring Wilkins presented none of the difficulties entwined with crediting Franklin because "we all *liked* him."

Soon enough, Jim Watson and I approached our final hour together. Over the period of a week, we discovered we genuinely liked each other and admired the other's work. Watson had my curriculum vitae on his desk during the entire time we were together and he asked me to send him a few of my books to read during his next overseas trip. I knew I had one last chance to get back to the gnawing question of whether he would finally agree that Franklin deserved a posthumous piece of the Nobel Prize.

This time, I began by asking how he came to receive his nickname, "Honest Jim," which was one of the titles he originally proposed for *The Double Helix*. After reading the book multiple times, I already knew the answer and could quote from memory a passage describing an Alpine climb during the summer of 1955. Watson saw a group of climbers "coming down" from a higher path; one of them was Willy Seeds, the physicist at King's College who worked with Wilkins on the optical properties of DNA fibers. Seeds saw Watson "and momentarily gave the impression that he might remove his rucksack and chat for a while. But all he said was, 'How's Honest Jim?' and quickly increasing the pace [he] was soon below me on the path."[23]

In the cat-and-mouse game of interviewer and subject, I had much more in mind when asking him this question. I wanted to hear him discuss his self-proclaimed virtue of absolute intellectual honesty. The crafty old man thought for a bit before replying, "Willy Seeds was a physicist at King's College. He was a pretty cynical person. He was the one who first called me 'Honest Jim' because I always said exactly what was on my mind." He neglected to add that many believe the moniker was also Seeds's expression of disgust at Watson for poaching the King's College DNA data and, as a result, becoming a world-famous Nobel laureate.[24]

Watson startled me for the second time in a few days by puffing out his geriatric chest as fully as he could and straightening up his tired, bent body to almost—but not quite—reach his full height of six feet one inches. He proudly declared, loud enough for his secretary to hear in the next room, "And I do! I still do! I say exactly what I think, no matter what!" I could see that more was coming and knew well enough to keep my mouth shut. He paused for a bit—maybe two or three seconds—but to me it seemed far longer. He pursed his lips, bulged his eyes, and finally, slowly, said, "You know? I was—when I was visiting King's College that afternoon and looked at the photograph [No. 51]—I was not honest." As he slowly resumed his seat, it was my pulse that raced as I prepared to hear the key that would unlock the story of DNA once and for all. Was it possible that he was about to confess to adopting the name "Honest Jim" as a subtle form of self-recrimination? Would he finally give Franklin her due? And then the words tumbled out of James Watson's mouth: "I think I was honest. Perhaps that's the wrong word. I think I was honest but . . . you wouldn't say I was exactly honorable." I repeated his response, just as I was taught to do in medical school when interviewing a difficult patient. "So, you were not acting in an honorable manner when you looked at Photograph No. 51?" I asked. And then he swerved sharply away from Franklin and her lovely picture, like an X-ray bouncing off an atom, veering, instead, toward Wilkins, the man everybody liked and everyone could beat: "Well, in the sense that even though Wilkins said we could work on DNA, I was following him and not trying to beat him. But—once I did see it, it was so obvious and clear that I *had* to run with it."

HE *HAD* "TO RUN WITH IT." He *had* to meet his scientific destiny of solving the riddle of DNA first. He *had* to become James Watson, the Nobel laureate who changed how the world understood life itself, no matter what the cost to himself or to others. As he accommodated the legend he wanted to become, he *had* to obscure Rosalind Franklin's role in the landmark discovery.

Legends die hard. But as Franklin once admonished Crick, "facts are facts." During their long, lionized lives, James Watson, Francis Crick, and Maurice Wilkins enjoyed many temporal victories from the legends they spun. But Rosalind Franklin—a woman blessed with exquisite experimental skills and an indomitable resolve to find the facts underpinning DNA's structure—won a victory for the ages.

Acknowledgments

I began thinking about DNA in the spring of 2016, after a group of eager medical students at the University of Michigan asked me to design a course on "Great Papers in the History of Medicine and Science." Each month for a full two terms, we met, lunched, and read an important paper that changed medical practice or scientific knowledge. The course started by discussing Watson and Crick's brief yet extraordinarily powerful paper, "A Structure for Deoxyribose Nucleic Acid," from the April 25, 1953, issue of *Nature*. The students' excitement that afternoon inspired me to uncover the story behind this seminal study. The real work of plotting out this book commenced in October 2017, during a wonderful month of scholarship and collegiality at the Rockefeller Foundation's Bellagio Center in Bellagio, Italy, and, over the next two years, on subsequent research trips to Cambridge, London, Cold Spring Harbor, Philadelphia, Naples, Baltimore, New York, and Stockholm.

At Cold Spring Harbor Laboratory, I am greatly indebted to James D. Watson, who patiently tolerated my questions during a series of oral history interviews in late July 2018. I am also grateful to Elizabeth Watson, Ludmilla Polluck, Peter Tarr, Jan Witkowski, Alexander Gann, Bruce Stillman, Stephanie Satalino, and Maureen Berejka, who all helped to make my stay at Cold Spring Harbor so pleasant and so productive.

At Cambridge University, I had the pleasure of reviewing the Rosalind Franklin papers as well as those of John Randall, J. D. Bernal, Aaron Klug, and Max Perutz. This work was aided by the wonderful team of Allen Packwood, Julia Schmidt, and Natasha Swainston at the Churchill Archives Centre, Churchill College, University of Cambridge. I also thank Frank Bowles, of the Cambridge University Library Archives; and Jude Brimmer, of Clare College Archives, University of Cambridge, who helped me locate the rooms where Watson lived in 1952.

Professor Malcolm Longair, of the Cavendish Physics Laboratory, University of Cambridge, generously showed me around the Austin Wing only a few weeks before it was ripped down so that I could explore Room 103, where Watson and Crick once worked. I am especially grateful to Jenifer and Ian Glynn and Adrian Poole of Trinity College, University of Cambridge. Jenifer is Rosalind Franklin's younger sister (by nine years), an accomplished historian in her own right, and the author of a superb memoir of Rosalind. Her recollections of her sister's life were invaluable in helping me to depict the character of a remarkable woman.

Jeff Karr and Lindsey Loeper, at Special Collections, Albin O. Kuhn Library, University of Maryland at Baltimore County went above and beyond the call of duty in gaining me access to the Anne Sayre papers, a rich collection of interviews and letters that were essential to writing this book.

Charles Greifenstein, David Gary, Tracey de Jong, and Michael Miller at the American Philosophical Society, Philadelphia, were instrumental in my work on the Horace Judson papers and the Erwin Chargaff papers.

Geoff Browell, Katrina DiMuro, Diana Manipud, Kate O'Brien, Frances Pattman, and Cathy Williams, at King's College Archives, University of London, helped me to navigate the Maurice Wilkins papers.

I am also grateful to Charlotte New, at the Royal Institution of Great Britain, who helped me as I reviewed the William Lawrence Bragg papers; Sarah Hall and Emma Illingworth, at Birkbeck College, University of London, who located materials from Rosalind Franklin's tenure there; Chris Petersen at the Special Collections and Archives Research Center of the Oregon State University Libraries, which holds the Linus and Ava Helen Pauling Papers and Linus Pauling digital collections; Daniel DeMellier at the Louis Pasteur Institute Archives, Paris, for the

Nobel Prize nomination written by François Jacob; Anna Petre, at Special Collections, Weston Library, Bodleian Libraries, Oxford University, which holds the John Kendrew papers; Timothy Horning, at Van Pelt Library, University of Pennsylvania Archives, Philadelphia, which holds the Jerry Donohue papers; Peter Collopy and Loma Karklins at the California Institute of Technology Archives; and Claudia di Somma and Christiane Groeben at the Stazione Zoologica Anton Dohrn Archives in Naples, Italy, who so generously helped me document a critical and rarely discussed episode in the DNA story—when Watson first heard Wilkins discuss using X-ray crystallography to determine the structure of DNA.

In Stockholm, I am indebted to Ann-Mari Dumaski at the Nobel Forum, Karolinska Institutet, Stockholm; Professors Karl Grandin and Erling Norrby of the Center for the History of Science, Royal Swedish Academy of Sciences, Stockholm; and Madeline Engström Broberg at the Nobel Library of the Swedish Academy, Stockholm.

I also thank the staffs of the Wellcome History of Medicine Library, London, which has electronically digitized the papers of Francis Crick, James Watson, Rosalind Franklin, and Maurice Wilkins; the University of Michigan Libraries, Ann Arbor; the New York Public Library, New York City; the National Library of Medicine, Bethesda; and the scores of historians who have studied and analyzed the history of molecular biology, and whose work is cited in the notes of this book.

In Ann Arbor, beginning with the record-breaking sub-zero temperatures of late January 2018 and closing at the very end of the plague year of 2020, I composed many drafts of the book you have just read. Several colleagues generously read various versions and, in so doing, saved me from a committing a long list of howling errors. What mistakes remain are my own and for those I apologize. At my home institution, the University of Michigan, I thank my supportive and inspiring colleagues Michael Schoenfeldt, J. Alexander Navarro, Heidi Mueller, Leslie Atzmon, David Bloom, Francis Blouin, David Ginsberg, Michael Imperiale, Arthur Vander, and Thomas Gelehrter. I was also fortunate to rely on the wisdom of David Oshinsky at New York University, who invited me to give several seminars on DNA to his medical students; and to Harvey Fineberg of the Betty and Gordon Moore Foundation, the biographer Eric Lax, and Bruce Alberts of the University of California, San Francisco.

At Michigan, my work has benefitted significantly from the generous gifts endowed by the late Dr. George E. Wantz. Much of the research for this book was supported by the George E. Wantz, MD Distinguished Professorship, and the George E. Wantz History of Medicine Research Fund. George, a brilliant surgeon and historian of medicine who died in 2000, would have enjoyed reading this book.

My literary agents, Glen Hartley and Lynn Chu of Writers Representatives, remain the best advocates and critical readers an author could hope for. We have worked together for over two decades and I remain ever grateful to them.

At W. W. Norton and Company, I was blessed by the wisdoms and superb editorial skills of John Glusman, vice president and editor-in-chief, as well as the professionalism of his colleagues, assistant editor Helen Thomaides, proofreader Mary Kanable, and project editor Dassi Zeidel.

I have many family members to thank but most of all I appreciate the incredible support of Drs. Sheldon and Geraldine Markel.

Throughout the writing of this book—particularly when describing the life and career of Rosalind Franklin—I thought a great deal about my two daughters, Samantha, age sixteen, and Bess, nearly twenty-one, Markel. I hope this volume inspires them to take bold, brave paths in their lives, no matter what obstacles they encounter on their way.

Howard Markel
Ann Arbor, Michigan
December 31, 2020

Notes

The following abbreviations are used for references to archival collections:

JDWP   James D. Watson Collection, Cold Spring Harbor Laboratory Archives, Cold Spring Harbor, NY. All citations quoted with permission of James D. Watson.

WFAT   Watson Family Asset Trust, Cold Spring Harbor Laboratory Archives, Cold Spring Harbor, NY. All citations quoted with permission of James D. Watson.

FCP   Francis Crick Papers, Wellcome Library, London.

RFP   Rosalind Franklin Papers, Churchill College Archives Centre, University of Cambridge.

LAHPP Linus and Ava Helen Pauling Papers, Oregon State University, Corvallis, OR.

MDP   Max Delbrück Papers, Archives and Special Collections, California Institute of Technology, Pasadena, CA.

MWP   Maurice Hugh Frederick Wilkins Collection, King's College, London.

JRP   Sir John Randall Papers, Churchill College Archives Centre, University of Cambridge.

WLBP   William Lawrence Bragg Papers, Archives of the Royal Institution, London.

AKP   Aaron Klug Papers, Churchill College Archives Centre, University of Cambridge.

ECP   Erwin Chargaff Papers, American Philosophical Society, Philadelphia, PA.

HFJP   Horace Freeland Judson Papers, American Philosophical Society, Philadelphia, PA.

ASP   Anne Sayre Papers, American Society of Microbiology Collection, University of Maryland at Baltimore County.

MPP   Max Ferdinand Perutz Papers, Churchill College Archives Centre, University of Cambridge.

Part I: Prologue

1. Voltaire, *Jeannot et Colin* (1764), in *Œuvres complètes de Voltaire* (Paris: Garnier, 1877), vol. 21, 235–42, quote is on 237.

2. Foreign Affairs, House of Commons Debate, 23 January 1948, vol. 446, 529–622, https://api.parliament.uk/historic-hansard/commons/1948/jan/23/foreign-affairs #S5CV0446P0_19480123_HOC_45.

Chapter 1: Opening Credits

1. Francis Crick, *What Mad Pursuit: A Personal View of Scientific Discovery* (New York: Basic Books, 1988), 35, 62.

2. James D. Watson, *The Double Helix: A Personal Account of the Discovery of the Structure of DNA* (New York: Atheneum, 1968); for the remainder of the notes, I use the Norton Critical Edition, edited by Gunther Stent (New York: Norton, 1980), this first quote is from p. 9.

3. Monthly Weather Report of the Meteorological Office, Summary of Observations Compiled from Returns of Official Stations and Volunteer Observers, 1953; 70:2 (London: Her Majesty's Stationery Office, 1953).

4. Watson, *The Double Helix*, 115.

5. "Of that I have no recollection," Crick said many times in his long life. See Francis Crick, "How to Live with a Golden Helix," *The Sciences* 19 (September 1979): 6–9.

6. The historian Robert Olby has argued that the revelation of DNA's structure on April 28, 1953, was only quietly received in the press; Robert Olby, "Quiet Debut for the Double Helix," *Nature* 421 (2003): 402–5. Yves Gingras, on the other hand, has argued against this "quiet debut" and documents, using bibliometric data and citation analysis, the immediate and long-term impact of the announcement; Yves Gingras, "Revisiting the 'Quiet Debut' of the Double Helix: SA Bibliometric and Methodological Note on the 'Impact' of Scientific Publications," *Journal of the History of Biology* 43, no. 1 (2010): 159–81.

7. George Johnson, "Murray Gell-Mann, Who Peered at Particles and Saw the Universe, Dies at 89," *New York Times*, May 25, 2019, B12.

8. Daniel J. Kevles, *The Physicists: The History of a Scientific Community in Modern America* (New York: Knopf, 1978); Richard Rhodes, *The Making of the Atomic Bomb* (New York: Simon and Schuster, 1986), 113–17, 127–29, 131–33.

9. Abraham Pais, *Niels Bohr's Times in Physics, Philosophy, and Polity* (Oxford: Clarendon Press, 1991), 176–210, 267–94; John Gribbin, *Erwin Schrödinger and the Quantum Revolution* (Hoboken, NJ: John Wiley and Sons, 2013); George Gamow, *Thirty Years That Shook Physics: The Story of Quantum Theory* (New York: Dover, 1966).

10. Rhodes, *The Making of the Atomic Bomb*; Andrew Hodges, *Alan Turing: The Enigma* (Princeton: Princeton University Press, 2014); Kai Bird and Martin J. Sherwin, *American Prometheus: The Triumph and Tragedy of J. Robert Oppenheimer* (New York: Knopf, 2005).

11. Author interview with James D. Watson (no. 4), July 26, 2018.

12. John Gribbin, *In Search of Schrödinger's Cat: Quantum Physics and Reality* (New York: Bantam, 1984).

13. Schrödinger shared the Nobel Prize in Physics with Paul A. M. Dirac in 1933. See "The Nobel Prize in Physics 1933," https://www.nobelprize.org/prizes/physics/1933/summary/.

14. Erwin Schrödinger, *What Is Life? The Physical Aspect of the Living Cell, with Mind and Matter and Autobiographical Sketches* (Cambridge: Cambridge University Press, 1992).

15. N. W. Timofeeff-Ressovsky, K. G. Zimmer, and M. Delbrück, "Uber die Natur der Genmutation und der Genstruktur: Nachrichten von der Gessellschaft der Wissenschaften zu Gottingen" (On the Nature of Gene Mutation and Structure), *Biologie, Neue Folge* 1, no. 13 (1935): 189–245. Among the scientists who disagreed with Delbrück's and, by extension, Schrödinger's concept of the "aperiodic crystal" were Linus Pauling and Max Perutz. See Linus Pauling, "Schrödinger's Contribution to Chemistry and Biology," and Max Perutz, "Erwin Schrödinger's *What Is Life?* and Molecular Biology," in C. W. Kilmister, ed., *Schrödinger: Centenary Celebration of a Polymath* (Cambridge: Cambridge University Press, 1987), 225–33 and 234–51.

16. J. T. Randall, "An Experiment in Biophysics," *Proceedings of the Royal Society of London, Series A, Mathematical and Physical Sciences* 208, no. 1092 (1951): 1–24; Horace Freeland Judson, *The Eighth Day of Creation: Makers of the Revolution in Biology* (Cold Spring Harbor, NY: Cold Spring Harbor Laboratory Press, 2013), 77; Robert Olby, *The Path to the Double Helix* (Seattle: University of Washington Press, 1974), 326–33.

17. Lily E. Kay, *The Molecular Vision of Life: Caltech, the Rockefeller Foundation, and the Rise of the New Biology* (New York: Oxford University Press, 1993); Robert E. Kohler, *Partners in Science: Foundations and Natural Scientists, 1900–1945* (Chicago: University of Chicago Press, 1991).

18. Watson, *The Double Helix*.

19. Matthew Cobb, "Happy 100th Birthday, Francis Crick (1916–2004)," *Why Evolution Is True* blog, https://whyevolutionistrue.wordpress.com/2016/06/08/happy-100th-birthday-francis-crick-1916-2004/.

20. Author interview with James D. Watson (no. 1), July 23, 2018.

21. Vilayanur S. Ramachandran, "The Astonishing Francis Crick," *Perception* 33 (2004): 1151–54; Rupert Shortt, "Idle Components: An Argument Against Richard Dawkins," *Times Literary Supplement*, no. 6089 (December 13, 2019): 12–13.

22. Howard Markel, "Who's On First?: Medical Discoveries and Scientific Priority," *New England Journal of Medicine* 351 (2004): 2792–94.

Chapter 2: The Monk and the Biochemist

1. Charles Darwin, *On the Origin of Species by Means of Natural Selection, or the Preservation of Favoured Races in the Struggle for Life* (London: John Murray, 1859), 13.

2. There may have been two gardens: the smaller one described above and another one on the southern side of the courtyard gate near the service entrance. Robin Marantz Henig, *The Monk in the Garden: The Lost and Found Genius of Gregor Mendel, the Father of Modern Genetics* (Boston: Houghton Mifflin, 2009), 21–36.

3. The Punnett square was developed by the British geneticist Reginald C. Punnett. It is a square diagram used to predict the genotypes of a cross-breeding experiment. F. A. E. Crew, "Reginald Crundall Punnett 1875–1967," *Biographical Memoirs of Fellows of the Royal Society* 13 (1967): 309–26.

4. Curriculum vitae, Gregor Mendel. Mendel Museum, Masarykova Univerzita, https://mendelmuseum.muni.cz/en/g-j-mendel/zivotopis.

5. Gregor Mendel, "Versuche über Plflanzenhybriden," *Verhandlungen des naturforschenden Vereines in Brünn, Bd. IV für das Jahr 1865, Abhandlungen* (Experiments in Plant Hybridization. Read at the February 8 and March 8, 1865, Meetings of the Brünn Natural History Society) (1866), 3–47; William Bateson and Gregor Mendel, *Mendel's Principles of Heredity: A Defense, with a Translation of Mendel's Original Papers on Hybridisation* (New York: Cambridge University Press, 2009).

6. Charles E. Rosenberg, "The Therapeutic Revolution: Medicine, Meaning, and Social Change in Nineteenth-Century America," in Morris J. Vogel and Charles E. Rosenberg, eds., *The Therapeutic Revolution: Essays in the Social History of American Medicine* (Philadelphia: University of Pennsylvania Press, 1979), 3–25.

7. Gunther S. Stent, "Prematurity and Uniqueness in Scientific Discovery," *Scientific American* 227, no. 6 (1972): 84–93.

8. Even this finding has been contested; some historians claim von Schermak did not fully understand Mendel's work and Spillman is often left off even the "parentheses" list. See Augustine Brannigan, "The Reification of Mendel," *Social Studies of Science* 9, no. 4 (1979): 423-54; Malcolm Kottler, "Hugo De Vries and the Rediscovery of Mendel's Laws," *Annals of Science* 36 (1979): 517–38; Randy Moore, "The Re-Discovery of Mendel's Work," *Bioscene* 27, no. 2 (2001): 13–24.

9. R. A. Fisher, "Has Mendel's Work Been Rediscovered?," *Annals of Science* 1 (1936): 115–37; Bob Montgomerie and Tim Birkhead, "A Beginner's Guide to Scientific Misconduct," *ISBE Newsletter* 17, no. 1 (2005): 16–21; Daniel L. Hartl and Daniel J. Fairbanks, "Mud Sticks: On the Alleged Falsification of Mendel's

Data," *Genetics* 175 (2007): 975–79; Allan Franklin, A. W. F. Edwards, Daniel J. Fairbanks, Daniel L. Hartl, and Teddy Seidenfeld, eds., *Ending the Mendel–Fisher Controversy* (Pittsburgh: University of Pittsburgh Press, 2008); Gregory Radick, "Beyond the 'Mendel–Fisher Controversy,'" *Science* 350, no. 6257 (2015): 159–60.

10. "Wilhelm His, Sr. (1831–1904), Embryologist and Anatomist," editorial, *Journal of the American Medical Association* 187, no. 1 (January 4, 1964): 58; Elan D. Louis and Christian Stapf, "Unraveling the Neuron Jungle: The 1879–1886 Publications by Wilhelm His on the Embryological Development of the Human Brain," *Archives of Neurology* 58, no. 11 (2001): 1932–35.

11. The weave structure of gauze includes pairs of weft yarns that are crossed before and after each warp yarn, to keep the weft in place. Interestingly, this arrangement looks a great deal like the double helix of DNA. A. Klose, "Victor von Bruns und die sterile Verbandswatte," ("Victor Bruns and the Sterile Cotton Wool"), *Ausstellungskatalog des Stadtsmuseums Tübinger Katalogue* 77 (2007): 36–46; D. J. Haubens, Victor von Bruns (1812–1883) and his contributions to plastic and reconstructive surgery," *Plastic and Reconstructive Surgery* 75, no. 1 (January 1985): 120–27.

12. Ralf Dahm, "Discovering DNA: Friedrich Miescher and the Early Years of Nucleic Acid Research," *Human Genetics* 122 (2008): 565–81; Ralf Dahm, "Friedrich Miescher and the Discovery of DNA," *Developmental Biology* 278, no. 2 (2005): 274–88; Ralf Dahm, "The Molecule from the Castle Kitchen," *Max Planck Research*, 2004, 50–55; Ulf Lagerkvist, *DNA Pioneers and Their Legacy* (New Haven: Yale University Press, 1998), 35–67.

13. Horace W. Davenport, "Physiology, 1850–1923: The View from Michigan," *Physiologist* 25, suppl. 1 (1982): 1–100.

14. Friedrich Miescher, "Ueber die chemische Zusammensetzung der Eiterzellen" (On the Chemical Composition of Pus Cells), *Medicinisch-chemische Untersuchungen* 4 (1871): 441–60; Felix Hoppe-Seyler, "Ueber die chemische Zusammensetzung des Eiter" (On the Chemical Composition of Pus), *Medicinisch-chemische Untersuchungen* 4 (1871): 486–501.

15. S. B. Weineck, D. Koelblinger, and T. Kiesslich, "Medizinische Habilitation im deutschsprachigen Raum: Quantitative Untersuchung zu Inhalt und Ausgestaltung der Habilitationsrichtlinien" (Medical Habilitation in German-Speaking Countries: Quantitative Assessment of Content and Elaboration of Habilitation Guidelines), *Der Chirurg* 86, no. 4 (April 2015): 355–65; Theodor Billroth, *The Medical Sciences in the German Universities: A Study in the History of Civilization* (New York: Macmillan, 1924).

16. Freidrich Miescher, "Die Spermatozoen einiger Wirbeltiere: Ein Beitrag zur Histochemie" (The Spermatazoa of Some Vertebrates: A Contribution to Histochemistry), *Verhandlungen der naturforschenden Gesellschaft in Basel* 6 (1874):

138–208; Dahm, "Discovering DNA"; Ulf Lagerkvist, *DNA Pioneers and Their Legacy* (New Haven: Yale University Press, 1998), 35–67.

17. Dahm, "Discovering DNA," 574.

Chapter 3: Before the Double Helix

1. Adolf Hitler, *Mein Kampf,* (My Struggle) translated by James Murphey (Munich: ZentralVerlag der NSDAP, Franz Eher Nachfolger, 1940), 149.

2. Much of the historical description on eugenics is drawn from one of my earlier books: Howard Markel, *The Kelloggs: The Battling Brothers of Battle Creek* (New York: Pantheon, 2017), 298–321.

3. Galton also coined the term "nurture vs. nature." He and Charles Darwin were both grandsons of the same Birmingham physician, Erasmus Darwin. See Francis Galton, *Inquiries into Human Faculty and its Development* (London: Macmillan, 1883), 17, 24–25, 44; Francis Galton, *Hereditary Genius: An Inquiry into its Laws and Consequences* (London: Macmillan, 1869); Francis Galton, "On Men of Science: Their Nature and Their Nurture," *Proceedings of the Royal Institution of Great Britain* 7 (1874): 227–36.

4. Howard Markel, *Quarantine: East European Jewish Immigrants and the New York City Epidemics of 1892* (Baltimore: Johns Hopkins University Press, 1997), 179–82; Howard Markel, *When Germs Travel: Six Major Epidemics That Invaded America Since 1900 and the Fears They Unleashed* (New York: Pantheon, 2004), 34–36; Kenneth M. Ludmerer, *Genetics and American Society: A Historical Appraisal* (Baltimore: Johns Hopkins University Press, 1972), 87–119.

5. Public Law 68-139, enacted by the 68th U.S. Congress; John Higham, *Strangers in the Land: Patterns of American Nativism, 1860–1925* (New York: Atheneum, 1963), 152; Barbara M. Solomon, *Ancestors and Immigrants: A Changing New England Tradition* (Cambridge, MA: Harvard University Press, 1956); Markel, *Quarantine,* 1–12, 66–67, 75–98, 133–52, 163–78, 181–85; Markel, *When Germs Travel,* 9–10, 35–36, 56, 87–89, 96–97, 102–3.

6. Charles E. Rosenberg, "Charles Benedict Davenport and the Irony of American Eugenics," in *No Other Gods: On Science and American Social Thought* (Baltimore: Johns Hopkins University, Press, 1976), 89–97; Garland E. Allen, "The Eugenics Record Office at Cold Spring Harbor, 1910–1940: An Essay in Institutional History," *OSIRIS* (second series) 2 (1986): 225–64; Oscar Riddle, "Biographical Memoir of Charles B. Davenport, 1866–1944," *Biographical Memoirs,* vol. 25 (Washington, DC: National Academy of Sciences of the United States of America, 1947).

7. Over the years, James Watson has made many racist (and public) statements that blacks are intrinsically and genetically not as intelligent as whites despite the absence of any scientific evidence. Most recently, these repugnant views were

presented in a PBS *American Masters* episode. See Amy Harmon, "For James Watson, the Price Was Exile," *New York Times*, January 1, 2019, D1; "Decoding Watson," *American Masters*, PBS, January 2, 2019, http://www.pbs.org/wnet/americanmasters/american-masters-decoding-watson-full-film/10923/?button=fullepisode.

8. Rosenberg, *No Other Gods*, 91.

9. Charles B. Davenport, "Report of the Committee on Eugenics," *American Breeders Magazine* 1 (1910): 129.

10. Letter from C. B. Davenport to Madison Grant, April 7, 1922, Charles B. Davenport Papers, American Philosophical Society, Philadelphia, cited in Rosenberg, *No Other Gods*, 95–96.

11. Madison Grant, *The Passing of the Great Race, or The Racial Basis of European History* (New York: Charles Scribner's Sons, 1916); Jacob H. Landman, *Human Sterilization: The History of the Sexual Sterilization Movement* (New York: Macmillan, 1932); Harry H. Laughlin, *Eugenical Sterilization in the United States* (Chicago: Municipal Court of Chicago, 1932); Paul Lombardo, *Three Generations, No Imbeciles: Eugenics, the Supreme Court, and Buck v. Bell* (Baltimore: Johns Hopkins University Press, 2010); Adam Cohen, *Imbeciles: The Supreme Court, American Eugenics and the Sterilization of Carrie Buck* (New York: Penguin, 2016); Daniel Kevles, *In the Name of Eugenics: Genetics and the Uses of Human Heredity* (New York: Knopf, 1985), 96–112. Harder to calculate are the many gay, handicapped, "gypsies," and other "so-called" defectives killed in Hitler's Final Solution. See U.S. Holocaust Museum, "Documenting the Numbers of Victims of the Holocaust and Nazi Persecution," https://encyclopedia.ushmm.org/content/en/article/documenting-numbers-of-victims-of-the-holocaust-and-nazi-persecution.

12. Archibald Garrod, *Garrod's Inborn Factors in Disease: Including an annotated facsimile reprint of The Inborn Factors in Disease* (New York: Oxford University Press, 1989); Thomas Hunt Morgan, "The Theory of the Gene," *American Naturalist* 51 (1917): 513–44; T. H. Morgan, A. H. Sturtevant, H. J. Muller, and C. B. Bridges, *The Mechanism of Mendelian Heredity*, revised ed. (New York: Henry Holt, 1922); T. H. Morgan, "Sex-linked Inheritance in Drosophila," *Science* 32, no. 812 (1910): 120–22; T. H. Morgan and C. B. Bridges, *Sex-linked Inheritance in Drosophila* (Washington, DC: Carnegie Institution of Washington/Press of Gibson Brothers, 1916). For an example of population genetics of this era, see Raymond Pearl, *Modes of Research in Genetics* (New York: Macmillan, 1915).

13. Matt Ridley, *Francis Crick: Discoverer of the Genetic Code* (New York: Harper Perennial, 2006), 33.

14. George W. Corner, *A History of the Rockefeller Institute, 1901–1953: Origins and Growth* (New York: Rockefeller Institute Press, 1964); E. R. Brown, *Rockefeller Medicine Men: Medicine and Capitalism in America* (Berkeley: University of California Press, 1979).

15. Howard Markel, "The Principles and Practice of Medicine: How a Textbook, a Former Baptist Minister, and an Oil Tycoon Shaped the Modern American Medical and Public Health Industrial–Research Complex," *Journal of the American Medical Association* 299, no. 10 (2008): 1199–201; Ron Chernow, *Titan: The Life of John D. Rockefeller* (New York: Random House, 1998), 470–79.

16. René Dubos, *The Professor, the Institute and DNA* (New York: Rockefeller University Press, 1976), 10, 161–79.

17. Robert D. Grove and Alice M. Hetzel, *Vital Statistics in the United States, 1940–1960*, U.S. Department of Health, Education and Welfare, Public Health Service, National Center for Health Statistics (Washington, DC: Government Printing Office, 1968), 92.

18. Frederick Griffith, "The Significance of *Pneumococcal* Types," *Journal of Hygiene* 27, no. 2 (1928): 113–59.

19. M. H. Dawson, "The transformation of *pneumococcal* types. I. The Conversion of R forms of *Pneumococcus* into S forms of the homologous type," *Journal of Experimental Medicine* 51, no. 1 (1930): 99–122; M. H. Dawson, "The Transformation of *Pneumococcal* Types. II. The interconvertibility of type-specific S *pneumococci*," *Journal of Experimental Medicine* 51, no. 1 (1930): 123–47; M. H. Dawson and R. H. Sia, "*In vitro* transformation of *Pneumococcal* types. I. A technique for inducing transformation of *Pneumococcal* types *in vitro*," *Journal of Experimental Medicine* 54, no. 5 (1931): 681–99; M. H. Dawson and R. H. Sia, "*In vitro* transformation of *Pneumococcal* types. II. The nature of the factor responsible for the transformation of *Pneumococcal* types," *Journal of Experimental Medicine* 54, no. 5 (1931): 701–10; J. L. Alloway, "The transformation *in vitro* of R *Pneumococci* into S forms of different specific types by the use of filtered *Pneumococcus* extracts," *Journal of Experimental Medicine* 55 No. 1 (1932): 91–99; J. L. Alloway, "Further observations on the use of *Pneumococcus* extracts in effecting transformation of type *in vitro*," *Journal of Experimental Medicine* 57, no. 2 (1933): 265–78.

20. Avery was nominated thirteen times, in 1932, 1933, 1934, 1935, 1936, 1937, 1938, 1939, 1942, 1945, 1946, 1947, and 1948, to no avail. See "List of Individuals Proposing Oswald Avery and others for the Nobel Prize (1932–1948)," Oswald Avery Collection, Profiles in Science, U.S. National Library of Medicine, https://profiles.nlm.nih.gov/ps/access/CCAAFV.pdf#xml=https://profiles.nlm.nih.gov:443/pdfhighlight?uid=CCAAFV&query=%28Nobel%2C%20Avery%29.

21. Dubos, *The Professor, the Institute and DNA*, 139.

22. Dubos, *The Professor, the Institute and DNA*, 66; Matthew Cobb, "Oswald Avery, DNA, and the Transformation of Biology," *Current Biology* 24, no. 2 (2014): R55–R60; Maclyn McCarty, *The Transforming Principle: Discovering that Genes Are Made of DNA* (New York: Norton, 1985); Maclyn McCarty, "Discovering Genes are Made of DNA," *Nature* 421 (2003): 406; Horace Freeland Judson, "Reflec-

tions on the Historiography of Molecular Biology," *Minerva* 18, no. 3 (1980): 369–421; Alan Kay, "Oswald T. Avery," in Charles C. Gillespie, ed., *Dictionary of Scientific Biography*, vol. 1 (New York: Scribner's, 1970); Charles L. Vigue, "Oswald Avery and DNA," *American Biology Teacher* 46, no. 4 (1984): 207–11; Nicholas Russell, "Oswald Avery and the Origin of Molecular Biology," *British Journal for the History of Science* 21, no. 4 (1988): 393–400; M. F. Perutz, "Co-Chairman's Remarks: Before the Double Helix," *Gene* 135 (1993): 9–13.

23. In the 1950s, the terminology of DNA shifted from *desoxyribonucleic* acid to *deoxyribonucleic* acid. This letter was excerpted by René Dubos in *The Professor, the Institute and DNA*, 217–20, quote is on 218–19. The original fourteen-page letter from Oswald Avery to Roy Avery, dated May 26, 1943, can be found in the Oswald Avery Papers, Tennessee State Library and Archives, Nashville, and online at Oswald Avery Collection, Profiles in Science, U.S. National Library of Medicine, https://profiles.nlm.nih.gov/ps/retrieve/ResourceMetadata/CCBDBF.

24. O. T. Avery, C. M. Macleod, and M. McCarty, "Studies on the chemical nature of the substance inducing transformation of pneumococcal types: Induction of transformation by a desoxyribonucleic acid fraction isolated from *pneumococcus* Type II," *Journal of Experimental Medicine* 79, no. 2 (1944): 137–58.

25. M. McCarty and O. T. Avery, "Studies on the chemical nature of the substance inducing transformation of pneumococcal types. II. Effect of desoxyribosenucleic on the biological activity of the transforming substance," *Journal of Experimental Medicine* 83, no. 2 (1946): 89–96; M. McCarty and O. T. Avery, "Studies on the chemical nature of the substance inducing transformation of pneumococcal types. III. An improved method for the isolation of the transforming substance and its application to *pneumococcus* types II, III, and VI," *Journal of Experimental Medicine* 83, no. 2 (1946): 97–104.

26. Cobb, "Oswald Avery, DNA, and the Transformation of Biology"; "List of Those Attending or Participating in the [Cold Spring Harbor on Heredity and Variation in Microorganisms] Symposium for 1946," Oswald Avery Papers, Tennessee State Public Library and Archives, Nashville.

27. H. V. Wyatt, "When Does Information Become Knowledge?," *Nature* 235 (1972): 86–89; Gunther S. Stent, "Prematurity and Uniqueness in Scientific Discovery," *Scientific American* 227, no. 6 (1972): 84–93.

28. Letter from W. T. Astbury to F. B. Hanson, October 19, 1944, Astbury Papers, University of Leeds Special Collections, Brotherton Library, (MS419, Box E. 152), quoted in Kirsten T. Hall, *The Man in the Monkeynut Coat: William Astbury and the Forgotten Road to the Double Helix* (Oxford: Oxford University Press, 2014); Kirsten T. Hall, "William Astbury and the Biological Significance of Nucleic Acids, 1938–1951," *Studies in History and Philosophy of Biological and Biomedical Sciences* 42 (2011): 119–28.

29. Kalckar insisted that Avery should have won two Nobel Prizes, for discovering that antigens need not be proteins as well as for his pneumococcus work. Horace Judson interview with Herman Kalckar, September 1973, 484, HFJP.

30. Cobb, "Oswald Avery, DNA, and the Transformation of Biology." Quote is cited in Joshua Lederberg Papers, U.S. National Library of Medicine, https://profiles .nlm.nih.gov/ps/retrieve/Narrative/BB/p-nid/30.

31. Joshua Lederberg, "Reply to H. V. Wyatt," *Nature* 239, no. 5369 (1972): 234. Lederberg made these assertions several times in his correspondence; see also letter from Joshua Lederberg to Maurice Wilkins, undated ?1973, inquiring about Wilkins's perceptions of Avery in 1944, Oswald Avery Collection, U.S. National Library of Medicine, https://profiles.nlm.nih.gov/spotlight/cc/catalog/ nlm:nlmuid-101584575X263-doc.

32. Horace Judson interview with Max Delbrück, July 9, 1972, HFJP.

33. Delbrück shared the 1969 Nobel Prize in Physiology or Medicine with Salvador Luria and Alfred D. Hershey for their work on phage genetics. The italicized "stupid" appears in Judson, "Reflections on the Historiography of Molecular Biology," 386. See also Horace Judson interview with Max Delbrück, July 9, 1972, HFJP.

Part II: The Players' Club

1. Oscar Wilde, *De Profundis* (New York: G. P. Putnam's Sons, 1905), 63.

Chapter 4: Take Me to the Cavendish Laboratory

1. James D. Watson, *The Double Helix: A Personal Account of the Discovery of the Structure of DNA*, edited by Gunther Stent (New York: Norton, 1980), 9. Elsewhere, Watson told audiences that he was aiming for "a book as good as *The Great Gatsby*"; James D. Watson, *A Passion for DNA: Genes, Genomes, and Society* (Cold Spring Harbor, NY: Cold Spring Harbor Laboratory Press, 2001), 120.

2. Atheneum was founded by Alfred A. Knopf, Jr., Simon Michael Bessie, and Hiram Haydn in 1959. See Herbert Mitgang, "Atheneum Publishers Celebrates its 25th Year," *New York Times*, December 23, 1984, 36.

3. The controversy over the publication of *The Double Helix*, and its unorthodox cancellation by Harvard University Press thanks to the robust letter-writing campaign orchestrated by Crick and Wilkins, is well documented in the William Lawrence Bragg Papers, RI.MS.WLB 12/3-12/100. Bragg wrote the introduction for the original edition. Sales figures are estimated in Nicholas Wade, "Twists in the Tale of the Great DNA Discovery," *New York Times*, November 13, 2012, D2.

4. Information on Crick's early life is drawn from Francis Crick, *What Mad Pur-*

suit: A Personal View of Scientific Discovery (New York: Basic Books, 1988), 3–80; the quote is on 40. See also Robert Olby, *Francis Crick: Hunter of Life's Secrets* (Cold Spring Harbor, NY: Cold Spring Harbor Laboratory Press, 2009); Matt Ridley, *Francis Crick: Discoverer of the Genetic Code* (New York: Harper Perennial, 2006); Mark S. Bretscher and Graeme Mitchison, "Francis Harry Compton Crick, O.M., 8 June 1916–28 July 2004," *Biographical Memoirs of Fellows of the Royal Society* 63 (2017): 159–96.

5. Horace W. Davenport, "The Apology of a Second-Class Man," *Annual Review of Physiology* 47 (1985): 1–14.

6. Crick, *What Mad Pursuit*, 13.

7. "Of the 236,000 British mines laid in World War II, one-third of them were of the non-contact type, i.e., magnetic or acoustic": Olby, *Francis Crick*, 53–54. See also Science Museum, "Naval Mining and Degaussing: Catalogue of an Exhibition of British and German Material Used in 1939–1954 (London: His Majesty's Stationery Office, 1946), iv; and Crick, *What Mad Pursuit*, 15.

8. Ridley, *Francis Crick*, 13.

9. Olby, *Francis Crick*, 62; Crick, *What Mad Pursuit*, 15.

10. Quote is from Crick, *What Mad Pursuit*, 18. See also Linus Pauling, *The Nature of the Chemical Bond and the Structure of Molecules and Crystals: An Introduction to Modern Structural Chemistry* (Ithaca, NY: Cornell University Press, 1939); Cyril Hinshelwood, *The Chemical Kinetics of the Bacterial Cell* (Oxford: Clarendon Press, 1946); Edgar D. Adrian, *The Mechanism of Nervous Action: Electrical Studies of the Neurone* (Philadelphia: University of Pennsylvania Press, 1932). Hinshelwood won the 1956 Nobel Prize in Physiology or Medicine and Lord Adrian shared the 1932 Nobel Prize in Physiology or Medicine with Charles Sherrington.

11. Ridley, *Francis Crick*, 23.

12. Crick, *What Mad Pursuit*, 15.

13. V. V. Ogryzko, "Erwin Schrödinger, Francis Crick, and epigenetic stability," *Biology Direct* 3 (April 17, 2008): 15, doi:10.1186/1745-6150-3-15.

14. Crick, *What Mad Pursuit*, 19–23; Brenda Maddox, *Rosalind Franklin: The Dark Lady of DNA* (New York: HarperCollins, 2002), 105.

15. Francis Crick's Application for a Studentship for Training in Research Methods, July 7, 1947, Medical Research Council, Francis Crick Personal File, FD21/13, British National Archives; Olby, *Francis Crick*, 69–90; Ridley, *Francis Crick*, 26.

16. H. H. Dale, "Edward Mellanby, 1884–1955," *Biographical Memoirs of Fellows of the Royal Society* 1 (1955): 192–222.

17. Crick, *What Mad Pursuit*, 19.

18. Edward Mellanby, memorandum of a meeting with Francis Crick, July 7, 1947, Medical Research Council, Francis Crick Personal File, FD21/13, British National Archives; Olby, *Francis Crick*, 69.

19. Papers of the Strangeways Laboratory, Cambridge Research Hospital, 1901–

1999, PP/HBF, Honor Fell Papers, Wellcome Library, London; L. A. Hall, "The Strangeways Research Laboratory: Archives in the Contemporary Medical Archives Centre," *Medical History* 40, no. 2 (1996): 231–38.

20. Crick, *What Mad Pursuit*, 22; F. H. C. Crick and A. F. W. Hughes, "The Physical properties of cytoplasm. A Study by means of the magnetic particle method. Part I. Experimental," *Experimental Cell Research* 1 (1950): 3–90; F. H. C. Crick, "The Physical properties of cytoplasm. A Study by means of the magnetic particle method. Part II. Theoretical Treatment," *Experimental Cell Research* 1 (1950): 505–33.

21. Crick, *What Mad Pursuit*, 22.

22. Olby, *Francis Crick*, 147.

23. Francis Crick, "Polypeptides and proteins: X-ray studies," PhD dissertation, Gonville and Caius College, University of Cambridge, submitted on July 1953, FCP, PPCRI/F/2, https://wellcomelibrary.org/item/b18184534.

24. Crick, *What Mad Pursuit*, 40.

25. I am indebted to Professor Malcolm Longair of the University of Cambridge Cavendish Physics Laboratory, who took me through the Austin Wing on February 19, 2018, only weeks before it was torn down. For background on the critically important work done there, see Malcolm Longair, *Maxwell's Enduring Legacy: A Scientific History of the Cavendish Laboratory* (Cambridge: Cambridge University Press, 2016); J. G. Crowther, *The Cavendish Laboratory, 1874–1974* (New York: Science History Publications, 1974); Thomas C. Fitzpatrick, *A History of the Cavendish Laboratory, 1871–1910* (London: Longmans, Green and Co., 1910); Dong-Won Kim, *Leadership and Creativity: A History of the Cavendish Laboratory 1871–1919* (Dordrecht, The Netherlands: Kluwer Academic Publishers, 2002); John Finch, *A Nobel Fellow on Every Floor: A History of the Medical Research Council Laboratory of Molecular Biology* (Cambridge: MRC/LMB, 2008); Egon Larsen, *The Cavendish Laboratory: Nursery of Genius* (London: Franklin Watts, 1952); Alexander Wood, *The Cavendish Laboratory* (Cambridge: Cambridge University Press, 1946); Basil Mahon, *The Man Who Changed Everything: The Life of James Clerk Maxwell* (Chichester, UK: John Wiley and Sons, 2004).

26. Letter from James Clerk Maxwell to L. Campbell, quoted in Lewis Campbell and William Garnet, *The Life of James Clerk Maxwell, with a selection from his correspondence and occasional writings and a sketch of his contributions to science* (London: Macmillan, 1882), 178.

27. Mahon, *The Man Who Changed Everything*.

28. Longair, *Maxwell's Enduring Legacy*, 55–60.

29. "Onward Christian Soldiers," lyrics by Sabine Baring-Gould (1865), music by Arthur Sullivan (1872), in Ivan L. Bennett, ed., *The Hymnal Army and Navy* (Washington, DC: Government Printing Office, 1942), 414.

30. Longair, *Maxwell's Enduring Legacy*, 255–318.

31. William Henry Bragg held several positions, including Cavendish Professor of Physics at the University of Leeds 1909–18 and director of the Royal Institution in London 1923–42. The mineral braggite is named for his father and him. See A. M. Glazer and Patience Thomson, eds., *Crystal Clear: The Autobiographies of Sir Lawrence and Lady Bragg* (Oxford: Oxford University Press, 2015); John Jenkin, *William and Lawrence Bragg, Father and Son: The Most Extraordinary Collaboration in Science* (Oxford: Oxford University Press, 2008); André Authier, *Early Days of X-ray Crystallography* (Oxford: Oxford University Press/International Union of Crystallography Book Series, 2013); Anthony Kelly, "Lawrence Bragg's interest in the deformation of metals and 1950–1953 in the Cavendish—a worm's-eye view," *Acta Crystallographica* A69 (2013): 16–24; Edward Neville Da Costa Andrade and Kathleen Yardley Londsale, "William Henry Bragg, 1862–1942," *Biographical Memoirs of Fellows of the Royal Society* 4 (1943): 276–300; David Chilton Phillips, "William Lawrence Bragg, 31 March 1890–1 July 1971. Elected F.R.S. 1921," *Biographical Memoirs of Fellows of the Royal Society* 25 (1979): 75–142.

32. Chilton Phillips, "William Lawrence Bragg."

33. "Cavendish Laboratory, Cambridge, Benefaction by Sir Herbert Austin, K.B.E.," editorial, *Nature* 137, no. 3471 (May 9, 1936): 765–66; "Cavendish Laboratory: The Austin Wing," editorial, *Nature* 158, no. 4005 (August 3, 1946): 160; W. L. Bragg, "The Austin Wing of the Cavendish Laboratory," *Nature* 158, no. 4010 (September 7, 1946): 326–27. Bragg later solicited many other gifts, including £37,000 for a new cyclotron and another £100,000 for construction of a connector building between the Austin and the original wings.

34. Adam Smith interview with James D. Watson, December 10, 2012, https://old.nobelprize.org/nobel_prizes/medicine/laureates/1962/watson-interview.html.

35. Anne Sayre interview with Francis Crick, June 16, 1970, ASP, box 2, folder 9.

36. Angus Wilson, "Critique of the Prizewinners," typescript for article in *The Queen*, January 2, 1963, FCP, PP/CRI/I/2/4, box 102.

37. Olby, *Francis Crick*, 108–9.

38. Crick, *What Mad Pursuit*, 50.

39. Murray Sayle, "The Race to Find the Secret of Life," *Sunday Times*, May 5, 1968, 49–50. Bragg later denied much of this version of his relationship with Crick and called many of Watson's recollections "pure imagination." See Horace Judson interview with William Lawrence Bragg, January 28, 1971, HFJP.

40. Author interview with James D. Watson (no. 1), July 23, 2018.

Chapter 5: The Third Man

1. The phrase "the third man" comes from Wilkins's memoir, *The Third Man of the Double Helix* (Oxford: Oxford University Press, 2003). *The Third Man* (1949) is

a famous British film noir, directed by Carol Reed, written by Graham Greene, produced by David O. Selznick, and starring Joseph Cotten and Orson Welles. In that film, a mysterious murder is witnessed by three men but no one can recall who the third man was nor where he went. At the end of the movie, the third man is exposed as the villain Harry Lime, played by Orson Welles.

2. Horace Freeland Judson, *The Eighth Day of Creation: Makers of the Revolution in Biology* (Cold Spring Harbor, NY: Cold Spring Harbor Laboratory Press, 2013), 9.

3. Wilkins, *The Third Man of the Double Helix*, 112, 113, 150.

4. Anne Sayre interview with Maurice Wilkins, June 15, 1970, ASP, box 4, folder 32.

5. Steven Rose interview with Maurice Wilkins, "National Life Stories. Leaders of National Life. Professor Maurice Wilkins, FRS," C408/017 (London: British Library, 1990).

6. Anne Sayre interview with Maurice Wilkins, June 15, 1970.

7. Anne Sayre interview with Francis Crick, June 16, 1970, ASP, box 2, folder 9.

8. Wilkins, *The Third Man of the Double Helix*; Struther Arnott, T. W. B. Kibble, and Tim Shallice, "Maurice Hugh Frederick Wilkins, 15 December 1916–5 October 2004; Elected FRS 1959," *Biographical Memoirs of Fellows of the Royal Society* 52 (2006): 455–78; Steven Rose interview with Maurice Wilkins.

9. Wilkins, *The Third Man of the Double Helix*, 6–7.

10. Wilkins, *The Third Man of the Double Helix*, 16–17.

11. Wilkins, *The Third Man of the Double Helix*, 17–18.

12. Wilkins, *The Third Man of the Double Helix*, 19.

13. Edgar H. Wilkins, *Medical Inspection of School Children* (London: Balliere, Tindall and Cox, 1952).

14. Wilkins, *The Third Man of the Double Helix*, 31–32.

15. Eric Hobsbawm, "Bernal at Birkbeck," in Brenda Swann and Francis Aprahamian, eds., *J. D. Bernal: A Life in Science and Politics* (London: Verso, 1999), 235–54; Maurice Goldsmith, *Sage: A Life of J. D. Bernal* (London: Hutchinson, 1980); Andrew Brown, *J. D. Bernal: The Sage of Science* (Oxford: Oxford University Press, 2005).

16. Wilkins, *The Third Man of the Double Helix*, 41.

17. Wilkins, *The Third Man of the Double Helix*, 42.

18. Horace Judson interview with Maurice Wilkins, September 1975, 145, HFJP.

19. Steven Rose interview with Maurice Wilkins, 81.

20. Wilkins, *The Third Man of the Double Helix*, 44.

21. Wilkins, *The Third Man of the Double Helix*, 48.

22. Wilkins, *The Third Man of the Double Helix*, 48.

23. Wilkins, *The Third Man of the Double Helix*, 49.

24. M. H. F. Wilkins, "John Turton Randall, 23 March 1905–16 June 1984, Elected F.R.S. 1946," *Biographical Memoirs of Fellows of the Royal Society* 33 (1987): 493–535.

25. Wilkins, *The Third Man of the Double Helix*, 50, 100.

26. Wilkins, *The Third Man of the Double Helix*, 100.

27. Wilkins, *The Third Man of the Double Helix*, 101.

28. M. H. F. Wilkins, "Phosphorescence Decay Laws and Electronic Processes in Solids," PhD thesis, University of Birmingham, 1940; G. F. G. Garlick and M. H. F. Wilkins, "Short Period Phosphorescence and Electron Traps," *Proceedings of the Royal Society A: Mathematical, Physical and Engineering Sciences* 184, no. 999 (1945): 408–33; J. T. Randall and M. H. F. Wilkins, "Phosphorescence and Electron Traps. I. The Study of Trap Distributions," *Proceedings of the Royal Society A: Mathematical, Physical and Engineering Sciences* 184, no. 999 (1945): 365–89; J. T. Randall and M. H. F. Wilkins, "Phosphorescence and Electron Traps. II. The Interpretation of Long-Period Phosphorescence," *Proceedings of the Royal Society A: Mathematical, Physical and Engineering Sciences* 184, no. 999 (1945): 390–407; J. T. Randall and M. H. F. Wilkins, "The Phosphorescence of Various Solids," *Proceedings of the Royal Society A: Mathematical, Physical and Engineering Sciences* 184, no. 999 (1945): 347–64.

29. Wilkins, *The Third Man of the Double Helix*, 68.

30. Wilkins, *The Third Man of the Double Helix*, 65.

31. Wilkins, *The Third Man of the Double Helix*, 65.

32. Angela Hind, "The Briefcase 'That Changed the World'," *BBC News/Science*, February 5, 2007, http://news.bbc.co.uk/2/hi/science/nature/6331897.stm.

33. Wilkins, *The Third Man of the Double Helix*, 71–72.

34. Steven Rose interview with Maurice Wilkins, 81.

35. Wilkins, *The Third Man of the Double Helix*, 72.

36. "Secret Home Office Warrant from D. L. Stewart," August 7, 1953, MI5 file on M. H. F. Wilkins, allowing Wilkins's mail to be searched at his new address. His phone was also tapped. Reproduced in James D. Watson, *The Annotated and Illustrated Double Helix*, edited by Alexander Gann and Jan Witkowski (New York: Simon and Schuster, 2012), 123.

37. Wilkins, *The Third Man of the Double Helix*, 86.

38. Wilkins, *The Third Man of the Double Helix*, 86.

39. Letter from Maurice Wilkins to John Randall, August 2, 1945, JRP, RNDL File 3/3/4 "One Man's Science."

40. Steven Rose interview with Maurice Wilkins, 95.

41. Wilkins, *The Third Man of the Double Helix*, 84.

42. Steven Rose interview with Maurice Wilkins, 95.

43. Naomi Attar, "Raymond Gosling: The Man Who Crystalized Genes," *Genome Biology* 14 (2013): 402–14, quote is on 403.

44. The term "molecular biology" is said to have been coined in 1938 by Warren Weaver, the director of the Rockefeller Foundation's natural science division. See Warren Weaver, "Molecular Biology: Origins of the Term," *Science* 170 (1970): 591–92.

45. Wilkins, *The Third Man of the Double Helix*, 99.

46. "Engineering, Physics and Biophysics at King's College, London, New Building," editorial, *Nature* 170, no. 4320 (August 16, 1952): 261–63. The plans, acetate slides, papers, and publications for this unit can be found in the King's College, London, Department of Biophysics Records, Archives and Special Collections, KDBP 1/1–10; 2/1–8; 3/1–3; 4/1–71; 5/1–3.

47. "The Strand Quadrangle Redevelopment: History of the Quad," King's College, London, website, https://www.kclac.uk/aboutkings/orgstructure/ps/estates/quad-hub-2/history-of-the-quad.

48. Wilkins, *The Third Man of the Double Helix*, 111–12.

49. Wilkins, *The Third Man of the Double Helix*, 106.

50. Wilkins, *The Third Man of the Double Helix*, 101, 106.

51. Wilkins, *The Third Man of the Double Helix*, 106–7, 135, 142; Brenda Maddox, *Rosalind Franklin: The Dark Lady of DNA* (New York: HarperCollins, 2002), 156; Matthias Meili, "Signer's Gift: Rudolf Signer and DNA," *Chimia* 57, no. 11 (2003): 734–40; Tonja Koeppel interview with Rudolf Signer, September 30, 1986, Beckman Center for the History of Chemistry (Philadelphia: Chemical Heritage Foundation, Oral History Transcript no. 0056); Attar, "Raymond Gosling," 402.

Chapter 6: Like Touching the Fronds of a Sea Anemone

1. Letter from Anne Sayre to Muriel Franklin, February 5, 1970, ASP, box 2, folder 15.1.

2. Letter from James D. Watson to Jenifer Glynn, June 11, 2008. Quoted with permission of Jenifer Glynn.

3. Brenda Maddox, *Rosalind Franklin: The Dark Lady of DNA* (New York: HarperCollins, 2002); Anne Sayre, *Rosalind Franklin and DNA* (New York: Norton, 1975); J. D. Bernal, "Dr. Rosalind E. Franklin," *Nature* 182 (1958): 154; Jenifer Glynn, *My Sister Rosalind Franklin: A Family Memoir* (Oxford: Oxford University Press, 2012); Jenifer Glynn, "Rosalind Franklin, Fifty Years On," *Notes and Records of the Royal Society* 62 (2008): 253–55; Jenifer Glynn, "Rosalind Franklin, 1920–1958," in Edward Shils and Carmen Blacker, eds., *Cambridge Women: Twelve Portraits* (Cambridge: Cambridge University Press, 1996), 267–82; Arthur Ellis Franklin, *Records of the Franklin Family and Collaterals* (London: George Routledge and Sons, 1915, printed for private circulation); Muriel Franklin, "Rosalind," privately printed obituary pamphlet, RFP, "Articles and Obituaries," FRKN 6/6.

4. Author interview with James D. Watson (no. 3), July 25, 2018.

5. Franklin, *Records of the Franklin Family and Collaterals*, 4. The Franklin family banking firm, A. Keyser and Company, specialized in American rail bonds. It pur-

chased the publishing house of George Routledge in 1902 and in 1911 bought the publishers Kegan Paul. These firms employed many Franklin men over the years.

6. "The Golem" tells the story of a rabbi, based on Rabbi Löwe, who builds a man from clay and then cannot control his creation. In some versions, the Golem goes on a murderous rampage. See Friedrich Korn, *Der Jüdische Gil Blas* (Leipzig: Friese, 1834); Gustave Meyrink, *The Golem* (London: Victor Gollancz, 1928); Chayim Bloch, *The Golem: Legends of the Ghetto of Prague* (Vienna: John N. Vernay, 1925); Mary Shelley, *Frankenstein, or The Modern Prometheus* (London: Lackington, Hughes, Harding, Mavor and Jones, 1818).

7. Chaim Bermant, *The Cousinhood: The Anglo-Jewish Gentry* (New York: Macmillan, 1971), 1.

8. (The Right Honorable Viscount) Herbert Samuel, "The Future of Palestine," January 15, 1915, CAB (Cabinet Office Archives), British National Archives, 37/123/43; Bernard Wasserman, *Herbert Samuel: A Political Life* (Oxford: Clarendon Press, 1992).

9. Letter from Muriel Franklin to Anne Sayre, November 23, 1969, ASP, box 2, folder 15.1.

10. The five Franklin children were David, b. 1919; Rosalind, b. 1920; Colin, b. 1923; Roland, b. 1926; and Jenifer, b. 1929. See Helen Franklin Bentwich, *Tidings from Zion: Helen Bentwich's Letters from Jerusalem, 1919–1931* (London: I. B. Tauris and European Jewish Publication Society, 2000), 147; Helen Franklin Bentwich, *If I Forget Thee: Some Chapters of Autobiography, 1912–1920* (London: Elek for the Friends of the Hebrew University of Jerusalem, 1973); Maddox, *Rosalind Franklin*, 15. See also Norman Bentwich, *The Jews in Our Time: The Development of Jewish Life in the Modern World* (London: Penguin, 1960); Norman and Helen Bentwich, *Mandate Memories, 1918–1948: From the Balfour Declaration to the Establishment of Israel* (New York: Schocken, 1965).

11. Letter from Muriel Franklin to Anne Sayre, July 10, 1970, ASP, box 2, folder 15.1.

12. Letter from Colin Franklin to Jenifer Glynn, quoted in Glynn, *My Sister Rosalind Franklin*, 26.

13. Muriel Franklin, "Rosalind," 4.

14. Muriel Franklin, "Rosalind," 3.

15. Sayre, *Rosalind Franklin and DNA*, 39.

16. Maddox, *Rosalind Franklin*, 18.

17. J. F. C. Harrison, *A History of the Working Men's College, 1854–1954* (London: Routledge and Kegan Paul, 1954), 157, 164, 168.

18. Muriel Franklin, *Portrait of Ellis* (London: Willmer Brothers, 1964, printed for private circulation); Maddox, *Rosalind Franklin*, 5.

19. George Orwell, "Anti-Semitism in Britain," *Contemporary Jewish Record*, April 1945, reprinted in George Orwell, *Essays* (New York: Everyman's Library/Knopf, 2002), 847–56.

20. St. Paul's School was administered by the Worshipful Company of Mercers, a livery guild of the City of London. It was a trade association for general merchants, and especially for exporters of wool and importers of velvet, silk, and other luxurious fabrics. Not coincidentally, many Anglo-Jews were in the clothing and textile business. Maddox, *Rosalind Franklin*, 21–42; "Notes on the Opening of the Rosalind Franklin Workshop at St. Paul's Girls School, February 1988" and *Paulina* (St. Paul's Girls School yearbook), 1988, AKP, 2/6/2/4.

21. Maddox, *Rosalind Franklin*, 24.

22. Maddox, *Rosalind Franklin*, 33.

23. Elisabeth Leedham-Green, *A Concise History of the University of Cambridge* (Cambridge: Cambridge University Press, 1996).

24. Letter from Rosalind Franklin to Muriel and Ellis Franklin, January 20, 1939, ASP, box 3, folder 1; Maddox, *Rosalind Franklin*, 48.

25. Philippa Strachey, *Memorandum on the Position of English Women in Relation to that of English Men* (Westminster: London and National Society for Women's Service, 1935); Virginia Woolf, *Three Guineas* (New York: Harcourt, 1938), 30–31; Maddox. *Rosalind Franklin*, 44.

26. Virginia Woolf, *A Room of One's Own* (London: Hogarth Press, 1929), 6.

27. Letter from Rosalind Franklin to Muriel and Ellis Franklin, "Saturday, 7 Mill Road, undated," cited in Maddox, *Rosalind Franklin*, 72; Virginia Woolf, *To the Lighthouse* (London: Hogarth Press, 1927).

28. Woolf, *Three Guineas*, 17–18.

29. Letter from Rosalind Franklin to Muriel and Ellis Franklin, October 26, 1939, ASP, box 3, folder 1.

30. Letter from Rosalind Franklin to Muriel and Ellis Franklin, November 25, 1940, ASP, box 3, folder 1.

31. Letter from Rosalind Franklin to Muriel and Ellis Franklin, February 18, 1940, ASP, box 3, folder 1.

32. Letters from Rosalind Franklin to Muriel and Ellis Franklin, July 12, 1940, and February 7, 1941, ASP, box 3, folder 1.

33. Letter from Rosalind Franklin to Muriel and Ellis Franklin, December 8, 1940, ASP, box 3, folder 1.

34. Maddox, *Rosalind Franklin*, 65–66.

35. Letter from Rosalind Franklin to Muriel and Ellis Franklin, November 25, 1940, ASP, box 3, folder 1; see also, Jenifer Glynn, *My Sister Rosalind Franklin*, 56.

36. Maddox, *Rosalind Franklin*, 65.

37. Sayre, *Rosalind Franklin and DNA*, 45–46; Maddox, *Rosalind Franklin*, 94.

38. Muriel Franklin, "Rosalind," 5.

39. Letter from Rosalind Franklin to Ellis Franklin, undated, probably the summer of 1940, quoted in Glynn, *My Sister Rosalind Franklin*, 61–62; Glynn, "Rosalind Franklin, 1920–1958," 272; Maddox, *Rosalind Franklin*, 60–61.

40. Sayre, *Rosalind Franklin and DNA*, 45–46.

41. Letter from Muriel Franklin to Anne Sayre, July 24, 1974, ASP, box 2, folder 15.2.

42. Letter from Muriel Franklin to Anne Sayre, October 22, 1974, ASP, box 2, folder 15.2.

43. Letter from Anne Sayre to Muriel Franklin, October 30, 1974, ASP, box 2, folder 15.2.

44. Francis Crick, "How to Live with a Golden Helix," *The Sciences* 19, no 7 (September 1979): 6–9. A letter to the editor by Charlotte Friend of Mount Sinai Hospital in New York City, printed a few months later, complained, "Crick still feels the need to justify his condescension toward Rosalind Franklin": *The Sciences* 19, no. 3 (December 1979); Francis Crick, *What Mad Pursuit: A Personal View of Scientific Discovery* (New York: Basic Books, 1988), 68–69; author interview with James D. Watson (no. 3), July 25, 2018.

45. Anne Sayre interview with Gertrude "Peggy" Clark Dyche, May 31, 1977, ASP, box 7, "Post Publication Correspondence A–E"; Maddox, *Rosalind Franklin*, 306.

46. Glynn, *My Sister Rosalind Franklin*, 61. Glynn told me, "she was of infinite good company. A terrific sense of humor, quite loyal to her friends and very unforgiving of her enemies [but] trivial topics bored her and she did not well tolerate those who indulged in them when she thought they ought to be considering things of greater import, or, at least, what she felt were of greater import." Author interview with Jenifer Glynn, May 7, 2018.

47. Rosalind Franklin, "Notebook: X-ray Crystallography II," March 7, 1939, RFP; Maddox, *Rosalind Franklin*, 55–56.

48. Letter from Sir Frederick Dainton to Anne Sayre, November 8, 1976, ASP, box 7, "Post Publication Correspondence A–E."

49. Marion Elizabeth Rodgers, *Mencken and Sara: A Life in Letters* (New York: McGraw-Hill, 1987), 29; Maddox, *Rosalind Franklin*, 68.

50. Letter from Sir Frederick Dainton to Anne Sayre, November 24, 1976, ASP, box 7, "Post Publication Correspondence A–E."

51. Letter from Anne Sayre to Sir Frederick Dainton, November 14, 1976, ASP, box 7, "Post Publication Correspondence A–E."

52. Letter from Frederick Dainton to Anne Sayre, November 8, 1976, ASP, box 7, "Post Publication Correspondence A–E."

53. J. E. Carruthers and R. G. W. Norrish, "The polymerisation of gaseous formaldehyde and acetaldehyde," *Transactions of the Faraday Society* 32 (1936): 195–208. The society was named for Michael Faraday (1791–1867), who made many important contributions to electrochemistry and electromagnetism.

54. Glynn, *My Sister Rosalind Franklin*, 60.

55. Glynn, *My Sister Rosalind Franklin*, 61.

56. Letter from Rosalind Franklin to Ellis Franklin, June 1, 1942, ASP, box 3, folder 1.

57. Sayre, *Rosalind Franklin and DNA*, 203.

58. D. H. Bangham and Rosalind E. Franklin, "Thermal Expansion of Coals and Carbonized Coals," *Transactions of the Faraday Society* 42 (1946): B289–94.

59. Maddox, *Rosalind Franklin*, 87–107.

60. "The X-ray Crystallography that Propelled the Race for DNA: Astbury's Pictures vs. Franklin's Photo 51," *The Pauling Blog*, July 9, 2009, https://paulingblog .wordpress.com/2009/07/09/the-X-ray-crystallography-that-propelled-the-race -for-dna-astburys-pictures-vs-franklins-photo-51/.

61. Peter J. F. Harris, "Rosalind Franklin's Work on Coal, Carbon and Graphite," *Interdisciplinary Science Reviews* 26, no. 3 (2001): 204–9.

62. Letter from Vittorio Luzzati to Anne Sayre, May 17, 1968, ASP, box 4, folder 13.

63. Maddox, *Rosalind Franklin*, 96.

64. Maddox, *Rosalind Franklin*, 93.

65. Maddox suggested a flirtation between the married Mering and the "puritanical" Franklin that came close to, but never reached, consummation; Maddox, *Rosalind Franklin*, 85, 96–97. Franklin's sister, Jenifer Glynn, holds true to the narrative that Franklin never found the right man and that the stories about Mering are "pure fantasy." Author interview with Jenifer Glynn, May 7, 2018.

66. Maddox, *Rosalind Franklin*, 90.

67. Letter from Vittorio Luzzati to Anne Sayre, May 17, 1968, ASP, box 4, folder 13; Robert Olby, *Francis Crick: Hunter of Life's Secrets* (Cold Spring Harbor, NY: Cold Spring Harbor Laboratory Press, 2009), 212–13, 221.

68. Anne Sayre interview with Geoffrey Brown, May 12, 1970, ASP, box 2, folder 3.

69. Maddox, *Rosalind Franklin*, 174–75. Maddox interviewed Brown on February 10, 2000.

70. Letter from Rosalind Franklin to Muriel and Ellis Franklin, undated, March 1950, quoted in Glynn, *My Sister Rosalind Franklin*, 108.

71. Rosalind Franklin, "Résumé and Application for Fellowship," undated, early 1950, JRP, Franklin personnel file.

72. Quotes are from letter from I. C. M. Maxwell, Secretary I.C.I. and Turner and Newall Research Fellowships Committee to John Randall, July 7, 1950; letter from John Randall to Principal, King's College, June 19, 1950; letter from Principal, King's College, to John Randall, June 20, 1950, all in JRP, RNDL 3/1/6.

73. Louise Heller, a volunteer worker at King's during this period, was a graduate of Syracuse University and formerly a health physics employee at the U.S. Atomic Energy Facility at Oak Ridge, Tennessee. Letter from John Randall to Rosalind Franklin, December 4, 1950, JRP, RNDL 3/1/6.

74. Maurice Wilkins, *The Third Man of the Double Helix* (Oxford: Oxford University Press, 2003), 128.

75. Wilkins, *The Third Man of the Double Helix*, 129.

76. James D. Watson, *The Double Helix: A Personal Account of the Discovery of the Structure of DNA*, edited by Gunther Stent (New York: Norton, 1980), 14–15.

77. Anne Sayre interview with Maurice Wilkins, June 15, 1970, 18, ASP, box 4, folder 32.

78. Letter from Maurice Wilkins to Roy Markham, February 6, 1951, MWP (Letters to Roy Markham, supplied by Robert Olby), K/PP178/3/5/11.

79. Brenda Maddox interview with Maurice Wilkins, November 4, 2000, cited in Maddox, *Rosalind Franklin*, 130; Maurice Wilkins, "Origins of DNA Research at King's College, London," in Seweryn Chomet, ed., *D.N.A.: Genesis of a Discovery* (London: Newman–Hemisphere, 1995), 10–26; Wilkins, *The Third Man of the Double Helix*, 126–35.

80. Wilkins, *The Third Man of the Double Helix*, 148–49.

81. Wilkins, *The Third Man of the Double Helix*, 156.

82. Anne Sayre interview with Sir John Randall, May 18, 1970, ASP, box 4, folder 27.

Chapter 7: There Was No One Like Linus in All the World

1. The chapter title and the quote that follows are from the same passage, in James D. Watson, *The Double Helix: A Personal Account of the Discovery of the Structure of DNA*, edited by Gunther Stent (New York: Norton, 1980), 25.

2. Thomas Hager, *Force of Nature: The Life of Linus Pauling* (New York: Simon and Schuster, 1995), 207.

3. Warren Weaver, "Molecular Biology: Origin of the Term," *Science* 170 (1970): 581–82; Warren Weaver, "The Natural Sciences," in *Annual Report of the Rockefeller Foundation for 1938*, 203–51 (quote is on 203), https://assets.rockefellerfoundation.org/app/uploads/20150530122134/Annual-Report-1938.pdf.

4. Hager, *Force of Nature*, 214; Linus Pauling and E. Bright Wilson, *Introduction to Quantum Mechanics With Applications to Chemistry* (New York: McGraw-Hill, 1935).

5. Horace Freeland Judson, *The Eighth Day of Creation: The Makers of the Revolution in Biology* (Cold Spring Harbor, NY: Cold Spring Harbor Laboratory Press, 1996), 60; Horace Judson interviews with Linus Pauling, March 1, 1971, and December 23, 1975, HFJP.

6. Biographical information on Pauling is drawn from Hager, *Force of Nature*; Jack D. Dunitz, *A Biographical Memoir of Linus Carl Pauling, 1901–1994* (Washington, DC: National Academy of Sciences/National Academies Press, 1997), 221–61; Anthony Serafini, *Linus Pauling: A Man and His Science* (St. Paul, MN: Paragon House, 1989); Ted Goertzel and Ben Goertzel, *Linus Pauling: A Life in Science and Politics* (New York: Basic Books, 1995); Clifford Mead and Thomas Hager, eds., *Linus Pauling: Scientist and Peacemaker* (Corvallis: Oregon State University Press, 2001); Mina Carson, *Ava Helen Pauling: Partner, Activist, Visionary* (Corvallis: Oregon State University Press, 2013); Barbara Marinacci, ed., *Linus Pauling: In His Own Words* (New York: Touchstone Books/Simon and Schuster,

1995); Chris Petersen and Cliff Mead, eds., *The Pauling Catalogue: The Ava Helen and Linus Pauling Papers at Oregon State University*, 6 vols. (Corvallis: Valley Library Special Collections, Oregon State University, 2006); Lily E. Kay, *The Molecular Vision of Life: Caltech, the Rockefeller Foundation, and the Rise of the New Biology* (New York: Oxford University Press, 1993); Richard Severo, "Linus C. Pauling Dies at 93; Chemist and Voice for Peace," *New York Times*, August 21, 1994, 1A, 51B.

7. The best friend's name was Lloyd Jeffress. Irwin Abrams, *The Nobel Peace Prize and the Laureates: An Illustrated Biographical History, 1901–2001* (Nantucket: Science History Publications USA, 2001), 198.

8. Hager, *Force of Nature*, 68–71.

9. The California Institute of Technology was founded as a vocational and preparatory school by Amos G. Throop in 1891. It was named, successively, Throop University, Throop Polytechnic Institute (and Manual Training School), and Throop College of Technology. In 1921, under the presidency of Nobel laureate Robert Millikin, the institution was expanded into the California Institute of Technology. (The vocational school was disbanded and preparatory school spun off as a separate institution in 1907.) Pauling left Caltech in 1963 because he believed the institution ignored the occasion of his second Nobel Prize and was politically opposed to his outspoken antinuclear and leftist beliefs.

10. Initially, Guggenheim Fellows were "required to spend their terms outside of the United States . . . but eager to place as few restrictions as possible on the Fellows, the Foundation rescinded that requirement with the completion of 1941." "History of the Fellowship," John Simon Guggenheim Memorial Foundation, https://www.gf.org/about/history/.

11. In 1925, Pauling applied to work at both the Sommerfield and the Bohr institutes: "Sommerfield answered my letter but Bohr didn't." Linus Pauling oral history interview by John L. Greenberg, May 10, 1984, 11, Archives of the California Institute of Technology, Pasadena, CA.

12. Dunitz, *Biographical Memoir*, 226. The paper Pauling wrote as a Guggenheim Fellow is "The theoretical prediction of the physical properties of many electron atoms and ions: Mole refraction, diamagnetic susceptibility, and extension in space," *Proceedings of the Royal Society A: Mathematical, Physical and Engineering Sciences* 114, no. 767 (1927): 181–211. See also Linus Pauling, "The Nature of the Chemical Bond: Application of Results Obtained from the Quantum Mechanics and From a Theory of Paramagnetic Susceptibility to the Structure of Molecules," *Journal of the American Chemical Society* 53, no. 4 (1931): 1367–400; and Linus Pauling, *The Nature of the Chemical Bond and the Structure of Molecules and Crystals: An Introduction to Modern Structural Chemistry* (Ithaca, NY: Cornell University Press, 1939).

13. Apparently, Pauling spent very little time with Bohr, whose "mind was on larger

questions." He left after about a month there. Hager, *Force of Nature*, 131. See also Werner Heisenberg, "Preface," *The Physical Principles of the Quantum Theory*, translated by Carl Eckart and F. C. Hoyt (New York: Dover, 1950), iv.

14. Hager, *Force of Nature*, 161; Severo, "Linus C. Pauling Dies at 93."

15. Severo, "Linus C. Pauling Dies at 93."

16. W. T. Astbury and H. J. Woods, "The Molecular Weights of Proteins," *Nature* 127 (1931): 663–65; W. T. Astbury and A. Street, "X-ray studies of the structures of hair, wool and related fibers. I. General," *Philosophical Transactions of the Royal Society of London A 230* (March 1931): 75–101; W. T. Astbury, "Some Problems in the X-ray Analysis of the Structure of Animal Hairs and Other Protein Fibres," *Transactions of the Faraday Society* 29 (1933): 193–211; W. T. Astbury and H. J. Woods, "X-ray studies of the structures of hair, wool and related fibers. II. The molecular structure and elastic properties of hair keratin," *Philosophical Transactions of the Royal Society of London A* 232 (1934): 333–94; W. T. Astbury and W. A. Sisson, "X-ray Studies of the Structures of Hair, Wool and Related Fibres. III. The configuration of the keratin molecule and its orientation in the biological cell," *Philosophical Transactions of the Royal Society of London A* 150 (1935): 533–51.

17. Horace Judson interview with Linus Pauling, December 23, 1975, HFJP; see also Judson, *The Eighth Day of Creation*, 61–62.

18. L. C. Pauling, "The Structure of the Micas and Related Minerals," *Proceedings of the National Academy of Sciences* 16, no. 2 (February 1930): 123–29.

19. *Oxford English Dictionary*, 2nd edition, vol. 16 (Oxford: Oxford University Press, 1989), 730.

20. Pauling, *The Nature of the Chemical Bond*, 411.

21. Jack Dunitz, "The Scientific Contributions of Linus Pauling," in Clifford Mead and Thomas Hager, eds., *Linus Pauling: Scientist and Peacemaker* (Corvallis: Oregon State University Press, 2001), 78-97, quote is on 89.

22. Hager, *Force of Nature*, 282. In 1987, Pauling wrote, "It was, and still is, my opinion that Schrödinger made no contribution to our understanding of life"; Linus Pauling, "Schrödinger's Contribution to Chemistry and Biology," in C. W. Kilmister, ed., *Schrödinger: Centenary Celebration of a Polymath* (Cambridge: Cambridge University Press, 1987), 225–33.

23. Linus Pauling and Max Delbrück, "The Nature of the Intermolecular Operative in Biological Processes," *Science* 92, no. 2378 (1940): 77–99, quote is on 78. The typescript of this paper is in LAHPP, Manuscript Notes and Typescripts, The Race for DNA, http://scarc.library.oregonstate.edu/coll/pauling/dna/notes/1940a.5-03.html. See also Dunitz, "The Scientific Contributions of Linus Pauling," 8; Pascual Jordan, "Biologische Strahlenwirkung und Physik der Gene" (Biological Radiation Effects and Physics of Genes), *Physikalische Zeitschrift* 39 (1938): 345–66, 711; Pascual Jordan, "Problem der spezifischen Immunität" (Prob-

lem of Specific Immunity), *Fundamenta Radiologica* 5 (1939): 43–56; Richard H. Beyler, "Targeting the Organism: The Scientific and Cultural Context of Pascual Jordan's Quantum Biology, 1932–1947," *Isis* 87, no. 2 (1996): 248–73; Nils Roll-Hansen, "The Application of Complementarity to Biology: From Niels Bohr to Max Delbrück," *Historical Studies in the Physical and Biological Sciences* 30, no. 2 (2000): 417–42; Daniel J. McKaughan, "The Influence of Niels Bohr on Max Delbrück," *Isis* 96, no. 4 (2005): 507–29; Bernard S. Strauss, "A Physicist's Quest in Biology: Max Delbrück and "Complementarity," *Genetics* 206 (2017): 641–50; James D. Watson, "Growing Up in the Phage Group," JDWP, JDW/2/3/1/38.

24. Linus Pauling, *Molecular Architecture and Processes of Life: The 21st Annual Sir Jesse Boot Foundation Lecture* (Nottingham, UK: Sir Jesse Boot Foundation, 1948), 1–13, esp. 10; see also L. C. Pauling, "Molecular Basis of Biological Specificity," *Nature* 258, no. 5451 (1974): 769–71.

25. The National Institute of Health changed its name to the National Institutes of Health in 1948. Richard E. Marsh, *Robert Brainard Corey, 1897–1971: A Biographical Memoir* (Washington, DC: National Academies Press, 1997), 51-67; quote is on 55.

26. Beaumont Newhall, "The George Eastman Visiting Professorship at Oxford University," *American Oxonian* 52, no. 2 (April 1965): 65–69.

27. Francis Crick, *What Mad Pursuit: A Personal View of Scientific Discovery* (New York: Basic Books, 1988), 54.

28. Linus Pauling, *Vitamin C, the Common Cold and the Flu* (New York: W. H. Freeman, 1977).

29. Thomas Hager, *Linus Pauling and the Chemistry of Life* (New York: Oxford University Press, 1998), 86.

30. Hager, *Linus Pauling*, 323–24; see also Horace Judson interview with Linus Pauling, December 23, 1975, HFJP.

31. The sixth amino acid on the 147-amino-acid chain making up the β-chain of hemoglobin is glutamic acid; in sickle cell anemia, the mutation substitutes valine rather than glutamic acid. L. C. Pauling, H. A. Itano, S. J. Singer, and A. C. Wells, "Sickle Cell Anemia, a Molecular Disease," *Science* 110, no. 2865 (1949): 543–48. The same year, James Neel of the University Michigan also demonstrated that sickle cell anemia is an inherited disease; James V. Neel, "The Inheritance of Sickle Cell Anemia," *Science* 110, no. 2846 (1949): 64–66.

32. Linus Pauling, "Reflections on the New Biology," *UCLA Law Review* 15 (February 1968): 268–72.

33. Max F. Perutz, *Science is Not a Quiet Life: Unraveling the Atomic Mechanism of Haemoglobin* (Singapore: World Scientific, 1997), 41.

34. W. L. Bragg, J. C. Kendrew, and M. F. Perutz, "Polypeptide Chain Configurations in Crystalline Proteins," *Proceedings of the Royal Society of London A: Mathematical and Physical Sciences* 203, no. 1074 (October 10, 1950), 321–57.

35. David Eisenberg, "The discovery of the α-helix and β-sheet, the principle structural feature of proteins," *Proceedings of the National Academy of Sciences* 100, no. 20 (September 30, 2003): 11207–10. See also M. F. Perutz, "New X-ray Evidence on the Configuration of Polypeptide Chains: Polypeptide Chains in Poly-γ-benzyl-L-glutamate, Keratin and Hæmoglobin," *Nature* 167, no. 4261 (1951): 1053–54; Arthur S. Edison, "Linus Pauling and the Planar Peptide Bond," *Nature Structural Biology* 8, no. 3 (2001): 201–2; California Institute of Technology press release on Pauling and Corey's protein research, September 4, 1951, LAHPP, http://scarc.library.oregonstate.edu/coll/pauling/proteins/papers/1951n.7.html.

36. Quote is from Edison, "Linus Pauling and the Planar Peptide Bond." See also Linus Pauling, Robert B. Corey, and Herman R. Branson, "The structure of proteins; two hydrogen-bonded helical configurations of the polypeptide chain," *Proceedings of the National Academy of Sciences* 37, no. 4 (1951): 205–11; L. C. Pauling and R. B. Corey, "Atomic coordinates and structure factors for two helical configurations of polypeptide chains," *Proceedings of the National Academy of Sciences* 37, no. 5 (1951): 235–40; L. C. Pauling and R. B. Corey, "The structure of synthetic polypeptides," *Proceedings of the National Academy of Sciences* 37, no. 5 (1951): 241–50; L. C. Pauling and R. B. Corey, "The Pleated Sheet, A New Layer Configuration of Polypeptide Chains," *Proceedings of the National Academy of Sciences* 37, no. 5 (1951): 251–56; L. C. Pauling and R. B. Corey, "The structure of feather rachis keratin," *Proceedings of the National Academy of Sciences* 37, no. 5 (1951): 256–61; L. C. Pauling and R. B. Corey, "The Structure of Hair, Muscle, and Related Proteins," *Proceedings of the National Academy of Sciences* 37, no. 5 (1951): 261–71; L. C. Pauling and R. B. Corey, "The Structure of Fibrous Proteins of the Collagen–Gelatin Group," *Proceedings of the National Academy of Sciences* 37, no. 5 (1951): 272–81; L. C. Pauling and R. B. Corey, "The polypeptide-chain configuration in hemoglobin and other globular proteins," *Proceedings of the National Academy of Sciences* 37, no. 5 (1951): 282–85.

37. W. L. Bragg, "First Stages in the Analysis of Proteins," *Reports of Progress in Physics* 28 (1965): 1–16; quote is on 6–7. This is the text of his lecture to the X-ray Analysis Group, November 15, 1963.

Chapter 8: The Quiz Kid

1. Carl Sandburg, "Chicago Poems," *Poetry* 3, no. 4 (March 1914): 191–92.

2. The full first line is, "I am an American, Chicago born—Chicago that somber city—and go at things as I have taught myself, free-style, and will make the record in my own way: first to knock, first admitted." Saul Bellow, *The Adventures of Augie March* (New York: Viking, 1953), 1.

3. James D. Watson, *Avoid Boring People: Lessons from a Life in Science* (New York: Knopf, 2007), 4; author interview with James D. Watson (no. 1), July 23, 2018.

4. Watson, *Avoid Boring People*, 5.

5. Watson, *Avoid Boring People*, 5. More than twenty species of warblers, most notably the Kirkland warbler, migrated to Chicago's Jackson Park, near the Watson family home, each year. James D. Watson (Sr.), George Porter Lewis, Nathan F. Leopold, Jr., *Spring Migration Notes of the Chicago Area*, privately printed pamphlet, 1920, JDWP.

6. Friedrich Nietzsche, *Thus Spake Zarathustra*, translated by Thomas Common (New York: Modern Library/Boni and Liveright, 1917). The novel was originally published in Germany in four parts from 1883 to 1885. The title is now more commonly translated as *Thus Spoke Zarathustra*.

7. James D. Watson, ed., *Father to Son: Truth, Reason and Decency* (Cold Spring Harbor, NY: Cold Spring Harbor Laboratory Press, 2014), 53–87; Simon Baatz, *For the Thrill of It: Leopold, Loeb, and the Murder That Shocked Jazz Age Chicago* (New York: Harper Perennial, 2009).

8. Watson, ed., *Father to Son*, title page.

9. Watson, *Avoid Boring People*, 6.

10. Victor K. McElheny, *Watson and DNA: Making a Scientific Revolution* (New York: Perseus, 2003), 7.

11. James D. Watson, *Genes, Girls and Gamow: After the Double Helix* (New York: Knopf, 2002), 118.

12. Carolyn Hong, "Focus: Newsmakers: How Beautiful It Was, This Thing Called DNA," *New Straits Times* (Malaysia), December 1, 1995, 15.

13. David Ewing Duncan, "Discover Magazine Interview: Geneticist, James Watson," *Discover*, July 1, 2003, http://discovermagazine.com/2003/jul/featdialogue.

14. Watson, *Avoid Boring People*, 7.

15. McElheny, *Watson and DNA*, 6–7.

16. Lee Edson, "Says Nobelist James (Double Helix) Watson: 'To Hell With Being Discovered When You're Dead,'" *New York Times Magazine*, August 18, 1968, 26, 27, 31, 34.

17. Cowan later created *The $64,000 Question* television show and became the president of the CBS network. During the Second World War, he was the director of Voice of America. His wife, Pauline, was a major civil rights activist in Mississippi and Alabama from 1964 to 1965. They both died in 1976 in a fire in their apartment in the Westbury Hotel at 15 East Sixty-Ninth Street in New York City, caused by "smoking carelessness"; "Louis Cowan, Killed with Wife in a Fire; Created Quiz Shows," *New York Times*, November 19, 1976, 1. The original sponsor was Alka-Seltzer, made by Miles Laboratories; later, the show was sponsored by both Alka-Seltzer and another Miles Labs product, One-A-Day vitamin

tablets. The quizmaster on the show was Joe Kelly. See also Ruth Duskin Feldman, *Whatever Happened to the Quiz Kids: Perils and Profits of Growing Up Gifted* (Chicago: Chicago Review Press, 1982), 10.

18. Author interview with James D. Watson (no. 4), July 26, 2018. See also Larry Thompson, "The Man Behind the Double Helix: Gene-Buster James Watson Moves on to Biology's Biggest Challenge, Mapping Heredity," *Washington Post*, September 12, 1989, Z12; Feldman, *Whatever Happened to the Quiz Kids.*

19. McElheny, *Watson and DNA*, 8.

20. "Heads University at 30, Dean Hutchins of Yale Named U. of C. Chief, Youngest American College President," *Chicago Daily Tribune*, April 26, 1929, 1.

21. Nathaniel Comfort, "'The Spirit of the New Biology': Jim Watson and the Nobel Prize," in Christie's auction catalogue, *Dr. James Watson's Nobel Medal and Related Papers: Thursday 4 December 2014* (New York: Christie's, 2014), 11–19; quote is on 13.

22. McElheny, *Watson and DNA*, 7.

23. Robert Olby, *The Path to the Double Helix* (Seattle: University of Washington Press, 1974), 297. Olby interviewed Weiss for his book on April 25, 1973.

24. Interview with James D. Watson on *Talk of the Nation/Science Friday*, NPR, June 2, 2000, https://www.npr.org/templates/story/story.php?storyId=1074946. See also James D. Watson, "Values from a Chicago Upbringing," *Annals of the New York Academy of Sciences* 758 (1995): 194–97, reprinted in James D. Watson, *A Passion for DNA: Genes, Genomes and Society* (Cold Spring Harbor, NY: Cold Spring Harbor Laboratory Press, 2001), 3–5; this article was adapted from an after-dinner talk given on October 14, 1993, at a meeting on "The Double Helix: 40 Years Prospective and Perspective," sponsored by the University of Illinois at Chicago, the New York Academy of Sciences, and Green College, Oxford University. See also McElheny, *Watson and DNA*, 14–16.

25. Watson, "Values from a Chicago Upbringing."

26. Watson, *Avoid Boring People*, 49.

27. Sinclair Lewis, *Arrowsmith* (New York: Harcourt, Brace, 1925); Howard Markel, "Prescribing Arrowsmith," *New York Times Book Review*, September 24, 2000, D8.

28. Watson, "Values from a Chicago Upbringing," 5.

29. Erwin Schrödinger, *What Is Life?: The Physical Aspect of the Living Cell, with Mind and Matter and Autobiographical Sketches* (Cambridge: Cambridge University Press, 1992), 21.

30. Letter from James Watson to his parents, November 21, 1947, WFAT, "Letters to Family, Bloomington Sept. 1947–May 1948." See also William Provine, *Sewall Wright and Evolutionary Biology* (Chicago: University of Chicago Press, 1986).

31. James D. Watson, "Winding Your Way Through DNA," video of symposium, University of California, San Francisco, September 25, 1992 (Cold Spring Har-

bor, NY: Cold Spring Harbor Laboratory Press, 1992); quote appears in McElheny, *Watson and DNA*, 16.

32. Salvador Luria, *A Slot Machine, a Broken Test Tube: An Autobiography* (New York: Harper and Row, 1983), 41–43.

33. Thomas Hager, *Force of Nature: The Life of Linus Pauling* (New York: Simon and Schuster, 1995), 409.

34. McElheny, *Watson and DNA*, 17–29; Watson, *Avoid Boring People*, 38–54; William C. Summers, "How Bacteriophage Came to Be Used by the Phage Group," *Journal of the History of Biology* 26, no. 2 (1993): 255–67.

35. Letter from James Watson to his parents, undated, spring 1948, WFAT, "Letters to Family, Bloomington, September 1947–May 1948."

36. Howard Markel, "Happy Birthday, Renato Dulbecco, Cancer Researcher Extraordinaire," *PBS NewsHour*, February 22, 2014, https://www.pbs.org/newshour/health/happy-birthday-renato-dulbecco-cancer-researcher-extraordinaire.

37. Watson, *Avoid Boring People*, 40–41; James H. Jones, *Alfred Kinsey: A Public/Private Life* (New York: Norton, 1997); Jonathan Gathorne-Hardy, *Sex the Measure of All Things: A Life of Alfred C. Kinsey* (Bloomington: Indiana University Press, 1998).

38. The Hoosiers' 1947 season was pitiful; the team tied with Iowa for sixth place in the Big Nine Conference, which had recently decreased from its iconic number of Ten when the University of Chicago dropped out in 1946. The University of Chicago discontinued its football program in 1939. Jim much preferred the Indiana basketball matches, despite an eighth-place finish in the 1947–48 season. Watson, *Avoid Boring People*, 45; A few years later, while on a postdoctoral fellowship in Copenhagen, Watson wrote to his parents, "I miss the basketball games of Bloomington"; letter from James D. Watson to his parents, December 13, 1950, WFAT, "Letters to Family, Copenhagen, Fall–Dec. 1950."

39. Letter from James D. Watson to his parents, undated, fall 1947, WFAT, "Letters to Family, Bloomington Sept. 1947–May 1948." LaMont Cole was a prominent evolutionary biologist and ecologist at the University of Chicago, Indiana University, and, later, Cornell University. He was one of Watson's teachers at Indiana in 1947–48. See Gregory E. Blomquist, "Population Regulation and the Life History Studies of LaMont Cole," *History and Philosophy of the Life Sciences* 29, no. 4 (2007): 495–516.

40. Letter from James D. Watson to his parents, November 21, 1947, WFAT, "Letters to Family, Bloomington Sept. 1947–May 1948."

41. Letter from James D. Watson to his parents, November 21, 1947, WFAT, "Letters to Family, Bloomington Sept. 1947–May 1948."

42. James D. Watson, "Growing Up in the Phage Group," in John Cairns, Gunther S. Stent, and James D. Watson, eds., *Phage and the Origins of Molecular Biology* (1966; Cold Spring Harbor, NY: Cold Spring Harbor Laboratory Press,

2007), pp. 239–45, quote is on 239. (The article also appears in Watson, *A Passion for DNA*, 7–15.) See also James D. Watson, "Lectures on Microbial Genetics–Sonneborn (Fall Term, 1948)," JDWP, JDW/2/6/1/5.

43. Watson, *Avoid Boring People*, 42, 45.

44. Watson, *Avoid Boring People*, 46.

45. Luria and Delbrück shared the Nobel Prize in Physiology or Medicine with Alfred Hershey in 1969. Dulbecco shared his Nobel for Physiology or Medicine with David Baltimore and Howard Temin in 1976, and Watson shared his with Francis Crick and Maurice Wilkins in 1962. See also Watson, "Values from a Chicago Upbringing," and Watson, "Growing Up in the Phage Group."

46. John Kendrew, "How Molecular Biology Started," and Gunther Stent, "That Was the Molecular Biology That Was," in Cairns, Stent, and Watson, eds., *Phage and the Origins of Molecular Biology*, 343–47, 348–62.

47. Watson, "Growing Up in the Phage Group," 240; Ernst P. Fischer and Carol Lipson, *Thinking About Science: Max Delbrück and the Origins of Molecular Biology* (New York: Norton, 1988), 183, 196.

48. Letter from James D. Watson to his parents, July 5, 1948, WFAT, "Letters to Family, Cold Spring Harbor, June to September, 1948." Watson took the Long Island Railroad from Cold Spring Harbor into Manhattan and noted it was a fifty-three-minute trip.

49. Letter from Horace Judson to Alfred D. Hershey, August 27, 1976, HFJP.

50. Letters from James D. Watson to Elizabeth Watson, February 8 and March 6, 1950, and letter from James D. Watson to his parents, March 2, 1950, WFAT, "Letters to Family, Bloomington, Fall 1949–Spring 1950."

51. Letter from James D. Watson to his parents, March 12, 1950, WFAT, "Letters to Family, Bloomington, Fall 1949–Spring 1950." See also James D. Watson, 1950 Merck/NRC Fellowship Application Materials and Acceptance Letters, National Research Council, JDWP, JDW/2/2/12.

52. Letter from James D. Watson to his parents, March 24, 1950, WFAT, "Letters to Family, Bloomington, Fall 1949–Spring 1950."

53. Letter from James D. Watson to his parents, September 11, 1950, WFAT, "Letters to Family, Copenhagen, September 15, 1950–October 1, 1951."

54. Letter from James D. Watson to his parents, September 13, 1950, WFAT, "Letters to Family, Copenhagen, September 15, 1950–October 1, 1951." The music and lyrics of "Wonderful Copenhagen" were written by Frank Loesser in 195. (New York: Frank Music Corp, September 24, 1951); the song first appeared in the 1952 film *Hans Christian Andersen*, starring Danny Kaye; https://frankloesser.com/library/wonderful-copenhagen/.

55. "The Nobel Prize in Physics, 1922," Nobel Media AB 2019, https://www.nobelprize.org/prizes/physics/1922/summary/.

56. Fritz Kalckar obituary, *Nature* 141, no. 3564 (February 19, 1938): 319; Herman M. Kalckar, "40 Years of Biological Research: From oxidative phosphorylation to energy requiring transport regulation," *Annual Review of Biochemistry* 60 (1991): 1–37. Fritz Kalckar was working on a theory of nuclear reactions at the time of his death. The *Nature* obituary states that he died of heart failure, but Herman notes in the memoir cited here that his younger brother had epilepsy, in an era when there were no effective pharmacological treatments for seizures, and died during a spell of intractable seizures, or *status epilepticus*. Herman Kalckar dedicated his PhD thesis on oxidative phosphorylation in the kidney's cortex to Fritz's memory.

57. Paul Berg, "Moments of Discovery: My Favorite Experiments," *Journal of Biochemistry* 278, no. 42 (October 17, 2003): 40417–24, doi: 10.1074/jbc.X300004200; quotes are on 40419 and 40420. Berg was widely celebrated for his work on nucleic acid chemistry and recombinant DNA. He was also one of the key architects of the 1975 Asilomar conference on the potential hazards and ethics of the emerging field of biotechnology.

58. Berg, "Moments of Discovery," 40420–21; John H. Exton, *Crucible of Science: The Story of the Cori Laboratory* (New York: Oxford University Press, 2013), 21–28. See also Kalckar, "40 Years of Biological Research"; "Herman Kalckar, 83, Metabolism Authority," *New York Times* May 22, 1991, D25; James D. Watson, *The Double Helix: A Personal Account of the Discovery of the Structure of DNA*, edited by Gunther Stent (New York: Norton, 1980), 17–21.

59. Exton, *Crucible of Science*, 28.

60. Watson, *The Double Helix*, 19.

61. Watson, *The Double Helix*, 18.

62. Francis Crick, "The Double Helix: A Personal View," *Nature* 248, no. 5451 (April 26, 1974): 766–69.

63. Letter from James D. Watson to his parents, September 19, 1950, WFAT, "Letters to Family, Copenhagen, September 15, 1950–October 1, 1951."

64. Letter from James D. Watson to his parents, September 16, 1950, WFAT, "Letters to Family, Copenhagen, September 15, 1950–October 1, 1951." See also Eugene Goldwasser, *A Bloody Long Journey: Erythropoietin (Epo) and the Person Who Isolated It* (Bloomington, IN: Xlibris, 2011), 55–60. Goldwasser later became well-known for identifying erythropoietin, the hormone manufactured by the kidney that, upon sensing cellular hypoxia or lack of oxygen, stimulates the production of red blood cells.

65. Letter from James D. Watson to his parents, September 19, 1950, WFAT, "Letters to Family, Copenhagen, September 15, 1950–October 1, 1951."

66. Author interview with James D. Watson (no. 1), July 23, 2018.

67. As chance would have it, John Steinbeck won the Nobel Prize for Literature in

1962, the same year as Watson, Crick, and Wilkins won their prize. Letter from James D. Watson to his parents, January 14, 1951, WFAT, "Letters to Family, Copenhagen, September 15, 1950–October 1, 1951."

68. Letter from James D. Watson to Elizabeth Watson, February 4, 1951, WFAT, "Letters to Family, Copenhagen, September 15, 1950–October 1, 1951." *Sunset Boulevard* (1950) was directed by Billy Wilder, screenplay by Billy Wilder and Charles Brackett, and starred Gloria Swanson, William Holden, and Erich von Stroheim.

69. Watson, *The Double Helix*, 21.

70. Goldwasser, *A Bloody Long Journey*, 55–56.

71. Letter from James D. Watson to his parents, November 6, 1950, WFAT, "Letters to Family, Copenhagen, September 15, 1950–October 1, 1951."

72. Letter from James D. Watson to his parents, November 6, 1950, WFAT, "Letters to Family, Copenhagen, September 15, 1950–October 1, 1951."

73. Letter from James D. Watson to his parents, November 19, 1950, WFAT, "Letters to Family, Copenhagen, September 15, 1950–October 1, 1951." In the 1840s, Jacobsen became a great admirer of the scientific methods then being developed and applied them to the production of beer. See Carlsberg Foundation, "The Carlsberg Foundation's Home," https://www.carlsbergfondet.dk/en/About-the -Foundation/The-Carlsberg-Foundations%27s-home/Domicile.

74. Letter from James D. Watson to his parents, December 3, 1950, WFAT, "Letters to Family, Copenhagen, September 15, 1950–October 1, 1951."

75. Letters from James D. Watson to his parents, December 3 and 17, 1950, and January 1, 1951, WFAT, "Letters to Family, Copenhagen, September 15, 1950–October 1, 1951."

76. Letter from James D. Watson to his parents, December 21, 1950, WFAT, "Letters to Family, Copenhagen, September 15, 1950–October 1, 1951."

Part III: Tick-Tock, 1951

1. Sinclair Lewis, *Arrowsmith* (New York: Harcourt, Brace, 1925), 280–81.

Chapter 9: *Vide Napule e po' muore*

1. A literal translation of the proverb, which refers to the extraordinary beauty of the Bay of Naples and the view of Vesuvius in the horizon, is, "See Naples and then die"; more romantically, it is construed as "Nothing compares to the beauty of Naples, so you can die after you've seen it." Naples was a required visit on most eighteenth- and nineteenth-century "grand tours" of Europe. The phrase is often ascribed to Johann Wolfgang von Goethe, who made his grand tour

through Italy in 1786–88. See J. W. Goethe, *Italian Journey, 1786–1788*, translated by W. H. Auden and Elizabeth Meyer (London: Penguin, 1970), 189.

2. Letter from Herman Kalckar to Reinhard Dohrn, January 13, 1950 (sic, but probably 1951 as it was received January 18, 1951), Archives of the Naples Zoological Station, Correspondence, K:SZN, 1951, Naples, Italy.

3. Letter from Herman Kalckar to Reinhard Dohrn, January 13, "1950" and letter from Reinhard Dohrn to Herman Kalckar, January 21, 1951, Archives of the Naples Zoological Station, Correspondence, K:SZN, 1951. Dohrn was pleased to accept the Americans because he wanted to curry favor with his American funders. Neither Watson nor Wright required financial support from Dohrn's strapped budget, because their expenses were being covered by the National Research Council. Kalckar's colleague Heinz Holter, a cell physiologist who had a long history of working at the Stazione, also sent a letter of recommendation for Watson and Wright on January 18, 1951, which was answered by Dohrn on February 2, 1951. H:SZN, 1951. See also Jytte R. Nilsson, "In memoriam: Heinz Holter (1904–1993)," *Journal of Eukaryotic Microbiology* 41, no. 4 (1994): 432–33.

4. Letter from James D. Watson to Alberto Monroy, February 20, 1980, Archives of the Naples Zoological Station, uncatalogued.

5. Barbara Wright's father, Gilbert Munger Wright, was a writer and the son of one of America's best-selling writers of the day, Harold Bell Wright. Together, they wrote the best-selling 1932 science fiction tale *The Devil's Highway* (Gilbert using the pen name John Lebar) about a mad scientist who controls his victims' minds. Her mother, Leta Luella Brown Deery, was a physics major at Berkeley (class of 1919) and an English teacher in the California public school system. In addition to her scientific accomplishments, Wright was an able boater and, later in life, became an internationally ranked whitewater slalom kayaker. See obituary of Barbara Evelyn Wright, *The Missoulian* (Missoula, MT), July 14, 2016.

6. Letter from James D. Watson to his parents, August 15, 1949, WFAT, "Letters to Family, Pasadena, 1949."

7. Letter from James D. Watson to his parents, August 15, 1949, WFAT, "Letters to Family, Pasadena, 1949."

8. Letter from James D. Watson to his parents, August 15, 1949, WFAT, "Letters to Family, Pasadena, 1949." In *The Annotated and Illustrated Double Helix*, edited by Alexander Gann and Jan Witkowski (New York: Simon and Schuster, 2012), 20, In this volume, the editors claim that Watson and Wright were arrested by the sheriff, but Watson's letter written at the time makes no mention of this part of the adventure.

9. Letter from C. J. Lapp, National Research Council, to James D. Watson, December 14, 1950, JDWP, JDW/2/2/1284.

10. Letter from James D. Watson to Max Delbrück, March 22, 1951, MDP, box 23, folder 20.

11. James D. Watson, *The Double Helix: A Personal Account of the Discovery of the Structure of DNA*, edited by Gunther Stent (New York: Norton, 1980), 20; Eugene Goldwasser, *A Bloody Long Journey: Erythropoietin (Epo) and the Person Who Isolated It* (Bloomington, IN: Xlibris, 2011), 55–60.

12. All the quotes in this paragraph are drawn from letter from James D. Watson to Max Delbrück, March 22, 1951, MDP, box 23, folder 20.

13. Letter from James D. Watson to Max Delbrück, March 22, 1951, MDP, box 23, folder 20.

14. Wright and Kalckar married in the fall of 1951, just before the arrival of their baby girl, Sonia. They had two more children, a boy and a girl, Niels (after Niels Bohr) and Nina, but the scandal was too great to be contained within the tiny social circles of Copenhagen. The Cytophysiology Institute, which a wealthy donor endowed expressly to advance Kalckar's research, lost its funding, and in 1952, the Kalckars sailed to America, first for a job at the National Institutes of Health and, later, at Johns Hopkins University (1958), and then Massachusetts General Hospital and Harvard Medical School (1961). By 1963, Wright and Kalckar had divorced and in 1968 he married a former Copenhagen student, Agnete Fridericia.

15. Theodor Heuss, *Anton Dohrn: A Life for Science* (Berlin: Springer, 1991), 63; Christiane Groeben, ed., *Charles Darwin (1809–1882) –Anton Dohrn (1840–1909) Correspondence* (Naples: Macchiaroli, 1982); Christiane Groeben, "Stazione Zoologica Anton Dohrn," in *Encyclopedia of the Life Sciences* (Chichester, UK: John Wiley & Sons, 2013), doi.org/10.1002/9780470015902.a0024932.

16. In 1982, the name was formally changed to Stazione Zoologica Anton Dohrn. See Christiane Groeben, "The Stazione Zoologica Anton Dohrn as a Place for the Circulation of Scientific Ideas: Vision and Management," in K. L. Anderson and C. Thiery, eds., *Information for Responsible Fisheries: Libraries as Mediators. Proceedings of the 31st Annual Conference of the International Association of Aquatic and Marine Sciences, Rome, Italy, October 10–14, 2005* (Fort Pierce, FL: International Association of Aquatic and Marine Science Libraries and Information Centers, 2006); Christiane Groeben and Fabio de Sio, "Nobel Laureates at the Stazione Zoologica Anton Dohrn: Phenomenology and Paths to Discovery in Neuroscience," *Journal of the History of the Neurosciences* 15, no. 4 (2006): 376–95; Groeben, "Stazione Zoologica Anton Dohrn"; "Some Unwritten History of the Naples Zoological Station," *American Naturalist* 31, no. 371 (1897): 960–65 ("It is beyond question, the greatest establishment for research in the world," 960); Paul Gross, ed., "The Naples Zoological Station and the Woods Hole, Maine Marine Biological Laboratory: One Hundred Years of Biology," *Biological Bulletin* 168,

no. 3, supplement (June 1985): 1–207; M. H. F. Wilkins, "Essay," in Christiane Groeben, ed., *Reinhard Dohrn, 1880–1962: Reden, Briefe und Veroffentlichungen zum 100. Geburtstag* (Berlin: Springer, 1983), 5–10; Charles Lincoln Edwards, "The Zoological Station at Naples," *Popular Science Monthly* 77 (September 1910): 209–25; Giuliana Gemelli, "A Central Periphery: The Naples Stazione Zoologica as an 'Attractor,'" in William H. Schneider, ed., *Rockefeller Philanthropy and Modern Biomedicine: International Initiatives from World War I to the Cold War* (Bloomington: University of Indiana Press, 2002), 184–207.

17. Registration cards for laboratory tables at the Naples Zoological Station, for Herman Kalckar, 4/16/61–9[5]/25.51; Barbara Wright, 4/16/61–9[5]/25.51; and James D. Watson 4/16/51–5/26/51; Archives of the Naples Zoological Station.

18. Gemelli, "A Central Periphery." The 1949 symposium on genetics and mutagens featured a lecture by the Paris-based scientist Harriet E. Taylor, who worked with Oswald Avery on the "transforming principle" of pneumococcus and later married the molecular biologist Boris Ephrussi. See H. E. Taylor, "Biological Significance of the Transforming Principles of *Pneumococcus*," *Pubblicazioni della Stazione Zoologica di Napoli* 22, supplement (Relazioni Tenute al Convegno su Gli Agenti Mutageni, May 27–31, 1949), 65–77. Taylor also presented these data at the annual Cold Spring Harbor Symposium of 1946; see M. McCarty, H. E. Taylor, and O. T. Avery, "Biochemical Studies of Environmental Factors Essential in Transformation of *Pneumococcus* types," *Cold Spring Harbor Symposia* 11 (1946): 177–83. It is worth noting that the 1948 meeting on embryology and genetics included a paper on nucleic acids in the nuclei of bacteria, which discussed the importance of the Avery and Griffith pneumococci papers with respect to the nucleic acids; Luigi Califano, "Nuclei ed acidi nucleinici nei bacteri" (Nuclei and Nucleic Acid in Bacterium), *Pubblicazioni della Stazione Zoologica di Napoli* 21 (1949): 173–90.

19. Watson, *The Double Helix*, 22.

20. Letter from James D. Watson to his parents, April 17, 1951, WFAT, "Letters to Family, Naples, April–May 1951."

21. Registration cards for laboratory tables at the Naples Zoological Station; *Relazione sull'attivita della Stazione Zoologica di Napoli durante l'anno 1951* (annual report, 1951) lists Kalckar as working on "purine metabolism of sea-urchin eggs, *Paracentrotus*," Wright on "purine metabolism of sea-urchin eggs," and Watson on "bibliographic work," (4–6). Archives of the Naples Zoological Station.

22. Letter from James D. Watson to Elizabeth Watson, April 30, 1951, WFAT, "Letters to Family, Naples, April–May 1951."

23. Watson, *The Double Helix*, 22; *Relazione sull'attivita della Stazione Zoologica di Napoli durante l'anno 1952, 1953, 1954* (annual reports, 1952, 1953, 1954), 19–22. See also Biblioteca della Stazione Zoologica di Napoli, Report of Library Holdings for 1982, Archives of the Naples Zoological Station.

24. Frank Fehrenbach, "The Frescoes in the *Statione Zoologica* and Classical Ekphrasis," in Lea Ritter-Santini and Christiane Groeben, eds., *Art as Autobiography: Hans von Marées* (Naples: Pubblicazioni della Stazione Zoologica Anton Dohrn, 2008), 93–104, quote is on 98. See also Christiane Groeben, *The Fresco Room of the Stazione Zoologica Anton Dohrn: The Biography of a Work of Art* (Naples: Macchiaroli, 2000).

25. Watson, *The Double Helix*, 22.

26. Letter from James D. Watson to Max Delbrück enclosing manuscript of "The Transfer of Radioactive Phosphorus From Parental to Progeny Phage," April 22, 1951, MDP, box 23, folder 20; Victor K. McElheny, *Watson and DNA: Making a Scientific Revolution* (New York: Perseus, 2003), 28; Ole Maaløe and James D. Watson, "The Transfer of Radioactive Phosphorus from Parental to Progeny Phage," *Proceedings of the National Academy of Sciences* 37, no. 8 (1951): 507–13. For a more complete report, see: James D. Watson and Ole Maaløe, "Nucleic Acid Transfer from Parental to Progeny Bacteriophage," *Biochimica et Biophysica Acta* 10 (1953): 432–42. Here, they found that 40–50% of the radiolabeled phosphorus was transmitted from parental to phage progeny; only 5–10% stay associated with bacterial debris after lysis and the remaining 40% appear as non-sedimented material in the lysate.

27. Letter from James D. Watson to Local Draft Board No. 75, Chicago, March 13, 1951, and letter from James D. Watson to Max Delbrück, March 13, 1951, both in MDP, box 23, folder 20; letter from C. J. Lapp to James D. Watson, March 23, 1951, JDWP, JDW/2/2/1284; letter from James D. Watson to his parents, May 8, 1951, WFAT, "Letters to Family, Naples, April–May 1951"; S. E. Luria, *A Slot Machine, A Broken Test Tube: An Autobiography* (New York: Harper and Row, 1983), 88–90.

28. For biographical studies of William Astbury, see Kersten T. Hall, *The Man in the Monkeynut Coat: William Astbury and the Forgotten Road to the Double Helix* (Oxford: Oxford University Press, 2014); Kersten T. Hall, "William Astbury and the biological significance of nucleic acids, 1938–1951," *Studies in History and Philosophy of Biological and Biomedical Sciences* 42 (2011): 119–28; J. D. Bernal, "William Thomas Astbury, 1898–1961," *Biographical Memoirs of Fellows of the Royal Society* 9 (1963): 1–35; Robert Olby, *The Path to the Double Helix* (Seattle: University of Washington Press, 1974), 41–70. For Astbury's X-ray studies on nucleic acids, see W. T. Astbury, "X-ray Studies of Nucleic Acids," *Symposia of the Society for Experimental Biology* 1 (1947): 66–76; W. T. Astbury, "Protein and virus studies in relation to the problem of the gene," in R. C. Punnett, ed., *Proceedings of the Seventh International Congress on Genetics, Edinburgh, Scotland, August 20–23, 1939* (Cambridge: Cambridge University Press, 1941), 49–51; W. T. Astbury and F. O. Bell, "X-ray Study of Thymonucleic Acid," *Nature* 141 (1938): 747–48; W. T. Astbury and F. O. Bell, "Some Recent Developments in

the X-ray Study of Proteins and Related Structures," *Cold Spring Harbor Symposia on Quantitative Biology* 6 (1938): 109–18; W. T. Astbury, "X-ray Studies of the Structure of Compounds of Biological Interest," *Annual Review of Biochemistry* 8 (1939): 113–33; W. T. Astbury, "Adventures in Molecular Biology," Harvey Lecture for 1950, *Harvey Society Lectures* 46 (1950): 3–44.

29. Mansel Davies, "W. T. Astbury, Rosie Franklin, and DNA: A Memoir," *Annals of Science* 47 (1990): 607–18, quote is on 609; Hall, *The Man in the Monkeynut Coat*, 67–72, 91–102.

30. Astbury and Bell, "X-ray Study of Thymonucleic Acid." In 1951, a full year before Franklin took her famous Photograph No. 51, Elwyn Beighton, Astbury's research assistant, produced an X-ray photograph that showed a similar "Maltese cross" pattern. Astbury paid little attention to Beighton's picture; indeed, he could not "see" the helical form that would soon make Watson and Crick famous. Hall, "William Astbury and the biological significance of nucleic acids"; Davies, "W. T. Astbury, Rosie Franklin, and DNA."

31. Astbury, "X-ray Studies of Nucleic Acids," 68; Astbury and Bell, "X-ray Study of Thymonucleic Acid"; Horace Freeland Judson, *The Eighth Day of Creation: The Makers of the Revolution in Biology* (Cold Spring Harbor, NY: Cold Spring Harbor Laboratory Press, 1996), 93.

32. Watson recalled Astbury telling dirty jokes in author interview (no. 2) with James D. Watson, July 24, 2018. The letters of invitation from the Stazione and Astbury's responses and travel plans to Naples can be found in W. T. Astbury Papers, MS 419/File 4: Conference on Submicroscopic Structure of Protoplasm, May 22–25, 1951, University of Leeds; and letters from W. T. Astbury to Reinhold Dohrn regarding the conference, Archives of the Naples Zoological Station, Correspondence, A:SZN, 1951.

33. Astbury, "Protein and virus studies in relation to the problem of the gene"; Astbury, "X-ray Studies of the Structure of Compounds of Biological Interest"; Hall, *The Man in the Monkeynut Coat*, 100.

34. W. T. Astbury, "Some Recent Adventures Among Proteins," and H. M. Kalckar, "Biosynthetic aspects of nucleosides and nucleic acids," in *Pubblicazioni della Stazione Zoologica di Napoli* 23, supplement (1951): 1–18 and 87–103.

35. Author interview with James D. Watson (no. 1), July 23, 2018.

36. Hall, *The Man in the Monkeynut Coat*, 121–22.

37. Watson, *The Double Helix*, 23.

38. Letter from John Randall to Reinhard Dohrn, August 11, 1950 ("During the course of the visit I should like to spend two or three days in Naples collecting spermatozoa for our electron microscope research programme"), Archives of the Naples Zoological Station, Correspondence A.1950 (J–Z).

39. Maurice Wilkins, *The Third Man of the Double Helix* (Oxford: Oxford University Press, 2003), 135–39.

40. Maurice Wilkins, "The molecular configuration of nucleic acids," December 11, 1962, in *Nobel Lectures, Physiology or Medicine 1942–1962* (Amsterdam: Elsevier, 1964).

41. Anne Sayre interview with Raymond Gosling, May 18, 1970, ASP, box 4, folder 2.

42. Naomi Attar, "Raymond Gosling: The Man Who Crystallized Genes," *Genome Biology* 14 (2013): 402; Matthew Cobb, *Life's Greatest Secret: The Race to Crack the Genetic Code* (New York: Basic Books, 2015), 93.

43. Watson, *The Double Helix*, 23.

44. M. H. F. Wilkins, "I: Ultraviolet dichroism and molecular structure in living cells. II. Electron Microscopy of nuclear membranes," *Pubblicazioni della Stazione Zoologica di Napoli* 23, supplement (1951): 104–14. At lunch directly following Wilkins's presentation, Watson paired off with an Italian marine biologist named Elvezio Ghiradelli. All the while, Watson doodled onto a napkin versions of the X-ray photographs Wilkins had just projected and, when the meal ended, "he threw the napkin away!" Christiane Groeben, Archivist Emerita, Naples Zoological Station, email to the author, February 15, 2019.

45. Letter from Reinhard Dohrn to John Randall, May 31, 1951, Archives of the Naples Zoological Station, ASZN:R, Correspondence I–Z, 1951.

46. Wilkins, *The Third Man of the Double Helix*, 137.

47. Watson, *The Double Helix*, 23.

48. Pellegrino Claudio Sestieri, *Paestum: The City, the Prehistoric Necropolis in Contrada Gaudo, the Heraion at the Mouth of the Sele* (Rome: Istituto Poligrafico Dello Stato, 1967); Gabriel Zuchtriegel and Marta Ilaria Martorano, *Paestum: From Building Site to Temple* (Naples: Parco archeologico di Paestum minister dei beni e delle attività culturali, 2018); Paul Blanchard, *Blue Guide to Southern Italy* (New York: Norton, 2007), 271–79.

49. File "James D. Watson and his Sister's Tour of Europe," JDWP, JDW/1/1/30, which includes photographs of Betty and Jim Watson on trips to Salzburg, the Alps, Vienna, Paris, Bavaria, Munich, Brussels, Copenhagen, Florence, Rome, Bern, and Venice, including a shot of a young Jim standing in front of the Coliseum; letter from James D. Watson to Elizabeth Watson, January 8, 1951, regarding her plans to apply to Oxford and Cambridge, JDWP, "James D. Watson Letters" (1 of 5), JDW/2/2/1934.

50. Letter from Henri Chantrenne to Reinhold Dohrn, May 27, 1951, Archives of the Naples Zoological Station, H:SZN, 1951; H. Chantrenne, "Recherches sur le mécanisme de la synthèse des protéines" (Research on the mechanism of protein synthesis), *Pubblicazioni della Stazione Zoologica di Napoli* 23, supplement (1951), 70–86. Commenting on this paper, Astbury said, "I am particularly interested in the problem of the interplay between protein and nucleic acids in biogenesis . . . certain nucleoprotein combinations for which I have suggested the name 'viable growth complexes,' have the power, the minimum essential of reproduction, of

making exact copies of themselves given the right physico-chemical environment (whatever that may mean!) and we will not progress very far until we have found out the common structural principle underlying this property" (82).

51. Watson, *The Double Helix*, 23–24.

52. Watson, *The Double Helix*, 24.

53. Wilkins, *The Third Man of the Double Helix*, 139.

54. Attar, "Raymond Gosling."

55. James D. Watson, *The Annotated and Illustrated Double Helix*, edited by Alexander Gann and Jan Witkowski (New York: Simon and Schuster, 2012), 27.

56. Watson, *The Double Helix*, 31.

Chapter 10: From Ann Arbor to Cambridge

1. Letter from Salvador Luria to James D. Watson, October 20, 1951, WFAT, "DNA Letters."

2. "The Summer Symposium on Theoretical Physics at the University of Michigan," *Science* 83, no. 2162 (June 5, 1936): 544; "Calendar of Events," *Physics Today* 3, no. 6 (1950): 40; James Tobin, "Summer School for Geniuses," *Michigan Today*, November 10, 2010, https://michigantoday.umich.edu/2010/11/10/a7892/; Alaina G. Levine, "Summer Symposium in Theoretical Physics, University of Michigan, Ann Arbor, Michigan," APS Physics, https://www.aps.org/programs/outreach/history/historicsites/summer.cfm.

3. Sutherland later became director of Britain's National Physics Laboratory, 1956–64, and Master of Emmanuel College, Cambridge, 1964–77. See Norman Sheppard, "Gordon Brims Black McIvor Sutherland, 8 April 1907–27 June 1980," *Biographical Memoirs of Fellows of the Royal Society* 28 (1982): 589–626.

4. The biophysicists offered thirty-six lectures to "audiences composed of graduate students and staff members of the departments of Bacteriology, Biochemistry, Botany, Medicine, Physics, Public Health and Zoology": *The President's Report to the Board of Regents of the University of Michigan for the Academic Year 1951*, 191; *Proceedings of the Board of Regents of the University of Michigan, 1951–1954*: September 1951 meeting, 80, October 1951 meeting, 182; Sheppard, "Gordon Brims Black McIvor Sutherland"; Samuel Krimm, "On the Development of Biophysics at the University of Michigan," Michigan Physics, Histories of the Michigan Physics Department, https://michiganphysics.com/2012/06/24/development-of-biophysics-at-michigan/.

5. Sinclair Lewis, *Arrowsmith* (New York: Harcourt, Brace, 1925), 7. This was not Watson's first close encounter with the University of Michigan. He spent the summer of 1946 at its Biological Station on Douglas Lake, in northern Michigan. There, he waited tables to pay his tuition for two courses, "Systematic Botany" and "Advanced Ornithology," lived in a tented cabin, and briefly acquired the

unfortunate nickname Jimbo. James D. Watson, *Avoid Boring People: Lessons from a Life in Science* (New York: Knopf, 2007), 29.

6. Wilfred B. Shaw, *The University of Michigan: An Encyclopedic Survey,* vol. 1 (Ann Arbor: University of Michigan Press, 1942), 206.

7. Letter from James D. Watson to his parents, September 24, 1951, WFAT, "Letters to Family, Copenhagen, 1951"; George Santayana, *The Last Puritan: A Memoir in the Form of a Novel* (New York: Charles Scribner's Sons, 1936). The novel took Santayana forty-five years to complete and sold more copies than any book published in 1936 save Margaret Mitchell's *Gone With the Wind.*

8. James D. Watson, *The Double Helix: A Personal Account of the Discovery of the Structure of DNA,* edited by Gunther Stent (New York: Norton, 1980), 24–25; L. C. Pauling, R. B. Corey, and H. R. Branson, "The structure of proteins; two hydrogen-bonded helical configurations of the polypeptide chain," *Proceedings of the National Academy of Sciences* 37, no. 4 (1951): 205–11.

9. Watson, *The Double Helix,* 25.

10. Watson, *The Double Helix,* 24–25.

11. Letter from James D. Watson to his parents, July 12, 1951, WFAT, "Letters to Family, Copenhagen, 1951."

12. Letter from James D. Watson to Elizabeth Watson, July 14, 1951, JDWP, "James D. Watson Letters" (1 of 5), JDW/2/2/1934.

13. Horace Freeland Judson, *The Eighth Day of Creation: The Makers of the Revolution in Biology* (New York: Simon and Schuster, 1979), 97; Torbjörn Caspersson, "The Relations Between Nucleic Acid and Protein Synthesis," *Symposia of the Society for Experimental Medicine* 1 (1947): 127–51; R. Signer, T. Caspersson, and E. Hammarsten, "Molecular Shape and Size of Thymonucleic Acid," *Nature* 141 (1938): 122; G. Klein and E. Klein, "Torbjörn Caspersson, 15 October 1910–7 December 1997," *Proceedings of the American Philosophical Society* 147, no. 1 (2003): 73–75.

14. James D. Watson, Merck/National Research Council Fellowship correspondence, 1950–52, JDWP, JDW/2/2/1284.

15. Horace Judson interview with John Kendrew, November 11, 1975, HFJP.

16. Author interview with James D. Watson (no. 1), July 23, 2018.

17. Letter from James D. Watson to his parents, August 21, 1951, WFAT, "Letters to Family, Copenhagen, 1951."

18. Letter from James D. Watson to his parents, August 27, 1951, WFAT, "Letters to Family, Copenhagen, 1951."

19. James D. Watson, fellowship applications and correspondence with the National Foundation for Infantile Paralysis, 1951–53, JDWP, JDW/2/2/1276; letter from James D. Watson to his parents, August 27, 1951, WFAT, "Letters to Family, Copenhagen, 1951." See also Niels Bohr, "Medical Research and Natural Philosophy," Basil O'Connor, "Man's Responsibility in the Fight Against Disease," and

Max Delbrück, "Virus Multiplication and Variation," in International Poliomy-elitis Congress, *Poliomyelitis: Papers and Discussions Presented at the Second International Poliomyelitis Conference* (Philadelphia: J. B. Lippincott, 1952), xv–xviii, xix–xxi; 13–19. The conference was hosted by the Medicinsk–Anatomisk Institut, University of Copenhagen, September 3–7, 1951.

20. Howard Markel, "April 12, 1955: Tommy Francis and the Salk Vaccine," *New England Journal of Medicine* 352 (2005): 1408–10.

21. Watson, *The Double Helix*, 28.

22. Jane Smith, *Patenting the Sun: Polio and the Salk Vaccine* (New York: William Morrow, 1990), 171–72.

23. Letter from James D. Watson to his parents, September 15, 1951, WFAT, "Letters to Family, Copenhagen, 1951." In a letter dated September 29, he gives his new address as "Cavendish Laboratory, Cambridge England"; see Watson, *The Double Helix*, 28.

24. Letter from James D. Watson to C. J. Lapp, undated, early October 1951, WFAT, quoted in Watson, *The Annotated and Illustrated Double Helix*, 273.

25. Letter from Herman Kalckar to C. J. Lapp, October 5, 1951, JDWP, JDW/2/2/1284, "James Watson's Merck/National Research Council Fellowship Correspondence, 1950–1952."

26. Letter from James D. Watson to Elizabeth Watson, October 16, 1951, WFAT; quoted in Watson, *The Annotated and Illustrated Double Helix*, 275.

27. George H. F. Nuttall, "The Molteno Institute for Research in Parasitology, University of Cambridge, with an Account of How it Came to be Founded," *Parasitology* 14, no. 2 (1922): 97–126; S. R. Elsden, "Roy Markham, 29 January 1916–16 November 1979," *Biographical Memoirs of Fellows of the Royal Society* 28 (1982): 319–314; 319–45.

28. Letter from Salvador Luria to Paul Weiss, October 20, 1951, JDWP, JDW/2/2/1284, "James Watson's Merck/National Research Council Fellowship Correspondence, 1950–1952."

29. Watson, *The Annotated and Illustrated Double Helix*, 275.

30. Watson, *The Double Helix*, 30.

31. Letter from Paul Weiss to James D. Watson, October 22, 1951, JDWP, JDW/2/2/1284, "James Watson's Merck/National Research Council Fellowship Correspondence, 1950–1952."

32. Watson, *The Double Helix*, 30–31.

33. Letter from Catherine Worthingham, Director of Professional Education, NFIP, to James D. Watson, October 29, 1951, JDWP, JDW/2/2/1276, "James D. Watson Fellowship Applications and Correspondence to the National Foundation for Infantile Paralysis, 1951–1953." This letter was correctly addressed to Watson at the Cavendish Laboratory.

34. Letter from James Watson to Paul Weiss, November 13, 1951. A carbon copy

of this letter is dated November 14, 1951, but is otherwise the same. JDWP, JDW/2/2/1284, "James Watson's Merck/National Research Council Fellowship Correspondence, 1950–1952."

35. James D. Watson to C. J. Lapp, November 27, 1951 (see also C. J. Lapp to James D. Watson, November 21, 1951), JDWP, JDW/2/2/1284, "James Watson's Merck/National Research Council Fellowship Correspondence, 1950–1952"; quoted in Watson, *The Annotated and Illustrated Double Helix*, 277–78.

36. Letter from James D. Watson to his parents, November 28, 1951, WFAT, "Letters to Family, Cambridge, October 1951–August 1952."

37. Letter from James D. Watson to Elizabeth Watson, November 28, 1951, WFAT, "Letters to Family, Cambridge, October 1951–August 1952."

38. Letter from James D. Watson to Max Delbrück, December 9, 1951, MDP, box 23, folder 20.

39. Letter from James D. Watson to his parents, January 8, 1952, WFAT, "Letters to Family, Cambridge, October 1951–August 1952."

40. Letter from James D. Watson to his parents, January 18, 1951, WFAT, "Letters to Family, Cambridge, October 1951–August 1952."

41. National Research Council Merck Fellowship Board, minutes of meeting March 16, 1952, National Academy of Sciences Archives; quoted in Watson, *The Annotated and Illustrated Double Helix,* 279.

42. Letter from Salvador Luria to James D. Watson, March 5, 1952, JDWP, JDW 2/2/1284; see Watson, *The Annotated and Illustrated Double Helix,* 109 and 280.

43. Letter from James D. Watson to his parents, October 9, 1951, WFAT, "Letters to Family, Cambridge, October 1951–August 1952."

Chapter 11: An American in Cambridge

1. James D. Watson, *The Double Helix: A Personal Account of the Discovery of the Structure of DNA*, edited by Gunther Stent (New York: Norton, 1980), 31.

2. Horace Judson interview with John Kendrew, November 11, 1975, HFJP.

3. Georgina Ferry, *Max Perutz and the Secret of Life* (London: Chatto and Windus, 2007), 1–53; Max F. Perutz, "X-Ray Analysis of Hemoglobin," December 11, 1962, in *Nobel Lectures, Chemistry 1942–1962* (Amsterdam: Elsevier, 1964), 653–73; D. M. Blow, "Max Ferdinand Perutz, OM, CH, CBE. 19 May 1914–6 February 2002," *Biographical Memoirs of Fellows of the Royal Society* 50 (2004): 227–56; Alan R. Fersht, "Max Ferdinand Perutz, OM, FRS," *Nature Structural Biology* 9 (2002): 245–46.

4. Ferry, *Max Perutz and the Secret of Life,* 26; Blow, "Max Ferdinand Perutz."

5. Max F. Perutz, "True Science," review of *Advice to a Young Scientist* by P. B. Medawar, *London Review of Books*, March 19, 1981.

6. Max F. Perutz, "How the Secret of Life Was Discovered," *I Wish I'd Made You*

Angry Earlier: Essays on Science, Scientists and Humanity (Cold Spring Harbor, NY: Cold Spring Harbor Laboratory Press, 2003), 197–206, quote is on 204.

7. Watson, *The Double Helix*, 28.

8. Watson, *The Double Helix*, 28–29.

9. Watson, *The Double Helix*, 29.

10. Letter from James D. Watson to Elizabeth Watson, September 12, 1951, JDWP, JDW/2/2/1934.

11. K. C. Holmes, "Sir John Cowdery Kendrew, 24 March 1917–23 August 1997," *Biographical Memoirs of Fellows of the Royal Society* 47 (2001): 311–32; John C. Kendrew, *The Thread of Life: An Introduction to Molecular Biology* (Cambridge, MA: Harvard University Press, 1968); Soraya de Chadarevian, "John Kendrew and Myoglobin: Protein Structure Determination in the 1950s," *Protein Science* 27, no. 6 (2018): 1136–43.

12. Watson, *The Double Helix*, 29.

13. Author interview with James D. Watson (no. 1), July 23, 2018.

14. Watson, *The Double Helix*, 31.

15. Letter from James D. Watson to his parents, October 9, 1951, WFAT, "Letters to Family, Cambridge, October 1951–August 1952."

16. Watson, *The Double Helix*, 31.

17. The Kendrews divorced in 1956. Author interview with James D. Watson (no. 3), July 25, 2018; Paul M. Wasserman, *A Place in History: The Biography of John C. Kendrew* (New York: Oxford University Press, 2020), 130–36.

18. Watson, *The Double Helix*, 31.

19. Letter from James D. Watson to his parents, October 16, 1951, WFAT, "Letters to Family, Cambridge, October 1951–August 1952"; Denys Haigh Wilkinson, "Blood, Birds and the Old Road," *Annual Review of Nuclear Particle Science* 45 (1995): 1–39. Wilkinson was on the Cavendish staff from 1947 to 1957, before moving to Oxford. Interestingly, Sir William Lawrence Bragg was also an avid birdwatcher.

20. Author interview with James D. Watson (no. 2), July 24, 2018.

21. Sherwin B. Nuland, "The Art of Incision," *New Republic*, August 13, 2008, https://newrepublic.com/article/63327/the-art-incision.

22. Horace Judson interview with John Kendrew, November 11, 1975, HFJP.

23. Watson, *The Double Helix*, 31.

24. Francis Crick, *What Mad Pursuit: A Personal View of Scientific Discovery* (New York: Basic Books, 1988), 64.

25. Crick, *What Mad Pursuit*, 64.

26. Francis Crick interviewed on *The Prizewinners*, BBC Television, December 11, 1962; Horace Freeland Judson, *The Eighth Day of Creation: Makers of the Revolution in Biology* (Cold Spring Harbor, NY: Cold Spring Harbor Laboratory Press, 2013), 125.

27. Letter from James D. Watson to Max Delbrück, December 5, 1951, MDP, box 23, folder 20.

28. Erwin Chargaff, "A Quick Climb Up Mount Olympus," review of *The Double Helix* by James D. Watson, *Science* 159, no. 3822 (1968): 1448–49.

29. Crick, *What Mad Pursuit*, 65.

30. Matt Ridley, *Francis Crick: Discoverer of the Genetic Code* (New York: Harper Perennial, 2006), 50; email from Malcolm Longair to the author, June 12, 2020. When I visited Room 103 on February 19, 2018, just prior to the Austin Wing's demolition, it was a storeroom for the zoology department, filled to the ceiling with shelves and boxes containing the disarticulated skeletons of cows and other large creatures.

31. Letter from James D. Watson to his parents, November 4, 1951, WFAT, "Letters to Family, Cambridge, October 1951–August 1952."

32. Anne Sayre, *Rosalind Franklin and DNA* (New York: Norton, 1975), 131.

33. "The Race for the Double Helix," documentary television program, narrated by Isaac Asimov, *Nova*, PBS, March 7, 1976.

34. Watson, *The Double Helix*, 31–32.

35. Watson, *The Double Helix*, 13.

36. Watson, *The Double Helix,* 34.

37. Watson, *The Double Helix*, 34.

38. Watson, *The Double Helix*, 36.

39. Watson, *The Double Helix*, 37.

40. Crick, *What Mad Pursuit*, 65.

41. Watson, *The Double Helix*, 43.

42. Victor K. McElheny, *Watson and DNA: Making a Scientific Revolution* (New York: Perseus, 2003), 40.

43. Watson, *The Double Helix*, 37.

Chapter 12: The King's War

1. Horace Judson interview with Maurice Wilkins, March 12, 1976, HFJP.

2. Muriel Franklin, "Rosalind," privately printed obituary pamphlet, 16–17, RFP, "Articles and Obituaries," FRKN 6/6.

3. In the decade after the Second World War, there were approximately 400,000 Jews in Great Britain, compared to roughly 300,000 in 1933—the result of refugee migration. George Orwell, "Anti-Semitism in Britain," *Contemporary Jewish Record*, April 1945, reprinted in George Orwell, *Essays* (New York: Everyman's Library/ Knopf, 2002), 847–56; Eli Barnavi, *A Historical Atlas of the Jewish People: From the Time of the Patriarchs to the Present* (New York: Schocken, 1992); United States Holocaust Memorial Museum, "Jewish Population of Europe in 1933: Population Data by Country," *Holocaust Encyclopedia*, https://encyclopedia

.ushmm.org/content/en/article/jewish-population-of-europe-in-1933-population
-data-by-country.

4. Horace Judson interview with John Kendrew, November 11, 1975, HFJP.

5. Anne Sayre interview with Francis Crick, June 16, 1970, ASP, box 2, folder 9.

6. Anne Sayre interview with Geoffrey Brown, May 12, 1970, ASP, box 2, folder 3.

7. Horace Judson interview with Raymond Gosling, July 21, 1975, HFJP.

8. Raymond Gosling interview in "The Secret of Photo 51," documentary television program, *Nova*, PBS, April 22, 2003, https://www.pbs.org/wgbh/nova/transcripts/3009_photo51.html.

9. Anne Sayre interview with Raymond Gosling, May 18, 1970, ASP, box 4, folder 2.

10. Anne Sayre interview with Maurice Wilkins, June 15, 1970, 18, ASP, box 4, folder 32.

11. Letter from Maurice Wilkins to Horace Judson, July 12, 1976, HFJP.

12. Brenda Maddox, *Rosalind Franklin: The Dark Lady of DNA* (New York: HarperCollins, 2002), 146.

13. Author interview with Jenifer Glynn, May 7, 2018.

14. Horace Freeland Judson, *The Eighth Day of Creation: Makers of the Revolution in Biology* (Cold Spring Harbor, NY: Cold Spring Harbor Laboratory Press, 2013), 82–83; Maddox, *Rosalind Franklin*, 129.

15. Naomi Attar, "Raymond Gosling: The Man Who Crystallized Genes," *Genome Biology* 14 (2013): 402.

16. Raymond G. Gosling, "X-ray Diffraction Studies with Rosalind Franklin," in Seweryn Chomet, ed., *Genesis of a Discovery* (London: Newman Hemisphere, 1995), 43–73, quote is on 52.

17. Wilkins, *The Third Man of the Double Helix*, pp. 129–30.

18. Wilkins, *The Third Man of the Double Helix*, 130.

19. Wilkins, *The Third Man of the Double Helix*, 130.

20. Wilkins, *The Third Man of the Double Helix*, 132. He later claimed to have found the exchange amusing and the result of "not having had the advantage of living in post-war Paris where food was not rationed as it had been in Britain." In contrast to those living elsewhere, he had simply "forgotten what real cream was like." Rationing was not limited to Britain. Even as late as 1949 to 1950, "World War II still cast a shadow over France. Heat and hot water were scarce; baths were limited to once a week. Everyone . . . had a ration card for coffee and sugar"; Ann Mah, "After She Had Seen Paris," *New York Times*, June 30, 2019, TR1. See also Alice Kaplan, *Dreaming in French: The Paris Years of Jacqueline Bouvier Kennedy, Susan Sontag, and Angela Davis* (Chicago: University of Chicago Press, 2012), 7–80.

21. Wilkins, *The Third Man of the Double Helix*, 133. He recalled that at this time, "In any case my interests were fairly heavily occupied by Edel [Lange], whom I visited that summer at her family home in Berlin."

22. Judson, *The Eighth Day of Creation*, 626–27; see also Horace Judson interview with Sylvia Jackson, June 30, 1976, HFJP, "Women at King's College."

23. Anne Sayre, *Rosalind Franklin and DNA* (New York: Norton, 1975), 76–107; Maddox, *Rosalind Franklin*, 127–28, 134. Watson claims that Franklin was angry because "the women's combination room remained dingily pokey whereas money had been spent to make life agreeable for [Maurice] and his friends when they had their morning coffee": Watson, *The Double Helix*, 15.

24. Anne Sayre interview with Raymond Gosling, May 18, 1970, ASP, box 4, folder 2.

25. Margaret Wertheim, *Pythagoras's Trousers: God, Physics, and the Gender War* (New York: Norton, 1997), 12; Maddox, *Rosalind Franklin*, 134.

26. Letter from Maurice Wilkins to Horace Judson, April 28, 1976, HFJP.

27. As the King's MRC unit's Senior Biological Advisor, Fell "came in every week to give an experienced ear and council to each research team." Judson, *The Eighth Day of Creation*, 625–26; Horace Judson interview with Dame Honor Fell, January 28, 1977, HFJP, "Women at King's College."

28. They were Dr. E. Jean Hanson, Dr. Angela Martin Brown, Dr. Marjorie B. M'Ewan, Miss M. I. Pratt, Dr. Rosalind Franklin, Miss Pauline Cowan Harrison, Miss J. Towers, Dr. Mary Fraser, and a lab technician named Sylvia Fitton Jackson, who had published papers and later took a PhD degree at Randall's urging. Judson interviewed or corresponded with seven of these women; the transcripts of his interviews, and letters from the interviewees, are filed in HFJP under "Women at King's College" (Brown, Fell, Harrison, Jackson, North). Franklin and Hanson had died when he began his research, and he was unable to trace Miss Towers. He corresponded with M'Ewan but did not interview her in person. I am deeply indebted to Charles Griefenstein, chief archivist at the American Philosophical Society, for unsealing these critical documents for my review. See also Judson, *The Eighth Day of Creation*, 625–26; Maddox, *Rosalind Franklin*, 137; MRC Biophysics/Biophysics Research Unit, King's College London, PP/HBF/C.10, box 4, and MRC Biophysics/Biophysics Research Unit, King's College London, PP/HBF/C.11, box 4, Honor Fell Papers, Wellcome Library, London.

29. Judson, *The Eighth Day of Creation*, 626.

30. Sayre, *Rosalind Franklin and DNA*, 96–97. Sayre notes that, as of 1971, Wilkins had never directed a female PhD student at King's College (107).

31. Robert Olby, *The Path to the Double Helix* (Seattle: University of Washington Press, 1974), 331; W. E. Seeds and M. H. F. Wilkins, "A Simple Reflecting Microscope," *Nature* 164 (1949): 228–29; W. E. Seeds and M. H. F. Wilkins, "Ultraviolet Micrographic Studies of Nucleoproteins and Crystals of Biological Interest," *Discussions of the Faraday Society* 9 (1950): 417–23; M. H. F. Wilkins, R. G. Gosling and W. E. Seeds, "Physical Studies of Nucleic Acid," *Nature* 167 (1951): 759–60; M. H. F. Wilkins, W. E. Seeds, A. R. Stokes, H. R. Wilson, "Heli-

cal Structure of Crystalline Deoxypentose Nucleic Acid," *Nature* 172 (1953): 759–62.

32. Maddox, *Rosalind Franklin*, 160.

33. Maddox, *Rosalind Franklin*, 160, 256. Other of Seeds's nicknames included "Uncle" for Wilkins and "Aunty" for Honor Fell. He called Stokes "Archangel Gabriel."

34. Maddox, *Rosalind Franklin*, 288.

35. Maddox, *Rosalind Franklin*, 160, 288. Maddox notes that "she would accept 'Ros'" from close friends and family members. She also notes that the Reuters journalist Rosanna Groarke would "routinely, although no one else did, refer to her as 'Rosie.'"

36. Anne Sayre interview with Maurice Wilkins, June 15, 1970, ASP, box 4, folder 32.

37. Maddox, *Rosalind Franklin*, 160–61.

38. Maddox, *Rosalind Franklin*, 146.

39. Sayre, *Rosalind Franklin and DNA*, 102–3.

40. Anne Sayre interview with Raymond Gosling, May 18, 1970, ASP, box 4, folder 2.

41. Author interview with Jenifer Glynn, May 7, 2018.

42. Sayre, *Rosalind Franklin and DNA*, 105.

43. Letter from Mary Fraser to Horace Judson, August 22, 1978, HFJP, "Women at King's College."

44. Letter from Marjorie M'Ewan to Horace Judson, September 15, 1976, HFJP, "Women at King's College;" Judson, *The Eighth Day of Creation*, 625–26.

45. Anne Sayre interview with Raymond Gosling, May 18, 1970, ASP, box 4, folder 2; Sayre, *Rosalind Franklin and DNA*, 102–3.

46. Maddox, *Rosalind Franklin*, 145–47. She said to her former laboratory mate Vittorio Luzzati, of Wilkins, "He's so middle-class, Vittorio!"

47. In the coming months, after she obtained the proper camera setup, Franklin showed that "the lengthening of the DNA fibers was the same as the increase in the periodicity of the diffraction pattern," and "the change of fiber length resulted from the helices of DNA partially uncoiling and lengthening": Wilkins, *The Third Man of the Double Helix*, 134. See also Sayre, *Rosalind Franklin and DNA*, 103–4.

48. Sayre, *Rosalind Franklin and DNA*, 104.

49. Wilkins, *The Third Man of the Double Helix*, 134–35.

50. Maddox, *Rosalind Franklin*, 144.

51. Sayre, *Rosalind Franklin and DNA*, 104.

52. Anne Sayre interview with Maurice Wilkins, June 15, 1970, ASP, box 4, folder 32.

53. Wilkins, *The Third Man of the Double Helix*, 134–35.

54. Olby, *The Path to the Double Helix*, 341.

55. Wilkins, *The Third Man of the Double Helix*, 142.

56. Wilkins, *The Third Man of the Double Helix*, 142–43.

57. Letter from John Randall to Rosalind Franklin, December 4, 1950, JRP, RNDL 3/1/6.
58. Maddox, *Rosalind Franklin*, 150.
59. Wilkins, *The Third Man of the Double Helix*, 150–51.
60. Letter from Maurice Wilkins to Rosalind Franklin, July 1951, MWP, K/PP178/3/9.
61. Prime Minister Neville Chamberlain's disastrous policy of appeasement with Hitler was articulated in his infamous "Peace for our Time" speech, September 30, 1938. Anne Sayre interview with Maurice Wilkins, June 15, 1970, 11–12, ASP, box 4, folder 32.
62. Anne Sayre interview with Maurice Wilkins, June 15, 1970, 5, ASP, box 4, folder 32.
63. Wilkins, *The Third Man of the Double Helix,* 157–58.
64. Wilkins, *The Third Man of the Double Helix*, 156.
65. Letter from Muriel Franklin to Anne Sayre, November 23, 1969, ASP, box 2, folder 15.1.
66. Letter from Rosalind Franklin to Adrienne Weill, October 21, 1941, ASP, box 3, folder 1.

Chapter 13: The Lecture

1. James D. Watson, *The Double Helix: A Personal Account of the Discovery of the Structure of DNA*, edited by Gunther Stent (New York: Norton, 1980), 14. The term "bluestocking," for an intellectual or literary woman, originates with a mid- to late-eighteenth-century cohort of British feminists who were members of the Blue Stockings Society, which emphasized education, social cooperation, and the pursuit of intellectual accomplishments. Many of the women in the society were not wealthy enough to afford silk stockings or fancy clothes and, instead, wore worsted wool ones. See Gary Kelly, ed., *Bluestocking Feminism: Writings of the Bluestocking Circle, 1738–1785*, 6 vols. (London: Pickering & Chatto, 1999).
2. Letter from Muriel Franklin to Anne Sayre, undated, mid-April to early May 1970 (certainly after the publication of Watson's *The Double Helix*), ASP, box 2, folder 15.1.
3. Brenda Maddox, *Rosalind Franklin: The Dark Lady of DNA* (New York: Harper-Collins, 2002), 138.
4. Letter from Anne Sayre to Gertrude Clark Dyche, June 28, 1978, ASP, box 7, "Post-Publication Correspondence A–E"; Maddox. *Rosalind Franklin*, 52–53, 138–39.
5. Muriel Franklin, "Rosalind," 16, privately printed obituary pamphlet, RFP, "Articles and Obituaries," FRKN 6/6. Brenda Maddox in *Rosalind Franklin* (138) describes the flat as having four rooms; Jenifer Glynn noted in an interview with

the author, May 7, 2018, that there was one bedroom, a living/dining room, a full bathroom, and a kitchen.

6. Maddox, *Rosalind Franklin,* 139–140; Anne Sayre interview with Mrs. Simon Altmann, May 15, 1970, ASP, box 2, folder 2; Anne Sayre interview with Geoffrey Brown, May 12, 1970, ASP, box 2, folder 3.

7. Meteorological Office, United Kingdom, *British Rainfall, 1951. The 91st Annual Volume of the British Rainfall Organization. Report on the Distribution of Rain in Space and Time Over Great Britain and Northern Ireland During the 1951 as Recorded by About 5,000 Observers* (London: Her Majesty's Stationery Office, 1953), 17–18, 81–82.

8. Horace Judson interview with Raymond Gosling, July 21, 1975, HFJP.

9. Muriel Franklin, "Rosalind," 10; Maddox, *Rosalind Franklin,* 21.

10. Letter from Muriel Franklin to Anne Sayre, undated, probably mid-April to early May, 1970 (certainly after the publication of Watson's *The Double Helix*), ASP, box 2, folder 15.1.

11. King's College, London, "Strand Campus: Self-Guided Tour," pamphlet, 2; correspondence from Ben Barber, King's College, London, Archives, to the author, July 19, 2019. The building was designed by the architect Sir Robert Smirk, who also drew up plans for portions of the British Museum and the Royal Opera House at Covent Garden.

12. Maddox, *Rosalind Franklin,* 135, 255–56.

13. Maurice Wilkins, *The Third Man of the Double Helix* (Oxford: Oxford University Press, 2003), 163.

14. Robert Olby, *The Path to the Double Helix* (Seattle: University of Washington Press, 1974), 348.

15. Horace Judson interview with Alexander Stokes, August 11, 1976, HFJP. Wilkins recalled this lecture in his 2003 memoir even more hazily: "I don't think he attempted to link his work with Rosalind's new B pattern." This is a puzzling comment because if Stokes gave his talk before Franklin, it appears unlikely that he would go off point to discuss her data. See Wilkins, *The Third Man of the Double Helix,* 163.

16. François Jacob, *The Statue Within: An Autobiography* (Cold Spring Harbor, NY: Cold Spring Harbor Laboratory Press, 1995), 264.

17. Horace Freeland Judson, *The Eighth Day of Creation: Makers of the Revolution in Biology* (Cold Spring Harbor, NY: Cold Spring Harbor Laboratory Press, 2013), 97.

18. Letter from James D. Watson to his parents, November 20, 1951, WFAT, "Letters to Family, Cambridge, October 1951–August 1952."

19. Victor K. McElheny, *Watson and DNA: Making a Scientific Revolution* (New York: Perseus, 2003), 40.

20. Watson, *The Double Helix,* 44–45.

21. Watson, *The Double Helix*, 45.

22. Olby, *The Path to the Double Helix*, 316.

23. Watson, *The Double Helix*, 59.

24. Watson, *The Double Helix*, 45.

25. Wilkins, *The Third Man of the Double Helix*, 163–64.

26. Wilkins, *The Third Man of the Double Helix*, 164.

27. Anne Sayre interview with Maurice Wilkins, June 15, 1970, ASP, box 4, folder 32.

28. Letter from Maurice Wilkins to Robert Olby, December 18, 1972 (returning the author's manuscript chapters 19, 20, 21, with annotations), quoted in Olby, *The Path to the Double Helix*, 350. Olby is highly skeptical of this claim and, following the quote, writes, "But these speculations were very limited and they were supported by her realization that she had to do with near-hexagonal packing, indicative of cylindrical molecules. Surely she had every reason to refer to this feature and its significance in her talk?"

29. Letter from Maurice Wilkins to Horace Judson, April 28, 1976, HFJP.

30. Wilkins, *The Third Man of the Double Helix*, 163–64; see also letter from Maurice Wilkins to Robert Olby, December 18, 1972, quoted in Olby, *The Path to the Double Helix*, 350.

31. Horace Judson interview with Alexander Stokes, August 11, 1976, HFJP.

32. Rosalind Franklin, Colloquium, November 1951, RFP, FRKN 3/2; Rosalind Franklin, "Interim Annual Report: January 1, 1951–January 1, 1952," Wheatstone Laboratory, King's College, London, February 7, 1952, RFP, FRKN 4/3; Rosalind Franklin DNA research notebooks, September 1952–May 1953, RFP, FRKN 1/1.

33. Watson, *The Double Helix*, 45; Judson, *The Eighth Day of Creation*, 98.

34. Anne Sayre interview with Raymond Gosling, May 18, 1970, ASP, box 4, folder 2.

35. Franklin, Colloquium, November 1951.

36. Aaron Klug, "Rosalind Franklin and the Discovery of the Structure of DNA," *Nature* 219, no. 5156 (1968): 808–10, 843–44; Aaron Klug, "Rosalind Franklin and the Double Helix," *Nature* 248 (1974): 787–88.

37. C. Harry Carlisle, "Serving My Time in Crystallography at Birkbeck: Some Memories Spanning 40 Years." Unpublished lecture, partly delivered as a valedictory lecture at Birkbeck College, May 30, 1978. Birkbeck College, University of London Library and Repository Services. I am indebted to Sarah Hall and Emma Illingworth for helping me to find this manuscript.

38. Franklin, Colloquium, November 1951.

39. At this point in time, the terms "spiral" and "helical" were often used interchangeably. Franklin, Colloquium, November 1951.

40. Franklin, Colloquium, November 1951; Franklin, "Interim Annual Report."

41. Franklin, Colloquium, November 1951.

42. Franklin, Colloquium, November 1951; see also Sayre, *Rosalind Franklin and DNA*, 127–29; Judson, *The Eighth Day of Creation*, 98.

43. Franklin, "Interim Annual Report."

44. Judson, *The Eighth Day of Creation*, 100.

45. Horace Judson interview with Max Perutz, February 15, 1975, HFJP.

46. Horace Judson interview with Max Perutz, February 15, 1975, HFJP; see also Judson, *The Eighth Day of Creation*, 101–2.

47. Horace Judson interview with Max Perutz, February 15, 1975, HFJP; see also Judson, *The Eighth Day of Creation*, 102.

48. Horace Judson interview with Max Perutz February 15, 1975, HFJP; see also Judson, *The Eighth Day of Creation*, 102–3.

49. Horace Judson interview with Max Perutz, February 15, 1975; see also Judson, *The Eighth Day of Creation*, 102–3.

50. Watson, *The Double Helix*, 45.

51. Eugene Fodor and Frederick Rockwell, *Fodor's Guide to Britain and Ireland, 1958* (New York: David McKay, 1958), 122; British Library Learning Timelines: Sources from History, "Chinese Food, 1950s: Oral History with Wing Yip, Asian Food Restaurateur in London, 1950s and 1960s," http://www.bl.uk/learning/timeline/item107673.html.

52. Watson, *The Double Helix*, 46.

53. Watson, *The Double Helix*, 46.

54. Watson, *The Double Helix*, 46.

55. Judson, *The Eighth Day of Creation*, 102–3.

56. Watson, *The Double Helix*, 46.

57. Watson, *The Double Helix*, 48.

Chapter 14: The Dreaming Spires of Oxford

1. Matthew Arnold, "Thyrsis: A Monody, to Commemorate the Author's Friend, Arthur Hugh Clough," https://www.poetryfoundation.org/poems/43608/thyrsis-a-monody-to-commemorate-the-authors-friend-arthur-hugh-clough.

2. Fleming discovered the mold *Penicillin notatum* in 1928, but it was thirteen years before Howard Florey, Ernst Chain, and their team at Oxford University were able to mass-produce the antibiotic for wide-scale use. The three shared the 1945 Nobel Prize in Physiology or Medicine. See Eric Lax, *The Mold in Dr. Florey's Coat: The Story of the Penicillin Miracle* (New York: Henry Holt, 2004); Howard Markel, "Shaping the Mold, from Lab Glitch to Life Saver," *New York Times*, April 20, 2004, D6.

3. Georgina Ferry, *Dorothy Hodgkin: A Life* (London: Granta, 1998); Guy Dodson, "Dorothy Mary Crowfoot Hodgkin, O.M., 12 May 1910–29 July 1994," *Biographical Memoirs of Fellows of the Royal Society* 48 (2002): 179–219; Dorothy Crowfoot, Charles W. Bunn, Barbara W. Rogers-Low, and Annette Turner-Jones, "X-ray crystallographic investigation of the structure of penicillin,"

in H. Y. Clarke, J. R. Johnson, and R. Robinson, eds., *The Chemistry of Penicillin* (Princeton: Princeton University Press, 1949), 310–67.

4. W. Cochran and F. H. C. Crick, "Evidence for the Pauling–Corey α-Helix in Synthetic Polypeptides," *Nature* 169, no. 4293 (1952): 234–35; W. Cochran, F. H. C. Crick, and V. Vand, "The structure of synthetic peptides. I. The transform of atoms on a helix," *Acta Crystallographica* 5 (1952): 581–86.

5. James D. Watson, *The Double Helix: A Personal Account of the Discovery of the Structure of DNA*, edited by Gunther Stent (New York: Norton, 1980), 48.

6. Vand received his doctorate in physics and astrophysics from Charles University in Prague. After a few industrial jobs, first for the Skoda automobile works and then Lever Brothers, he was a research fellow at Glasgow University and ultimately a professor of physics at Pennsylvania State. He died on April 4, 1968, at the age of fifty-seven. See "Vladimir Vand, Pennsylvania State Crystallographer Dies," *Physics Today* 21, no. 7 (July 1, 1968): 115.

7. Robert Olby interview with Francis Crick, March 8, 1968, HFJP; Watson, *The Double Helix*, 41.

8. Watson, *The Double Helix*, 41.

9. Watson, *The Double Helix*, 41.

10. Watson, *The Double Helix*, 43.

11. Robert Olby interview with Francis Crick, March 8, 1968, HFJP.

12. *Wine Tasting: Vintage 1949*, mimeographed announcement, October 31, 1951, FCP, PP/CRI/H/1/42/6, box 73.

13. Watson, *The Double Helix*, 43.

14. Watson, *The Double Helix*, 43; Cochran and Crick, "Evidence for the Pauling–Corey α-Helix in Synthetic Polypeptides"; Cochran, Crick, and Vand, "The structure of synthetic peptides."

15. Watson, *The Double Helix*, 43.

16. Horace Judson interview with Max Perutz, February 15, 1975, HFJP; Horace Freeland Judson, *The Eighth Day of Creation: Makers of the Revolution in Biology* (Cold Spring Harbor, NY: Cold Spring Harbor Laboratory Press, 2013), 100–3, quote is on 101.

17. Horace Judson interview with Max Perutz, February 15, 1975, HFJP; Judson, *The Eighth Day of Creation*, 101. With respect to DNA, the "spot" at 3.4 Ångstroms (the same smudge described by William Astbury in 1939 and which Rosalind Franklin detected in her 1952 photographs) sits at the outside edge of the molecule. It diffracts X-rays so well because the nucleotide bases—especially the phosphorus atoms, which are the heaviest elements of those nucleotides—repeat along the helix at that interval. In a voice that defined the geometric expression "Q.E.D.," (*quod erat demonstrandum*, or "that which was to be demonstrated"), Perutz added with a flourish what became crystal clear to Watson and Crick in February 1953: "It's the same principle as the 1.5 Ångstrom spot I found in the

alpha helix, beyond where people had looked before. The fact that in DNA the bases are stacked in the helix parallel to each other at that 3.4 Ångstrom distance makes the spots more intense."

18. Maurice Wilkins, *The Third Man of the Double Helix* (Oxford: Oxford University Press, 2003), 160; the description of the November 22, 1951, colloquium is on 160–64. See also Michael Fry, *Landmark Experiments in Molecular Biology* (Amsterdam: Academic Press, 2016), 181. Stokes told Wilkins about his insight and Wilkins shared it with Crick, around the same time Crick and Cochran came to their own conclusions. Stokes never published his theory, but the Cochran, Crick, and Vand paper acknowledges that the theory "was also derived independently and almost simultaneously by Dr. A. R. Stokes (private communication)." See Cochran, Crick, and Vand, "The structure of synthetic peptides," 582. See also James D. Watson, *The Annotated and Illustrated Double Helix*, edited by Alexander Gann and Jan Witkowski (New York: Simon and Schuster, 2012), 90.

19. Wilkins, *The Third Man of the Double Helix*, 161.

20. Ferry, *Dorothy Hodgkin*, 275. Dunitz was a research fellow in Hodgkin's laboratory at the time and later became a professor at the Swiss Federal Institute of Technology in Zurich.

21. Ferry, *Dorothy Hodgkin*, 275–76.

22. Author interview with James D. Watson (no. 2), July 24, 2018.

23. Watson, *The Double Helix*, 45.

24. Watson, *The Double Helix*, 49.

25. Watson, *The Double Helix*, 48.

26. Watson, *The Double Helix*, 49.

27. Watson, *The Double Helix*, 49.

28. Watson, *The Double Helix*, 49.

29. Peter Pauling, "DNA: The Race That Never Was?," *New Scientist* 58 (May 31, 1973): 558–60.

30. W. L. Bragg, J. C. Kendrew, and M. F. Perutz, "Polypeptide Chain Configurations in Crystalline Proteins," *Proceedings of the Royal Society of London A: Mathematical and Physical Sciences* 203, no. 1074 (October 10, 1950): 321–57; L. C. Pauling, R. B. Corey, and H. R. Branson, "The structure of proteins; two hydrogen-bonded helical configurations of the polypeptide chain," *Proceedings of the National Academy of Sciences* 37, no. 4 (1951): 205–11.

31. Watson, *The Double Helix*, 49.

32. Rosalind Franklin, Colloquium, November 1951. RFP, FRKN 3/2.

33. Watson, *The Double Helix*, 51.

34. Jenny Pickworth Glusker, "ACA Living History," *ACA [American Crystallographic Association] Reflections* 4 (Winter 2011): 6–10; Ian Hesketh, *Of Apes and Ancestors: Evolution, Christianity, and the Oxford Debate* (Toronto: University of Toronto Press, 2009).

35. Ferry, *Dorothy Hodgkin*, 63, 106.

36. Samanth Subramanian, *A Dominant Character: The Radical Science and Restless Politics of J. B. S. Haldane* (New York: Norton, 2020); Claude Gordon Douglas, "John Scott Haldane, 1860–1936," *Biographical Memoirs of Fellows of the Royal Society* 2, no. 5 (December 1, 1936): 115–39.

37. Watson, *The Double Helix*, 52.

38. Letter from James D. Watson to Elizabeth Watson, November 28, 1951, WFAT, "Letters to Family, Cambridge, October 1951–August 1952." "Apparently the family is very rich. They have a mansion in Scotland. There is a chance I may be invited for Christmas." Elizabeth was in Copenhagen in the weeks before Christmas and told her jealous brother she was being "pursued by a Dane," who was an actor; "sensing impending disaster," Watson asked Mitchison if she could come along as well. Watson, *The Double Helix*, 63; author interview with James D. Watson (no. 2), July 24, 2018.

39. Letter from James D. Watson to Max and Manny Delbrück, December 9, 1951, MDP, box 23, folder 20.

Chapter 15: Mr. Crick and Dr. Watson Build Their Dream Model

1. Horace Judson interview with Raymond Gosling, July 21, 1975, HFJP.

2. Author interview with James D. Watson (no. 3), July 25, 2018; James D. Watson, *The Double Helix: A Personal Account of the Discovery of the Structure of DNA*, edited by Gunther Stent (New York: Norton, 1980), 48.

3. Watson, *The Double Helix*, 52.

4. Watson, *The Double Helix*, 53.

5. Anne Sayre, *Rosalind Franklin and DNA* (New York: Norton, 1975), 131.

6. Letter from Dorothy Hodgkin to David Sayre, January 7, 1975, ASP, box 4, folder 7; Sayre, *Rosalind Franklin and DNA*, 134. Anne Sayre's husband, David, was a crystallographer who at one time worked with Hodgkin.

7. Sayre, *Rosalind Franklin and DNA*, 134.

8. Letter from Dorothy Hodgkin to David Sayre, January 7, 1975, ASP, box 4, folder 7; a slightly misquoted version of Hodgkin's observation appears in Brenda Maddox, *Rosalind Franklin: The Dark Lady of DNA* (New York: HarperCollins, 2002), 178–79.

9. Watson, *The Double Helix*, 53.

10. Watson, *The Double Helix*, 53.

11. Watson, *The Double Helix*, 53.

12. Watson, *The Double Helix*, 53.

13. Margaret Bullard, *A Perch in Paradise* (London: Hamish Hamilton, 1952); James D. Watson, *The Annotated and Illustrated Double Helix*, edited by Alexander Gann and Jan Witkowski (New York: Simon and Schuster, 2012), 82. Bertrand

Russell was one of the people characterized in Bullard's novel; see Kenneth Blackwell, "Two Days in the Dictation of Bertrand Russell," *Russell: The Journal of the Bertrand Russell Archives* 15 (new series, Summer 1995): 37–52.

14. Watson, *The Double Helix*, 53.

15. Watson, *The Double Helix*, 53.

16. Horace Freeland Judson, *The Eighth Day of Creation: Makers of the Revolution in Biology* (Cold Spring Harbor, NY: Cold Spring Harbor Laboratory Press, 2013), 118.

17. Watson, *The Double Helix*, 55.

18. Watson, *The Double Helix*, 56.

19. Watson, *The Double Helix*, 42, 57.

20. Francis Crick, *What Mad Pursuit: A Personal View of Scientific Discovery* (New York: Basic Books, 1988), 35.

21. Francis Crick and James D. Watson, "A Structure of Sodium Thymonucleate: A Possible Approach," 1951, FCP, PP/CRI/H/1/42/1, box 72. Thymonucleate is the sodium salt of DNA extracted from calves' thymus.

22. Watson, *The Double Helix*, 57.

23. Wilkins, *The Third Man of the Double Helix*, 164–65.

24. Wilkins, *The Third Man of the Double Helix*, 165.

25. Wilkins, *The Third Man of the Double Helix*, 165–66; J. M. Gulland, D. O. Jordan, and C. J. Threlfall, "212. Deoxypentose Nucleic Acids. Part I. Preparation of the Tetrasodium Salt of the Deoxypentose Nucleic Acid of Calf Thymus," *Journal of the Chemical Society* 1947: 1129–30; J. M. Gulland, D. O. Jordan, and H. F. W. Taylor. "213. Deoxypentose Nucleic Acids. Part II. Electrometric Titration of the Acidic and the Basic Groups of the Deoxypentose Nucleic Acid of Calf Thymus," *Journal of the Chemical Society* 1947: 1131–41; J. M. Creeth, J. M. Gulland, and D. O. Jordan, "214. Deoxypentose Nucleic Acids. Part III. Viscosity and Streaming Birefringence of Solutions of the Sodium Salt of the Deoxypentose Nucleic Acid Thymus," *Journal of the Chemical Society* 1947: 1141–45.

26. Sven Furberg, "An X-ray study of some nucleosides and nucleotides," PhD diss., University of London, 1949; Sven Furberg, "On the Structure of Nucleic Acids," *Acta Chemica Scandinavica* 6 (1952): 634–40.

27. Watson adds, "But not knowing the details of the King's College experiments, [Furberg] built only single-stranded structures, and so his structural ideas were never seriously considered in the Cavendish." *The Double Helix*, 54.

28. Wilkins, *The Third Man of the Double Helix*, 166.

29. Wilkins, *The Third Man of the Double Helix*, 166.

30. Rosalind Franklin, "Interim Annual Report: January 1, 1951–January 1, 1952," Wheatstone Laboratory, King's College, London, February 7, 1952. RFP, FRKN 4/3.

31. Wilkins, *The Third Man of the Double Helix*, 166; Rosalind Franklin, Colloquium, November 1951, RFP, FRKN 3/2; Franklin, "Interim Annual Report."

32. Watson, *The Double Helix*, 58.

33. Wilkins, *The Third Man of the Double Helix*, 171.

34. Watson, *The Double Helix*, 58.

35. Watson, *The Double Helix*, 58.

36. Watson, *The Double Helix*, 58.

37. Watson, *The Double Helix*, 59.

38. Robert Olby, *Francis Crick: Hunter of Life's Secrets* (Cold Spring Harbor, NY: Cold Spring Harbor Laboratory Press, 2009), 134. Olby uses the script of the 2003 BBC documentary *Double Helix: The DNA Story* as his source for Franklin's "tickled pink" mood and exclamation.

39. Gosling wrote these comments to the editors in a letter dated January 28, 2012; see Watson, *The Annotated and Illustrated Double Helix*, 91.

40. Watson, *The Double Helix*, 59.

41. Watson, *The Double Helix*, 59; Olby, *Francis Crick*, 135.

42. Robert Olby, *The Path to the Double Helix* (Seattle: University of Washington Press, 1974), 362.

43. Judson, *The Eighth Day of Creation*, 106–7.

44. "The Race for the Double Helix," documentary television program, narrated by Isaac Asimov, *Nova*, PBS, March 7, 1976.

45. Watson, *The Double Helix*, 59.

46. Watson, *The Double Helix*, 59; see also Wilkins, *The Third Man of the Double Helix*, 171–75; Crick, *What Mad Pursuit*, 65; Judson, *The Eighth Day of Creation*, 105–7; Olby, *The Path to the Double Helix*, 357–63.

47. Author interviews with James D. Watson (nos. 1 and 2), July 23 and 24, 2018.

48. Watson, *The Double Helix*, 60–61.

49. Wilkins, *The Third Man of the Double Helix*, 173–75.

50. Brenner shared the 2002 Nobel Prize in Physiology or Medicine with H. Robert Horvitz and John E. Sulston, "for their discoveries concerning genetic regulation of organ development and programmed cell death." I am grateful to professors Alexander Gann and Jan Witkowski at the Cold Spring Harbor Laboratory for their work in restoring these once lost letters to the historical record and their generosity in discussing them with me. See A. Gann and J. Witkowski, "The Lost Correspondence of Francis Crick," *Nature* 467, no. 7315 (September 30, 2010): 519–24. The thirty-four letters, which date from 1951 to 1964, can be found at in the Cold Spring Harbor Laboratory Archives Repository, Cold Spring Harbor, NY, SB/11/1/177, http://libgallery.cshl.edu/items/show/52125.

51. Letter from Maurice Wilkins to Francis Crick, December 11, 1951, Cold Spring Harbor Laboratory Archives Repository, SB/11/1/177. Quoted with permission.

52. Letter from Maurice Wilkins to Francis Crick, December 11, 1951, Cold Spring Harbor Laboratory Archives Repository, SB/11/1/177. Quoted with permission.

53. Letter from Francis Crick to Maurice Wilkins, December 13, 1951. Cold Spring Harbor Laboratory Archives Repository, SB/11/1/177. Quoted with permission.

54. Wilkins says this event occurred in early December 1951; see Wilkins, *The Third Man of the Double Helix*, 170–71.

55. Wilkins, *The Third Man of the Double Helix*, 171.

56. Anne Sayre interview with Francis Crick, June 16, 1970, ASP, box 2, folder 9.

57. After leaving King's College for J. D. Bernal's laboratory at Birkbeck College in the spring of 1953, Rosalind Franklin would make great strides in TMV and virology research. See Rosalind Franklin and K. C. Holmes, "The Helical Arrangement of the Protein Sub-Units in Tobacco Mosaic Virus," *Biochimica et Biophysica Acta* 21, no. 2 (1956): 405–6; Rosalind Franklin and Aaron Klug, "The Nature of the Helical Groove on the Tobacco Mosaic Virus," *Biochimica et Biophysica Acta* 19, no. 3 (1956): 403–16; J. G. Shaw, "Tobacco Mosaic Virus and the Study of Early Events in Virus Infections," *Philosophical Transactions of the Royal Society B: Biological Sciences* 354, no. 1383 (1999): 603–11; A. N. Craeger and G. J. Morgan, "After the Double Helix: Rosalind Franklin's Research on Tobacco Mosaic Virus," *Isis* 99, no. 2 (2008): 239–72.

58. Patricia Fara, "Beyond the Double Helix: Rosalind Franklin's work on viruses," *Times Literary Supplement*, July 24, 2020, https://www.the-tls.co.uk/articles/beyond-the-double-helix-rosalind-franklins-work-on-viruses/.

59. Watson, *The Double Helix*, 74.

60. Watson, *The Double Helix*, 74.

61. Watson, *The Double Helix*, 62.

62. Watson, *The Double Helix*, 67.

63. Watson, *The Double Helix*, 62.

64. Watson, *The Double Helix*, 62.

65. Letter from W. L. Bragg to A. V. Hill, January 18, 1952, A.V. Hill Papers, II 4/18, Churchill College Archives Centre, University of Cambridge.

Part IV: Moratorium, 1952

1. Horace Judson interview with William Lawrence Bragg, January 28, 1971, HFJP.

Chapter 16: Dr. Pauling's Predicament

1. This chapter title was taken from the title of an editorial that appeared in the *New York Times*, May 19, 1952, 16.

2. Linus Pauling, notarized statement, June 20, 1952, LAHPP, http://scarc.library.oregonstate.edu/coll/pauling/peace/papers/bio2.003.1-ts-19520620.html.

3. Thomas Hager, *Force of Nature: The Life of Linus Pauling* (New York: Simon and Schuster, 1995), 335–407, quote is on 358.

4. David Oshinsky, *A Conspiracy So Immense: The World of Joe McCarthy* (New York: Oxford University Press, 2005); Ellen Schrecker, *Many Are the Crimes: McCarthyism in America* (Princeton: Princeton University Press, 1999); Ellen Schrecker, *No Ivory Tower: McCarthyism and the Universities* (New York: Oxford University Press, 1986).

5. Hager, *Force of Nature*, 335–407; Victor Navasky, *Naming Names* (New York: Viking, 1980), 78–96, 169–78; "Statement by Prof. Linus Pauling, regarding clemency plea for Julius and Ethel Rosenberg," January 1953 (typescript), LAHPP; Helen Manfull, ed., *Additional Dialogue: Letters of Dalton Trumbo, 1942–1962* (New York: M. Evans/J. B. Lippincott, 1970), 172, 176, 191–92, 328.

6. James D. Watson, *The Double Helix: A Personal Account of the Discovery of the Structure of DNA*, edited by Gunther Stent (New York: Norton, 1980), 63.

7. Hager, *Force of Nature*, 357.

8. Robert Olby interview with Linus Pauling, November 1968, quoted in Robert Olby, *The Path to the Double Helix* (Seattle: University of Washington Press, 1974), 376–77; see also Hager, *Force of Nature*, 397.

9. Letter from John Randall to Linus Pauling, August 28, 1951, LAHPP, http://scarc .library.oregonstate.edu/coll/pauling/dna/corr/sci9.001.2-randall-lp-19510828 .html. Letter from Linus Pauling to John Randall, September 25, 1951, LAHPP, http://scarc.library.oregonstate.edu/coll/pauling/dna/corr/sci9.001.2-lp-randall -19510925.html.

10. *Life Story: Linus Pauling*, documentary film, BBC, 1997. Transcript and video clip in LAHPP, http://scarc.library.oregonstate.edu/coll/pauling/dna/audio/1997v .1-photos.html.

11. Olby, *The Path to the Double Helix*, 400.

12. Linus Pauling, "My Efforts to Obtain a Passport," *Bulletin of the Atomic Scientists* 8, no. 7 (October 1952): 253–56.

13. Ruth Bielaski married Frederick Shipley in 1909 and left government service when her husband was appointed a federal government administrator in the Panama Canal Zone. The couple had a son in 1911 but returned to Washington in 1914 after Fred contracted yellow fever and could no longer work. From 1912 to 1919, her brother, A. Bruce Bielaski, ran the U.S. Department of Justice's Bureau of Investigation, the forerunner to the FBI. It was through Bielaski's influence that Mrs. Shipley was given a job on the passport desk at the State Department. She declined the promotion to head of the passport office twice before finally accepting it in 1928. "Basic Passports," *Fortune* 32, no. 4 (October 1945): 123.

14. "Ogre," *Newsweek*, May 29, 1944, 38; "Sorry, Mrs. Shipley," *Time*, December 31, 1951, 15. The Subversive Activities Control Act of 1950, also known as the Internal

Security Act of 1950, can be accessed at https://www.loc.gov/law/help/statutes-at
-large/81st-congress/session-2/c81s2ch1024.pdf.

15. "Sorry, Mrs. Shipley." The *Time* cover that week featured a portrait of the come-
    dian Groucho Marx, with the caption "Trademark: effrontery."

16. Andre Visson, "Ruth Shipley: The State Department's Watchdog," *Reader's
    Digest*, October 1951, 73–74 (condensed and reprinted from *Independent Woman*,
    August 1951); Richard L. Strout, "Win a Prize—Get a Passport," *New Republic*,
    November 28, 1955, 11–13.

17. "Woman's Place Also in the Office, Finds Chief of the Nation's Passport Divi-
    sion," *New York Times*, December 24, 1939, 22. See also Hager, *Force of Nature*,
    335–407; Jeffrey Kahn, *Mrs. Shipley's Ghosts: The Right to Travel and Terrorist
    Watch Lists* (Ann Arbor: University of Michigan Press, 2013); Jeffrey Kahn, "The
    Extraordinary Mrs. Shipley: How the United States Controlled International
    Travel Before the Age of Terrorism," *Connecticut Law Review* 43 (February 2011):
    821–88; "Passport Chief to End Career; Mrs. Shipley Retiring After 47 Years in
    Government—Figured in Controversies," *New York Times*, February 25, 1955, 15;
    "Ruth B. Shipley, Ex-Passport Head, Federal Employee 47 Years Dies at 81 in
    Washington," *New York Times*, November 5, 1966, 29.

18. "Woman's Place Also in the Office."

19. "Mrs. Shipley Abdicates," editorial, *New York Times*, February 26, 1955, 14.

20. Luria, a leftist sympathizer, was refused a passport to travel to Oxford in April
    1952 for a Society for General Microbiology symposium, at which he had been
    asked to give a paper on the highly apolitical topic of bacteriophage multiplica-
    tion. His paper was read in absentia and included in the published proceedings
    of the meeting. Letter from James D. Watson to Elizabeth Watson, April 3, 1952,
    WFAT, "Letters to Family, Cambridge, October 1951–August 1952"; James D.
    Watson, *The Annotated and Illustrated Double Helix*, edited by Alexander Gann
    and Jan Witkowski (New York: Simon and Schuster, 2012), 121–24; S. E. Luria,
    "An Analysis of Bacteriophage Multiplication," in Paul Fieldes and W. E. Van
    Heyningen, eds., *The Nature of Virus Multiplication: Second Symposium for the
    Society of General Microbiology Held at Oxford University, April 1952* (Cam-
    bridge: Cambridge University Press, 1953). Luria also sent Watson a précis of the
    Hershey–Chase Waring blender experiment to be read at the Oxford conference;
    see letter from Horace Judson to Alfred D. Hershey, August 27, 1976, HFJP.

21. Letter from Ruth B. Shipley, U.S. State Department, to Linus Pauling, February
    14, 1952, LAHPP, http://scarc.library.oregonstate.edu/coll/pauling/dna/corr/bio2
    .002.5-shipley-lp-19520214.html.

22. Biographer Thomas Hager investigated Pauling's State Department file through
    a Freedom of Information Act application and quotes this passage and others in
    Force of Nature, 401–3.

23. Hager, *Force of Nature*, 401.

24. 77th Congress of the United States, Public Law 77-671, 56 Stat 662, S.2404, enacted July 20, 1942: "To Create the Decorations to be Known as the Legion of Merit, and the Medal for Merit."

25. Letter from Linus Pauling to President Harry Truman, February 29, 1952. The letter is included in Pauling's State Department file and is quoted in Hager, *Force of Nature*, 401.

26. Hager, *Force of Nature*, 402.

27. Graham Berry oral history interview with Edward Hughes, 1984, California Institute of Technology Archives; Hager, *Force of Nature*, 401–4.

28. "Passport is Denied to Dr. Linus Pauling; Scientist Assails Action as 'Interference'," *New York Times*, May 12, 1952, 8; "Passport Denial Decried: British Scientists Score U.S. Action on Prof. Linus Pauling," *New York Times*, May 13, 1952, 10; "Dr. Pauling's Predicament"; "Linus Pauling and the Race for DNA," documentary film, *Nova*, PBS and Oregon State University, 1977, available at http://osulibrary.oregonstate.edu/specialcollections/coll/pauling/dna/audio/1977v.66.html.

29. Robert Robinson, letter to the editor, *The Times*, May 2, 1952; Hager, *Force of Nature*, 405. Robinson's letter is dated May 1, which, he admitted, was "possibly an unfortunate choice of day" to invite an accused Communist to speak at the Royal Society.

30. "Second International Congress of Biochemistry (July 21–27, 1952)," *Nature* 170, no. 4324 (1952): 443–44; Hager, *Force of Nature*, 405.

31. *Tech* (magazine of the California Institute of Technology), May 15, 1952, 1.

32. Ruth B. Shipley, internal memorandum, May 16, 1952, quoted in Hager, *Force of Nature*, 406.

33. "Dr. Pauling Gets Limited Passport. State Department Reverses Its Stand in Cases of Famed Caltech Scientist," *Los Angeles Times*, July 16, 1952, 20.

34. "Linus Pauling Day-by-Day," July 1952, Linus Pauling Special Collections, Oregon State University, Corvallis, OR, http://scarc.library.oregonstate.edu/coll/pauling/calendar/1952/07/index.html.

35. Hager, *Force of Nature*, 414–15.

36. *Life Story: Linus Pauling*, documentary film, BBC, 1997. Transcript and video clip in LAHPP, http://scarc.library.oregonstate.edu/coll/pauling/dna/audio/1997v.1-photos.html.

Chapter 17: Chargaff's Rules

1. Erwin Chargaff, "Preface to a Grammar of Biology," *Science* 172, no. 3984 (May 14, 1971): 637–42, quote is on 639; Erwin Chargaff, *Heraclitean Fire: Sketches from a Life Before Nature* (New York: Rockefeller University Press, 1978), 81–82. The paper he refers to is O. T. Avery, C. M. Macleod, and M. McCarty, "Studies on the chemical nature of the substance inducing transformation of pneumococcal

types. Induction of transformation by a desoxyribonucleic acid fraction isolated from *pneumococcus* Type II," *Journal of Experimental Medicine* 79 (1944): 137–58 (DNA was still referred to as desoxyribonucleic acid rather than deoxyribonucleic acid); the Cardinal Newman book mentioned is John Henry Newman, *An Essay in Aid of the Grammar of Assent* (London: Burns, Oates, 1870).

2. Seymour S. Cohen, "Erwin Chargaff, 1905–2002," *Biographical Memoirs of the National Academy of Sciences* (Washington, DC: National Academy of Sciences, 2010), 5 (reprinted from *Proceedings of the American Philosophical Society* 148, no. 2 (2004): 221–28. See also Nicholas Wade, "Erwin Chargaff, 96, Pioneer in DNA Chemical Research," *New York Times*, June 30, 2002, 27; Nicole Kresge, Robert D. Simoni, and Robert L. Hill, "Chargaff's Rules: The Work of Erwin Chargaff," *Journal of Biological Chemistry* 280, no. 24 (2005): 172–74.

3. *Naturphilosophie* is a now obscure German theory of biology, nature, and mystical pantheism once adored by German academics. Chargaff, *Heraclitean Fire*, 15–16; Howard Markel, *An Anatomy of Addiction: Sigmund Freud, William Halsted, and the Miracle Drug, Cocaine* (New York: Pantheon, 2011), 21.

4. Chargaff, *Heraclitean Fire*, 16.

5. C. J. M., "Léon Charles Albert Calmette, 1863–1933," *Obituary Notices of Fellows of the Royal Society* 1 (1934): 315–25.

6. Chargaff, *Heraclitean Fire*, 52–54.

7. Chargaff first lived at 410 Central Park West: *Manhattan (New York) Telephone Directory, 1940* (New York: New York Telephone Co., 1939), 184. He later moved to 350 Central Park West: *National Academy of Sciences, National Academy of Engineering, Institute of Medicine, National Research Council: Annual Report, Fiscal Year, 1974–1975* (Washington, DC: National Academy of Sciences), 213.

8. Chargaff, *Heraclitean Fire*, 39–40.

9. Chargaff, *Heraclitean Fire*, 84–85.

10. Chargaff, *Heraclitean Fire*, 85.

11. Chargaff, "Preface to a Grammar of Biology," 639.

12. Cohen, "Erwin Chargaff, 1905–2002," 8.

13. Ernst Vischer and Erwin Chargaff, "The Separation and Quantitative Estimation of Purines and Pyrimidines in Minute Amounts," *Journal of Biological Chemistry* 176 (1948): 703–14; Erwin Chargaff, "On the nucleoproteins and nucleic acids of microorganisms," *Cold Spring Harbor Symposia of Quantitative Biology* 12 (1947): 28–34; Erwin Chargaff and Ernst Vischer, "Nucleoproteins, nucleic acids, and related substances," *Annual Review of Biochemistry* 17 (1948): 201–26; Erwin Chargaff, "Chemical Specificity of Nucleic Acids and Mechanism of Their Enzymatic Degradation," *Experientia* 6 (1950): 201–9; Erwin Chargaff, "Some Recent Studies of the Composition and Structure of Nucleic Acids," *Journal of Cellular and Comparative Physiology* 38, suppl. I (1951): 41–59. See also Erwin Chargaff and J. N. Davidson, eds., *The Nucleic Acids: Chemistry and Biology*, 2 vols.

(New York: Academic Publishers, 1955); Pnina Abir-Am, "From Biochemistry to Molecular Biology: DNA and the Acculturated Journey of the Critic of Science, Erwin Chargaff," *History and Philosophy of the Life Sciences* 2, no. 1 (1980): 3–60.

14. Chargaff, "Chemical Specificity of Nucleic Acid and Mechanism of Their Enzymatic Degradation."

15. Chargaff, *Heraclitean Fire*, 87.

16. Horace Freeland Judson, *The Eighth Day of Creation: Makers of the Revolution in Biology* (Cold Spring Harbor, NY: Cold Spring Harbor Laboratory Press, 2013), 75, see also 73–75, 117–21; Robert Olby, *Francis Crick: Hunter of Life's Secrets* (Cold Spring Harbor, NY: Cold Spring Harbor Laboratory Press, 2009), 140–43, 165–66.

17. Erwin Chargaff, "Amphisbaena," *Essays on Nucleic Acids* (New York: Elsevier, 1963), 174–99, quote is on 176; Chargaff, *Heraclitean Fire*, 140. The amphisbaena, from Greek mythology, is an ant-eating, double-headed serpent.

18. James D. Watson, *The Double Helix: A Personal Account of the Discovery of the Structure of DNA*, edited by Gunther Stent (New York: Norton, 1980), 74.

19. Watson, *The Double Helix*, 75.

20. Watson, *The Double Helix*, 75–76.

21. Watson, *The Double Helix*, 76.

22. Hermann Bondi and Thomas Gold, "The Steady State Theory of the Expanding Universe," *Monthly Notices of the Royal Astronomical Society* 109, no. 3 (1948): 252–70.

23. Watson, *The Double Helix*, 76. Self-replication, per se, was hardly a new hypothesis. Several scientists, including Pauling and Delbrück, had speculated about a process where a complementary, negative-shaped molecule or structure fit precisely into a positive one. This molecular arrangement also allowed for an extant negative structure to act as a mold or template for a new positive image. Not all scientists agreed with the notion of complementarity: Hermann Muller and the German theoretical physicist Pascual Jordan argued that "like attracts like." See L. C. Pauling and M. Delbrück, "The Nature of the Intermolecular Operative in Biological Processes," *Science* 92, no. 2378 (1940): 77–99; Pascual Jordan, "Biologische Strahlenwirkung und Physik der Gene" (Biological Radiation and Physics of Genes), *Physikalische Zeitschrift* 39 (1938): 345–66, 711; Pascual Jordan, "Problem der spezifischen Immunität" (Problem of Specific Immunity), *Fundamenta Radiologica* 5 (1939): 43–56.

24. John Griffith—the nephew of Frederick Griffith, who conducted some of the first transforming principle experiments on pneumococcus—was outraged by Watson's "uncalled-for remarks" in *The Double Helix* (77). See John Lagnado, "Past Times: From Pablum to Prions (via DNA): A Tale of Two Griffiths," *Biochemist* 27, no. 4 (August 2005): 33–35, http://www.biochemist.org/bio/02704/0033/027040033.pdf.

25. Watson, *The Double Helix*, 77.

26. Watson claimed this meeting occurred in July, but Chargaff dates it to May 24–27, 1952, which is far more likely. Watson, *The Double Helix*, 77–78; Chargaff, *Heraclitean Fire*, 100.

27. Chargaff had aspirations for an endowed chair in Switzerland at this time, but it came to naught. Horace Freeland Judson, "Reflections on the Historiography of Molecular Biology," *Minerva* 18, no. 3 (1980): 369–421.

28. Chargaff titles this chapter "Gullible's Troubles." Chargaff, *Heraclitean Fire*, 100–103.

29. Chargaff, *Heraclitean Fire*, 100.

30. Chargaff, "Preface to a Grammar of Biology," 641.

31. Watson, *The Double Helix*, 78.

32. Watson was actually twenty-four when he first met Chargaff; Chargaff, *Heraclitan Fire*, 100–2.

33. Erwin Chargaff, "Building the Tower of Babble," *Nature* 248 (April 26, 1974): 776–79, quote is on 776–77.

34. Watson, *The Double Helix*, 78.

35. Robert Olby, *The Path to the Double Helix* (Seattle: University of Washington Press, 1974), 385–423, quote is on 388; Olby, *Francis Crick*, 139–44; Royal Society interviews with Crick in Cambridge, conducted by Robert Olby, March 8, 1968, and August 7, 1972, Collections of the Royal Society, London.

36. Watson, *The Double Helix*, 77–78.

37. "And the king [Solomon] said: 'Fetch me a sword.' And they brought a sword before the king. And the king said: 'Divide the living child in two, and give half to the one, and half to the other.'" 1 Kings 3:24, 25.

38. Chargaff and Wilkins met during the annual Gordon Conference on Nucleic Acids and Proteins, held in New Hampton, NH, on August 27–31, 1951. The Chargaff–Wilkins letters, from late December 1951 through the end of 1953, are in ECP, box 59, Mss. B.C37. I am grateful to Charles Greifenstein, Associate Librarian and Curator of Manuscripts of the American Philosophical Society, for introducing me to these documents. See also Wilkins, *The Third Man of the Double Helix*, 151–54. *Bacillus coli*, or *B. coli*, is the antiquated name for *Escherichia coli*, or *E. coli*.

39. As late as 2000, he was still complaining that Franklin took the Signer DNA from him. Brenda Maddox, *Rosalind Franklin: The Dark Lady of DNA* (New York: HarperCollins, 2002), 195, 343. Maddox interviewed Wilkins on November 4, 2000.

40. Letter from Maurice Wilkins to Erwin Chargaff, January 6, 1952, ECP, box 59, Mss.B.C37.

41. Letter from Maurice Wilkins to Erwin Chargaff, January 6, 1952. "Nerk" is a British slang word of the era, referring to a foolish or objectionable person or activity.

42. Chargaff's mood only hardened further after enduring "shabby" treatment by his administrative superiors at Columbia. When he retired after forty years of service, the university refused "to endorse new grant applications, chang[ed] the locks on his old laboratory, strand[ed] him with a pension of 30 per cent of his salary." Judson, "Reflections on the Historiography of Molecular Biology."

43. Horace Freeland Judson, "No Nobel Prize for Whining," op-ed, *New York Times*, October 20, 2003, A17. Ironically, Chargaff was invited many times to nominate people for the Nobel Prize, a task he could not have completed with joy. ECP, Nobel Prize correspondence, box 121, Mss.B.C37.

44. Chargaff, *Heraclitan Fire*, 103; Erwin Chargaff, review of *The Path to the Double Helix* by Robert Olby, *Perspectives in Biology and Medicine* 19 (1976): 289–90.

Chapter 18: Paris and Royaumont

1. James D. Watson, *The Double Helix: A Personal Account of the Discovery of the Structure of DNA*, edited by Gunther Stent (New York: Norton, 1980), 80.

2. "Second International Congress of Biochemistry [July 21-27, 1952]," *Nature* 170, no. 4324 (1952): 443–44; Linus Pauling's annotated program from the Second International Congress of Biochemistry, Paris, July 21–27, 1952, LAHPP, http://scarc.library.oregonstate.edu/coll/pauling/proteins/papers/1952s.9-program.html.

3. This section of the forest was known as the Glade of the Armistice. Adolf Hitler had the railway carriage in which the First World War armistice was signed transported from Paris to Compiègne. It served as a symbol of Germany's humiliating defeat and the harsh treaty Hitler felt his nation was forced to sign in 1918 and his revenge in the form of the Third Reich's conquest of France. See William Shirer, *The Rise and Fall of the Third Reich* (New York: Simon and Schuster, 1960), 742.

4. Watson, *The Double Helix*, 79.

5. In 1622, Cardinal Richelieu was elected the *proviseur*, or principal, of the Sorbonne. Erwin Chargaff, "Building the Tower of Babble," *Nature* 248 (April 26, 1974): 776–79, quote is on 776.

6. Watson, *The Double Helix*, 79. Watson incorrectly recalled that Pauling presented his lecture during "the session at which [Max] Perutz spoke." But Perutz's session was the "first symposium," on the biochemistry of haemopoesis, and he covered the structure of hemoglobin. Pauling's archival papers, however, indicates that he spoke at the "second symposium," on the biogenesis of protein.

7. Commemorative dinner menu, International Congress of Biochemistry, Paris, July 26, 1952, LAHPP, http://scarc.library.oregonstate.edu/coll/pauling/proteins/pictures/1952s.9-menu.html.

8. Maurice Wilkins, *The Third Man of the Double Helix* (Oxford: Oxford University Press, 2003), 186.

9. Phage Conference, International (Summary of the Proceedings of the Conference), July 1952, JDWP, JDW/2/7/3/3.

10. Watson, *The Double Helix*, 80.

11. Frederick W. Stahl, ed., *We Can Sleep Later: Alfred D. Hershey and the Origins of Molecular Biology* (Cold Spring Harbor, NY: Cold Spring Harbor Laboratory Press, 2000).

12. Allen Campbell and Franklin W. Stahl, "Alfred D. Hershey," *Annual Review of Genetics* 32 (1998): 1–6.

13. Alfred Hershey and Martha Chase, "Independent Functions of Viral Protein and Nucleic Acid in Growth of Bacteriophage," *Journal of General Physiology* 36, no. 1 (1952): 39–56; see also "The Hershey–Chase Experiment," in Jan Witkowski, ed., *Illuminating Life: Selected Papers from Cold Spring Harbor, 1903–1969* (Cold Spring Harbor, NY: Cold Spring Harbor Laboratory Press, 2000), pp. 201–22; Stahl, ed., *We Can Sleep Later*, 171–207; Alfred D. Hershey, "The Injection of DNA into Cells by Phage," in John Cairns, Gunther S. Stent, and James D. Watson, eds., *Phage and the Origins of Molecular Biology* (Cold Spring Harbor, NY: Cold Spring Harbor Laboratory Press, 1966), 100–9. Unlike Watson's 1951 Copenhagen study, which only tagged the phosphorus of DNA, the 1952 Hershey study used radiolabeled tags for both protein and DNA, produced far better yields, and was considered by many to be the definitive study.

14. "The Hershey–Chase Experiment," 201; H. V. Wyatt, "How History Has Blended," *Nature* 249, no. 5460 (June 28, 1974): 803–4. Hershey shared the Nobel Prize with Luria and Delbrück ("two enemy aliens and one social misfit," as Hershey referred to the trio), "for their discoveries concerning the replication mechanism and the genetic structure of viruses." Unlike Oswald Avery, who never won a Nobel, these three men had the advantage of being associated with the Cold Spring Harbor Laboratory, where the scientists saw great value in writing, revising, and widely disseminating the literature of genetics in their own image.

15. James D. Watson, "The Lives They Lived: Alfred D. Hershey: Hershey Heaven," *New York Times Magazine*, January 4, 1998, 16; a longer version of this essay appears as "Alfred Day Hershey 1908–1997," in *Cold Spring Harbor Laboratory Annual Report 1997*, ix–x, http://repository.cshl.edu/id/eprint/36676/1/CSHL_AR_1997.pdf.

16. Thomas Hager, *Force of Nature: The Life of Linus Pauling* (New York: Simon and Schuster, 1995), 408.

17. Watson, *The Double Helix*, 80.

18. Letter from James D. Watson to Max Delbrück, May 20, 1952, MDP, box 23, folder 21.

19. Letter from Max Delbrück to James D. Watson, June 4, 1952, MDP, box 23, folder 21. Incidentally, Rosalind Franklin was not invited to Pauling's 1953 Pasadena Conference on Protein Structure at Caltech (September 21–25), although

Wilkins, Randall, Bragg, Kendrew, Perutz, Watson, and Crick were. See "Linus Pauling Day-by-Day," September 21, 1952, Linus Pauling Special Collections, Oregon State University, Corvallis, OR, http://scarc.library.oregonstate.edu/coll/pauling/calendar/1953/09/21.html.

20. Watson, *The Double Helix*, 81.

21. Watson, *The Double Helix*, 81. Watson confirmed his ingratiating approach to Mrs. Pauling to the author in interview no. 2, July 24, 2018.

22. Peter Pauling, "DNA: The Race That Never Was?," *New Scientist*, May 31, 1973, 558–60, quote is on 558.

23. Pauling, "DNA: The Race That Never Was?," 558; Horace Judson interview with Peter Pauling, February 1, 1970, HFJP.

24. Watson, *The Double Helix*, 81.

25. Photographs from the conference amply demonstrate Watson's odd attire. See JDWP, "Meeting at Royaumont, France," JDW/1/6/1, and "Bacteriophage Conference at Royaumont France," JDW/1/11/2.

26. Letter from James D. Watson to Francis Crick, August 11, 1952, FCP, PP/CRI/H/1/42/3, box 72. See also JDWP, "Italian Alps, 1952," JDW/1/15/2.

27. Letter from Jean Mitchell Watson to James D. Watson, Sr., June 18, 1952, WFAT, JDW/2/2/1947/55.

28. Watson, *The Double Helix*, 8.

29. Letter from James D. Watson to Francis and Odile Crick, August 11, 1952, FCP, PP/CRI/H/1/42/3, box 72.

30. Pauling, "DNA: The Race That Never Was?"

31. Author interview with James D. Watson (no. 2), July 24, 2018.

Chapter 19: A Haphazard Summer

1. Horace Judson interview with Francis Crick, July 3, 1975, HFJP.

2. Maurice Wilkins, *The Third Man of the Double Helix* (Oxford: Oxford University Press, 2003), 181.

3. Carlos Chagas, "Nova tripanozomiaze humana: estudos sobre a morfolojia e o ciclo evolutivo do *Schizotrypanum cruzi n. gen., n. sp.*, ajente etiolojico de nova entidade morbida do homem," (Human nova trypanossomia: studies on the morphology and evolutionary cycle of Schistrypanum cruzi (new genus, new species), etiological agent of a new morbid entity in man). *Memórias do Instituto Oswaldo Cruz* 1, no. 2 (1908): 158–218.

4. Letter from Maurice Wilkins to Francis Crick, undated, "on train, Innsbruck to Zurich," FCP, PP/CRI/H/1/42/4, box 72.

5. Wilkins, *The Third Man of the Double Helix*, 185–95, quote is on 194.

6. Wilkins, *The Third Man of the Double Helix*, 194.

7. Wilkins, *The Third Man of the Double Helix*, 195.

8. Wilkins, *The Third Man of the Double Helix*, 195.

9. Anne Sayre interview with Geoffrey Brown, May 12, 1970, ASP, box 2, folder 3.

10. Letter from Rosalind Franklin to Anne and David Sayre, March 1, 1952, ASP, box 2, folder 15.1.

11. Letter from Rosalind Franklin to Anne and David Sayre, March 1, 1952.

12. Letter from Rosalind Franklin to Anne and David Sayre, June 2, 1952, ASP, box 3, folder 1.

13. Letter from Rosalind Franklin to J. D. Bernal, June 19, 1952, RFP, personnel file, FRKN 2/31; Horace Freeland Judson, *The Eighth Day of Creation: Makers of the Revolution in Biology* (Cold Spring Harbor, NY: Cold Spring Harbor Laboratory Press, 2013), 114.

14. Brenda Maddox, *Rosalind Franklin: The Dark Lady of DNA* (New York: Harper-Collins, 2002), 183.

15. Randall approved the request on July 3 with the recommendation that Franklin transfer out of King's to Birkbeck on January 1, 1953. I. C. Maxwell, the chairman of the fellowship committee, echoed the recommendation on July 21. See I. C. Maxwell, Chair of the Turner and Newall Fellowships, to John Randall, July 1, 1952. JRP, RNDL 3/1/6; letter from Rosalind Franklin to J. D. Bernal, June 19, 1952, RFP, personnel file, FRKN 2/31; Rosalind Franklin, "Annual Report, 1 January 1954–1 January 1955," Birkbeck College, 1955, RFP, FRKN 1/4. See also Maddox, *Rosalind Franklin*, 183.

16. Maddox, *Rosalind Franklin*, 168–69. Both Crick and Wilkins said that Franklin pursued the cumbersome Patterson analysis based on Luzzati's advice. See Anne Sayre interview with Francis Crick, June 16, 1970, ASP, box 2, folder 9; Anne Sayre interview with Maurice Wilkins, June 15, 1970, ASP, box 4, folder 32. In a letter to Horace Judson, Luzzati complicates historical matters by stating that he did not see Franklin's B picture until after it was published, and therefore he did not push her in the direction of model building. He described his role as a minor one; while he taught her how to use the Beevers and Lipson's strips, he did not recall "seeing even a beginning of an application of Patterson superpositions, or any other of my pet ideas, to DNA." Letter from Vittorio Luzzati to Horace Judson, September 21, 1976, HFJP.

17. Raymond G. Gosling, "X-ray diffraction studies with Rosalind Franklin," in Seweryn Chomet, ed., *Genesis of a Discovery* (London: Newman Hemisphere, 1995), 43–73, esp. 47–48.

18. Judson, *The Eighth Day of Creation*, 128.

19. M. F. Perutz and J. C. Kendrew, "The Application of X-ray crystallography to the study of biological macromolecules," in F. J. W. Roughton and J. C. Kendrew, eds., *Haemoglobin: The Joseph Barcroft Memorial Conference* (London: Butterworths, 1949), 171.

20. Francis Crick, "The height of the vector rods in the three-dimensional Patterson

of haemoglobin," unpublished typescript (no. 1), signed by Crick and dated July 1951, and another typescript (no. 2) returned with editorial marks and figures following acceptance for publication in *Acta Crystallographica* 5 (1952): 381–86. FCP, PPCRI/H/1/4. Box 68.

21. Author interview with James D. Watson (no. 2), July 24, 2018.

22. Gosling, "X-ray diffraction studies with Rosalind Franklin," 66.

23. Rosalind Franklin, laboratory notebooks 1951–52, RFP, FRKN 1/1. When Franklin told Crick these findings, while in a tea queue at a conference in the Sedgwick Zoology Laboratory in July 1952, he condescendingly advised her to "scrutinize the evidence" she gathered, which appeared to be anti-helical, "very carefully." See Robert Olby, *Francis Crick: Hunter of Life's Secrets* (Cold Spring Harbor, NY: Cold Spring Harbor Laboratory Press, 2009), 152–53.

24. Postcard sent by Franklin and Gosling, "Announcing the Death of the DNA Helix, July 18, 1952." See Wilkins, *The Third Man of the Double Helix*, 182–83; Judson, *The Eighth Day of Creation*, 121; Maddox, *Rosalind Franklin*, 184–85. "Besselised" refers to the mathematical formula called a Bessel function, which is used in helical diffraction theory. According to Gosling, the "death notice" was given only to Wilkins and Stokes; Gosling preserved his copy. James D. Watson, *The Annotated and Illustrated Double Helix*, edited by Alexander Gann and Jan Witkowski (New York: Simon and Schuster, 2012), 179.

25. Maddox, *Rosalind Franklin*, 184; Jenifer Glynn, *My Sister Rosalind Franklin: A Family Memoir* (Oxford: Oxford University Press, 2012), 129; email from Jenifer Glynn to the author, August 27, 2020.

26. Gosling, "X-ray diffraction studies with Rosalind Franklin," 68.

27. Description of the Second European Symposium on Microbial Genetics, 1952, at Pallanza, by John Fincham, professor of genetics at Edinburgh and later Cambridge, JDWP, JDW/2/1/29; letters from Luca Cavalli-Sforza to James D. Watson, September–October 1952, JDWP, JDW/2/2/304; "Pallanza Italy Meeting," photographs of the attendees, JDWP, JDW/1/11/1; photographs of friends and colleagues at Cold Spring Harbor, 1946, and of attendees at the Pallanza conference, Guido Pontecorvo Papers, UGC198/10/1/1/11, Glasgow University Archive Services; Guido Pontecorvo, "Somatic recombination in genetics analysis without sexual reproduction in filamentous fungi," paper read at the conference, Guido Pontecorvo Papers, UGC198/7/3/3.

28. Watson, *The Double Helix*, 83.

29. J. Lederberg and E. L. Tatum, "Gene Recombination in *Escherichia coli*," *Nature* 158, no. 4016 (1946): 558; E. L. Tatum and J. Lederberg, "Gene Recombination in the Bacterium *Escherichia coli*," *Journal of Bacteriology* 53, no. 6 (1947): 673–84; J. Lederberg and N. D. Zinder, "Genetic Exchange in Salmonella," *Journal of Bacteriology* 64, no. 5 (1952): 679–99; J. Lederberg, L. L. Cavalli, and E. M. Lederberg, "Sex Compatibility in *Escherichia coli*," *Genetics* 37 (1952): 720–31; J. Lederberg,

"Genetic Recombination in Bacteria: A Discovery Account," *Annual Review of Genetics* 21 (1987): 23–46.

30. Watson, *The Double Helix*, 83.

31. Watson, *The Double Helix*, 83. Watson's ridicule was contagious. Lederberg's elaborate lecture and terminology was later "spoofed" in a joke letter to the editor of *Nature* about the "possible future importance of cyberkinetics at the bacterial level"; the editors of *Nature* did not realize it was a joke and published the letter. Boris Ephrussi, James Watson, Jean Weigle, and Urs Leopold, "Terminology in Bacterial Genetics," *Nature* 171, no. 4355 (April 18, 1953): 701. The letter ran in *Nature* only a week before Watson and Crick's famous DNA paper.

32. Watson, *The Double Helix*, 83–84; William Hayes, "Recombination in *B. coli*-12. Unidirectional transfer of genetic material," *Nature* 169 (1952): 118–19; William Hayes, "Observations on a transmissible agent determining sexual differentiation in *B. coli*," *Journal of General Microbiology* 8 (1953): 72–88; P. Broada and B. Holloway, "William Hayes, 19 January 1913–7 January 1994," *Biographical Memoirs of Fellows of the Royal Society* 42 (1996): 172–89; Roberta Bivins, "Sex Cells: Gender and the Language of Bacterial Genetics," *Journal of the History of Biology* 33, no. 1 (Spring 2000): 113–39; R. Jayaraman, "Bill Hayes and his Pallanza Bombshell," *Resonance*, October 2011, 911–21, https://www.ias.ac.in/article/fulltext/reso/016/10/0911-0921.

33. Letter from James D. Watson to Elizabeth Watson, October 27, 1952, WFAT, JDW/1/1/22. He uses a similar turn of phrase in a letter to Max Delbrück, September 23, 1952, MDP, box 23, folder 21.

34. Watson, *The Double Helix*, 84.

35. Watson, *The Double Helix*, 84.

36. Thomas Hager, *Force of Nature: The Life of Linus Pauling* (New York: Simon and Schuster, 1995), 413–15; letter from Linus Pauling to Arne Tiselius, October 17, 1952, LAHPP, http://scarc.library.oregonstate.edu/coll/pauling/calendar/1952/10/17.htmlNo.corr407.5-lp-tiselius-19521017.tei.xml.

37. Hager, *Force of Nature*, 413; W. Cochran and F. H. C. Crick, "Evidence for the Pauling–Corey α-Helix in Synthetic Polypeptides," *Nature* 169, no. 4293 (1952): 234–35; W. Cochran, F. H. C. Crick, and V. Vand, "The structure of synthetic peptides. I. The transform of atoms on a helix," *Acta Crystallographica* 5 (1952): 581–86.

38. Francis Crick, *What Mad Pursuit: A Personal View of Scientific Discovery* (New York: Basic Books, 1988), 60–61; Horace Judson interview with Francis Crick, July 3, 1975, HFJP. In this interview, Crick recalled teaching Watson helical diffraction theory around this time and how hard Watson worked to master it, "better than Max and John at that time, you see. Because he kept at it. I don't think he'd ever have learnt it by himself, he had to be taught."

39. Watson, *The Double Helix*, 9, 86.

40. Hager, *Force of Nature*, 414.

41. L. C. Pauling and R. B. Corey, "Compound Helical Configurations of Polypeptide Chains: Structure of Proteins of the α-Keratin Type," *Nature* 171, no. 4341 (January 10, 1953): 59–61.

42. F. H. C. Crick, "Is α-Keratin a Coiled Coil?," *Nature* 170, no. 4334 (November 22, 1952): 882–33; see also F. H. C. Crick, "The Packing of α-helices. Simple Coiled-Coils," *Acta Crystallographica* 6 (1953): 689–97.

43. On November 19, 1952, Pauling wrote to Donohue that "Crick had asked me if I had thought about the possibility of alpha helixes twisting around each other, and I said that I had—I don't remember that we said any more about the matter." Letter from Jerry Donohue to Linus Pauling, November 19, 1952, LAHPP, http://scarc.library.oregonstate.edu/coll/pauling/calendar/1952/11/index.html.

    In another letter, dated December 19, 1952, Donohue wrote of Crick's embarrassment over "the sloppy timing of the publication" of his and Pauling's α-keratin papers; LAHPP, http://scarc.library.oregonstate.edu/coll/pauling/dna/corr/sci9.001.14-donohue-lp-19521215-transcript.html. See also letter from Peter Pauling to Linus Pauling, January 13, 1953, and letter from Linus Pauling to Max Perutz, March 29, 1953, LAHPP, Quoted in: James Watson *The Annotated and Illustrated Double Helix*, 152, 325.

44. Hager, *Force of Nature*, 415–16.

Part V: The Home Stretch, November 1952–April 1953

1. Horace Judson interview with William Lawrence Bragg, January 28, 1971, HFJP.

2. "Nature Conference: Thirty Years of DNA," *Nature* 302 (April 21, 1983): 651–54, quote is on 652.

Chapter 20: Linus Sings

1. A Linos (Λίνος) song, or "Linus song," was a dirge sung to memorialize those who died young and to commemorate the end of summer. See Homer, *The Iliad*, translated by Robert Fagles (New York: Penguin, 1990), 586 (Book 18, lines 664–69).

2. Thomas Hager, *Force of Nature: The Life of Linus Pauling* (New York: Simon and Schuster, 1995), 416–21, quotes are on 417. It should be noted that in some viruses RNA, rather than DNA, carries genetic information.

3. Hager, *Force of Nature*, 417.

4. James D. Watson, *The Double Helix: A Personal Account of the Discovery of the Structure of DNA*, edited by Gunther Stent (New York: Norton, 1980), 33. Alexander Todd, an organic chemist from Scotland, would go on to win the 1957 Nobel Prize in Chemistry "for his work on nucleotides and nucleotide co-enzymes." See

Alexander R. Todd and Daniel M. Brown, "Nucleotides. Part 10. Some observations on the structure and chemical behavior of the nucleic acids," *Journal of the Chemical Society* 1952: 52–58; Daniel M. Brown and Hans Kornberg, "Alexander Robertus Todd, O.M., Baron Todd of Trumpington, 2 October 1907–10 January 1997," *Biographical Memoirs of Fellows of the Royal Society* 46 (2000): 515–32; Alexander Todd, *A Time to Remember: The Autobiography of a Chemist* (Cambridge: Cambridge University Press, 1983), 83–91; letter from Linus Pauling to Henry Allen Moe, December 19, 1952, LAHPP, http://scarc.library.oregonstate.edu/coll/pauling/dna/corr/sci14.014.7-lp-moe-19521219-01.html.

5. Linus Pauling, "A Proposed Structure for the Nucleic Acids" (70 pp. manuscript, 2 pp. typescript, 7 pp. notes), November–December 1952, and "Atomic Coordinates for Nucleic Acid, December 20, 1952," LAHPP, http://scarc.library.oregonstate.edu/coll/pauling/dna/notes/1952a.22.html.

6. "The Triple Helix," Narrative 19 in "Linus Pauling and the Race for DNA," documentary film, *Nova*, PBS and Oregon State University, 1977, LAHPP, http://scarc.library.oregonstate.edu/coll/pauling/dna/narrative/page19.html.

7. Pauling and Corey, "A Proposed Structure for the Nucleic Acids" and "Atomic Coordinates for Nucleic Acid, December 20, 1952."

8. Hager, *Force of Nature*, 418.

9. Hager, *Force of Nature*, 419.

10. Letter from Linus Pauling to E. Bright Wilson, December 4, 1952, cited in Hager, *Force of Nature*, 419.

11. Letter from Linus Pauling to Alexander Todd, December 19, 1952, LAHPP, http://scarc.library.oregonstate.edu/coll/pauling/dna/corr/sci9.001.16-lp-todd-19521219.html.

12. Hager, *Force of Nature*, 354–56, 420–21; "Budenz to Lecture on Communist Peril," *New York Times*, October 13, 1945, 5; Louis F. Budenz, *This Is My Story* (New York: McGraw-Hill, 1947); Louis F. Budenz, *Men Without Faces: The Communist Conspiracy in the U.S.A.* (New York: Harper, 1950); Robert M. Lichtman, "Louis Budenz, the FBI, and the 'List of 400 Concealed Communists': An Extended Tale of McCarthy-era Informing," *American Communist History* 3, no. 1 (2004): 25–54; "Louis Budenz, McCarthy Witness, Dies," *New York Times*, April 28, 1972, 44.

13. Louis F. Budenz, "Do Colleges Have to Hire Red Professors," *American Legion* 51, no. 5 (November 1951): 11–13, 40–43.

14. *Hearings Before the Select Committee to Investigate Tax-Exempt Foundations and Comparable Organizations, U.S. House of Representatives, 82nd Congress, Second Session on H.R. 561, December 23, 1952* (Washington, DC: Government Printing Office, 1953), 715–27, quote is on 723.

15. Linus Pauling, memorandum without address or title regarding allegations by

Louis Budenz of Pauling's Communist affiliations, December 23, 1952, LAHPP, http://scarc.library.oregonstate.edu/coll/pauling/peace/notes/1952a.21.html.

16. L. C. Pauling and R. B. Corey, "A Proposed Structure for the Nucleic Acids," *Proceedings of the National Academy of Sciences* 39 (1953): 84–97. The short "preview version" was published as "Structure of the Nucleic Acids," *Nature* 171 (February 21, 1953): 346.

17. Hager, *Force of Nature*, 421.

18. Pauling and Corey, "A Proposed Structure for the Nucleic Acids."

19. Letter from Linus Pauling to John Randall, December 31, 1952, LAHPP, http://scarc.library.oregonstate.edu/coll/pauling/calendar/1952/12/31-xl.html. By the end of 1952, Alexander Rich, a superb American crystallographer, was working with Pauling in Pasadena on getting better X-ray photographs of DNA.

20. Pauling and Corey, "Structure of the Nucleic Acids."

21. Horace Freeland Judson, *The Eighth Day of Creation: Makers of the Revolution in Biology* (Cold Spring Harbor, NY: Cold Spring Harbor Laboratory Press, 2013), 131–35; Hager, *Force of Nature*, 420–22; Pauling and Corey, "A Proposed Structure for the Nucleic Acids"; Pauling and Corey, "Structure of the Nucleic Acids."

Chapter 21: A Stomach Ache in Clare College

1. Walt Whitman, "Manly Health and Training, With Off-Hand Hints Toward Their Conditions," *Walt Whitman Quarterly Review* 33 (2016): 184–310, quote is on 210. The emphases are Whitman's. These essays were originally published in the *New York Atlas*, in serial form on successive Sundays from September 12 to December 26, 1858, under the pseudonym Mose Velsor.

2. Letter from L. M. Harvey, Secretary, Board of Research Studies, Assistant Registrary, to James Watson, November 17, 1952, JDWP, JDW/2/2/1862. The Registrary is a senior academic officer of the University of Cambridge; the archaic spelling of "registrar" is unique in its use to Cambridge University.

3. James D. Watson, *The Double Helix: A Personal Account of the Discovery of the Structure of DNA*, edited by Gunther Stent (New York: Norton, 1980), 87.

4. Watson, *The Double Helix*, 87. The physicist Denis Wilkinson, a Fellow of Jesus College (and later Professor of Experimental Physics at Oxford), was the point person for Watson's possible matriculation to Jesus.

5. Watson, *The Double Helix*, 87–88.

6. Watson's room was number 5 on R stairwell. "Room Assignments: Lent Term, 1953, Easter Term, 1953; both in Clare College Archives, University of Cambridge; Clare College, Cambridge, Extensions, 1951: Layout of typical bedroom and bed sitting rooms. Architects, Sir Giles Gilbert Scott and Son"; October Term, 1952; See also JDWP, "Receipts and Correspondence, 1953–1956, Clare College, Cambridge" (1 of 2), JDW/2/2/338, and "Correspondence 1967–1986, Clare College,

Cambridge" (2 of 2). I am indebted to Jude Brimmer at Clare College Archives, who helped me locate the rooms where Watson lived in 1952.

7. Letter from James D. Watson to Elizabeth Watson, October 8, 1952, JDW/2/2/1934, JDWP.

8. Watson, *The Double Helix*, 87. In 1944, Hammond commanded the Allied military mission that supported the Greek resistance in Thessaly and Macedonia. He was an author of many books on classical Greece and Rome. Nicholas Hammond (Obituary), *The Guardian*, April 4, 2001. Accessed on December 13, 2020 at: https://www.theguardian.com/news/2001/apr/05/guardianobituaries1.

9. Letter from James D. Watson to Elizabeth Watson, October 18, 1952, JDWP, JDW/2/2/1934.

10. The cost of 42 pence in 1952 would equal about £5, or $6.50, today. Watson, *The Double Helix*, 88. For the Whim restaurant ("In Cambridge, All Roads Lead to the Whim"), see http://www.iankitching.me.uk/history/cam/whim.html?LMCL=PkVbfy.

11. Author interview with James D. Watson (no. 4), July 26, 2018.

12. Watson, *The Double Helix*, 88.

13. The English-Speaking Union was founded as an international trust in 1918 by Sir John Evelyn Wrench, popular journalist and editor of the *Spectator*. Its purpose was to bring together students of different cultures in the belief that "the peace of the world and the progress of mankind can be largely helped by a unity of purpose of the English-speaking democracies." See "Creed," *Landmark* 1, no. 4 (April 1919): ix.

14. Author interview with James D. Watson (no. 4), July 26, 2018.

15. Watson, *The Double Helix*, 88; Howard Markel, *The Kelloggs: The Battling Brothers of Battle Creek* (New York: Pantheon, 2017); James C. Whorton, *Inner Hygiene: Constipation and the Pursuit of Health in Modern Society* (New York: Oxford University Press, 2000).

16. Watson, *The Double Helix*, 88.

17. S. C. Roberts, *Adventures with Authors* (Cambridge: Cambridge University Press, 1966), 144.

18. Watson, *The Double Helix*, 88–89. On October 8, 1952, Watson wrote to his sister, Elizabeth, "I have started taking private French lessons from the famed Mrs. Camille Prior who runs the 'high class' boarding house for young Continental girls. They should be rather pleasant as well as instructive"; JDWP, JDW/2/2/1934.

19. In London, this event came to be known as the Great Smog of 1952. Before the thick, grimy, particle-carrying wave of sulfur dioxide dispersed, at least 4,000 Londoners died (recent epidemiological analyses peg the mortality rate at more than 12,000 deaths); in the months that followed, 6,000 or more succumbed to

respiratory illnesses and more than 100,000 Britons fell ill. M. L. Bell, D. L. Davis, and T. Fletcher, "A retrospective assessment of mortality from the London smog episode of 1952: the role of influenza and pollution," *Environmental Health Perspectives* 112, no. 1 (2004): 6–8. This event led to the passage of some of the first air pollution laws in England, including the Clean Air Act of 1956; see Peter Hennessy, *Having It So Good: Britain in the Fifties* (London: Penguin, 2006), 117–18, 120–22.

20. Watson, *The Double Helix*, 89.

21. Francis Crick, "On Protein Synthesis," typescript of a lecture delivered on September 19, 1957, at a Society for Experimental Biology Symposium on the Biological Replication of Macromolecules, held at University College, London) Sydney Brenner Collection, SB/11/5/4, Cold Spring Harbor Laboratory Archives, Cold Spring Harbor, NY, published as F. H. C. Crick, "On Protein Synthesis," *The Symposia of the Society for Experimental Biology* 12 (1958): 138–63; F. H. C. Crick, "The Central Dogma of Molecular Biology," *Nature* 227 (August 8, 1970): 561–63; Matthew Cobb, "60 Years Ago, Francis Crick Changed the Logic of Biology," *PLoS Biology* 15, no. 9 (2017): e2003243, doi.org/10.1371/journal.pbio.2003243.

22. Watson, *The Double Helix*, 89; author interview with James D. Watson (no. 4), July 26, 2018.

23. Watson, *The Double Helix*, 89.

24. Watson, *The Double Helix,* 89–90.

25. Taslima Khan, "A Visit to Abergwenlais Mill," *The Pauling Blog*, https://paulingblog.wordpress.com/tag/abergwenlais-mill/; Peter Pauling, "DNA: The Race That Never Was?," *New Scientist*, May 31, 1973, 558–60.

26. Watson, *The Double Helix*, 91.

27. Thomas Hager, *Force of Nature: The Life of Linus Pauling* (New York: Simon and Schuster, 1995), 420.

28. Watson, *The Double Helix*, 91.

29. Watson, *The Double Helix*, 91.

Chapter 22: Peter and the Wolf

1. Peter Pauling, "DNA: The Race That Never Was?," *New Scientist*, May 31, 1973, 558–60, quote is on 559.

2. James D. Watson, *The Double Helix: A Personal Account of the Discovery of the Structure of DNA*, edited by Gunther Stent (New York: Norton, 1980), 92.

3. Pauling, "DNA: The Race That Never Was?," 559. Linus Pauling wrote to Jerry Donohue around the same time stating that he was "hoping soon to complete a short paper on nucleic acids"; Thomas Hager, *Force of Nature: The Life of Linus Pauling* (New York: Simon and Schuster, 1995), 420.

4. Linus Pauling sent the manuscript to Peter and to Bragg on January 21, 1952; it was received on January 28. Victor K. McElheny, *Watson and DNA: Making a Scientific Revolution* (New York: Perseus, 2003), 49–50.

5. Cynthia Sanz, "Brooklyn's Polytech: A Storybook Success," *New York Times*, January 5, 1986, 26.

6. Erwin Chargaff, "A Quick Climb Up Mount Olympus," review of *The Double Helix* by James D. Watson, *Science* 159, no. 3822 (1968): 1448–49.

7. Pauling, "DNA: The Race That Never Was?," 559.

8. Watson, *The Double Helix*, 93; Peter Pauling recalled merely giving the manuscript to Watson and Crick; see Pauling, "DNA: The Race That Never Was?," 559.

9. Horace Freeland Judson, *The Eighth Day of Creation: Makers of the Revolution in Biology* (Cold Spring Harbor, NY: Cold Spring Harbor Laboratory Press, 2013), 133. Judson does a superb job of explaining Pauling's errors in his triple helix paper on 133–35; see also Thomas Hager, *Force of Nature: The Life of Linus Pauling* (New York: Simon and Schuster, 1995), 416–25.

10. L. C. Pauling and R. B. Corey, "A Proposed Structure for the Nucleic Acids," *Proceedings of the National Academy of Sciences* 39 (1953): 84–97.

11. Howard Markel, "Science Diction: The Origin of Chemistry," *Science Friday/Talk of the Nation*, NPR, August 26, 2011, https://www.npr.org/2011/08/26/139972673/science-diction-the-origin-of-chemistry.

12. Judson, *The Eighth Day of Creation*, 135.

13. Watson, *The Double Helix*, 94.

14. Chemically speaking, an acid contains a hydrogen atom bonded to a negatively charged atom and when placed in water, that bond is broken. This facilitates a chemical process called dissociation wherein the acid (HA) releases the hydrogen ion (a proton or positive charge, H+) that binds with water to yield a conjugate base (H3O+) and a conjugate acid (A-). Watson, *The Double Helix*, 94.

15. Watson, *The Double Helix*, 93.

16. Watson, *The Double Helix*, 94.

17. Watson, *The Double Helix*, 94.

18. Watson, *The Double Helix*, 94.

19. Watson, *The Double Helix*, 94.

20. Judson, *The Eighth Day of Creation*, 135.

21. Watson, *The Double Helix*, 94.

22. Watson, *The Double Helix*, 95.

23. Defense of the Realm (No. 2) Regulations, 1914, s. 4. *London Gazette (Supplement)*, September 1, 1914, 6968–69.

24. Author interview with James D. Watson (no. 1), July 23, 2018.

25. Watson, *The Double Helix*, 95.

Chapter 23: Photograph No. 51

1. Steven Rose interview with Maurice Wilkins, "National Life Stories. Leaders of National Life. Professor Maurice Wilkins, FRS," C408/017 (London: British Library, 1990), 111.

2. James D. Watson, address at the inauguration of the Center for Genomic Research, Harvard University, September 30, 1999, quoted in "Linus Pauling and the Race for DNA," documentary film, PBS and Oregon State University, 1977, LAHPP, http://scarc.library.oregonstate.edu/coll/pauling/dna/quotes/rosalind_franklin.html.

3. Steven Rose interview with Maurice Wilkins.

4. Maurice Wilkins, *The Third Man of the Double Helix* (Oxford: Oxford University Press, 2003), 196.

5. Wilkins, *The Third Man of the Double Helix*, 196–98, quote is on 198.

6. Brenda Maddox interview with Raymond Gosling, c. 2000, cited in *Rosalind Franklin: The Dark Lady of DNA* (New York: HarperCollins, 2002), 196, 343; Raymond G. Gosling, "X-ray Diffraction Studies of Desoxyribose Nucleic Acid," PhD thesis, University of London, 1954.

7. "The Secret of Photo 51," documentary television program, *Nova*, PBS, April 22, 2003, https://www.pbs.org/wgbh/nova/transcripts/3009_photo51.html.

8. James D. Watson, *The Annotated and Illustrated Double Helix*, edited by Alexander Gann and Jan Witkowski (New York: Simon and Schuster, 2012), 182.

9. Author interview with Jenifer Glynn, May 7, 2018.

10. Maddox, *Rosalind Franklin*, 190–206, quote is on 190. Rosalind Franklin, laboratory notes for January 1953, Rosalind Franklin, laboratory notebooks, September 1951–May 1953, RFP, FRKN 1/1; Aaron Klug, "Rosalind Franklin and the Discovery of the Double Helix," *Nature* 219, no. 5156 (1968): 808–10 and 843–44; Aaron Klug, "Rosalind Franklin and the Double Helix," *Nature* 248 (1974): 787–88.

11. A. Gann and J. Witkowski, "The Lost Correspondence of Francis Crick," *Nature* 467 (2010): 519–24, quote is on 522.

12. Wilkins, *The Third Man of the Double Helix*, 203–4.

13. Wilkins, *The Third Man of the Double Helix*, 200–1.

14. Wilkins, *The Third Man of the Double Helix*, 200–3.

15. Herbert R. Wilson, "The Double Helix and All That," *Trends in Biochemical Sciences* 13, no. 7 (1988): 275–78; see also Herbert R. Wilson, "Connections," *Trends in Biochemical Sciences* 26, no. 5 (2000): 334–37; Maddox, *Rosalind Franklin*, 192.

16. Klug, "Rosalind Franklin and the Discovery of the Double Helix."

17. Horace Freeland Judson, *The Eighth Day of Creation: Makers of the Revolution in*

Biology (Cold Spring Harbor, NY: Cold Spring Harbor Laboratory Press, 2013), 145–52; Watson, *The Double Helix*, 95–99.

18. Watson, *The Double* Helix, 95.

19. Anne Sayre interview with André Lwoff, c. early October 1970, ASP, box 4, folder 14.

20. Author interview with James D. Watson (no. 4), July 26, 2018.

21. Maddox, *Rosalind Franklin*, 194.

22. Author interview with Jenifer Glynn, May 7, 2018; Jenifer Glynn, *My Sister Rosalind Franklin: A Family Memoir* (Oxford: Oxford University Press, 2012), 156.

23. Watson, *The Double Helix*, 95.

24. Judson, *The Eighth Day of Creation*, 136.

25. Klug, "Rosalind Franklin and the Discovery of the Double Helix."

26. Watson, *The Double Helix*, 96.

27. Watson, *The Double Helix*, 96.

28. Watson, *The Double Helix*, 96.

29. Watson, *The Double Helix*, 96.

30. Anne Sayre interview with Maurice Wilkins, June 15, 1970, ASP, box 4, folder 32.

31. "The Race for the Double Helix," documentary television program, narrated by Isaac Asimov, *Nova*, PBS, March 7, 1976.

32. Jenifer Glynn, email to the author, August 13, 2019.

33. Author interview with James D. Watson (no. 1), July 23, 2018.

34. Watson, *The Double Helix*, 97.

35. James D. Wilson, *Genes, Girls and Gamow: After the Double Helix* (New York: Knopf, 2002), 10.

36. Watson, *The Double Helix*, 98.

37. Watson, *The Double Helix*, 98. The many drafts of the book are preserved in the Watson Family Asset Trust. Suffice it to say, Watson worked long and hard to perfect his now famous narrative.

38. Anne Sayre interview with Maurice Wilkins, June 15, 1970, ASP, box 4, folder 32.

39. Letter from Francis Crick to Brenda Maddox, April 12, 2000, cited in Maddox, *Rosalind Franklin*, 343.

40. Wilkins, *The Third Man of the Double Helix*, 218–19.

41. Author interview with Jenifer Glynn, May 7, 2018.

42. Watson, *The Double Helix*, 99.

43. Watson, *The Double Helix*, 99.

44. Watson, *The Double Helix*, 98.

45. Watson, *The Double Helix*, 99.

46. Watson, *The Double Helix*, 99.

47. Watson, *The Double Helix*, 99.

Chapter 24: The Mornings After

1. Anne Sayre interview with Francis Crick, June 16, 1970, ASP, box 2, folder 9; see also Anne Sayre, *Rosalind Franklin and DNA* (New York: Norton, 1975), 214, n. 21.

2. James D. Watson, *The Double Helix: A Personal Account of the Discovery of the Structure of DNA*, edited by Gunther Stent (New York: Norton, 1980), 105.

3. Watson, *The Double Helix*, 61; J. G. Crowther, *The Cavendish Laboratory, 1874–1974* (New York: Science History Publications, 1974), 283.

4. Watson, *The Double Helix*, 100.

5. Watson, *The Double Helix*, 100.

6. Horace Freeland Judson, *The Eighth Day of Creation: Makers of the Revolution in Biology* (Cold Spring Harbor, NY: Cold Spring Harbor Laboratory Press, 2013), 139.

7. W. S. Gilbert and Arthur Sullivan, *H.M.S. Pinafore*, in *The Complete Plays of Gilbert and Sullivan* (New York: Modern Library, 1936), 99–137; "For He Is an Englishman," 131. See also Thomas Hager, *Force of Nature: The Life of Linus Pauling* (New York: Simon and Schuster, 1995), 424.

8. Watson, *The Double Helix*, 100.

9. Watson, *The Double Helix*, 100.

10. Watson, *The Double Helix*, 100.

11. Watson, *The Double Helix*, 100–1.

12. Watson insistently told Crick, "the meridional reflection at 3.4 Å was much stronger than any other reflection . . . [meaning that] the 3.4 Å-thick purine and pyrimidine bases were stacked on top of each other in a direction perpendicular to the helical axis. In addition, we could feel sure from both electron-microscope and X-ray evidence that the helix diameter was about 20 Å." Watson, *The Double Helix*, 101.

13. Watson, *The Double Helix,* 101.

14. Rosalind Franklin, laboratory notebooks, September 1951–May 1953, RFP, FRKN 1/1; Brenda Maddox, *Rosalind Franklin: The Dark Lady of DNA* (New York: HarperCollins, 2002), 197–98; Judson, *The Eighth Day of Creation*, 139–41.

15. Rosalind Franklin, laboratory notebooks, September 1951–May 1953, RFP, FRKN 1/1.

16. By this point, Franklin was so close to figuring out the helical structure that she was even consulting Crick's helical theory paper. Rosalind Franklin, laboratory notebooks, September 1951–May 1953, RFP, FRKN 1/1; W. Cochran, F. H. C. Crick, and V. Vand, "The structure of synthetic peptides. I. The transform of atoms on a helix," *Acta Crystallographica* 5 (1952): 581–86.

17. Judson, *The Eighth Day of Creation*, 627.

18. Judson, *The Eighth Day of Creation*, 627.

19. Aaron Klug, "Rosalind Franklin and the Discovery of the Double Helix," *Nature* 219, no. 5156 (1968): 808–10, 843–44.

20. Rosalind Franklin, laboratory notebooks, September 1951–May 1953, RFP, FRKN 1/1; Klug, "Rosalind Franklin and the Discovery of the Double Helix"; Aaron Klug, "Rosalind Franklin and the Double Helix," *Nature* 248 (1974): 787–88; Judson, *The Eighth Day of Creation*, 148.

21. Anne Sayre interview with Maurice Wilkins, June 15, 1970, ASP, box 4, folder 32.

22. Author interview with James D. Watson (no. 2), July 24, 2018.

23. Letter from Peter Pauling to Linus Pauling, January 13, 1953, and letter from Linus Pauling to Peter Pauling, February 4, 1953, both in LAHPP, http://scarc .library.oregonstate.edu/coll/pauling/dna/corr/bio5.041.6-peterpauling-paulings -19530113.html and http://scarc.library.oregonstate.edu/coll/pauling/dna/corr/ sci9.001.24-lp-peterpauling-19530204.html; Robert Olby, *The Path to the Double Helix* (Seattle: University of Washington Press, 1974), 382–83; "A Very Pretty Model," Narrative 25 in "Linus Pauling and the Race for DNA," documentary film, PBS and Oregon State University, 1977, LAHPP, http://scarc.library .oregonstate.edu/coll/pauling/dna/narrative/page25.html.

24. Letter from Linus Pauling to Peter Pauling, February 18, 1953, LAHPP, http:// scarc.library.oregonstate.edu/coll/pauling/dna/corr/sci9.001.26-lp-peterpauling -19530218.html.

25. Olby, *The Path to the Double Helix*, 383.

26. Watson, *The Double Helix*, 102.

27. Watson, *The Double Helix*, 102.

28. Watson, *The Double Helix*, 103.

29. Watson, *The Double Helix*, 103.

30. Watson, *The Double Helix*, 103.

31. Francis Crick, *What Mad Pursuit: A Personal View of Scientific Discovery* (New York: Basic Books, 1988), 70.

32. Watson, *The Double Helix*, 103.

33. Watson, *The Double Helix*, 103.

34. Francis Crick, "Polypeptides and proteins: X-ray studies," PhD dissertation, Gonville and Caius College, University of Cambridge, submitted on July 1953, FCP, PPCRI/F/2, https://wellcomelibrary.org/item/b18184534.

35. Watson, *The Double Helix*, 103.

36. Letter from Maurice Wilkins to Francis Crick, dated "Thursday," probably written on February 5, 1953, and received on Saturday, February 7, 1953, FP, PPCRI/H/1/42/4. See also Judson, *The Eighth Day of Creation*, 140, 664; Wilkins, *The Third Man of the Double Helix*, 203.

37. Watson, *The Double Helix*, 103.

38. Wilkins, *The Third Man of the Double Helix*, 203–5.

39. Watson, *The Double Helix*, 104.

40. Wilkins, *The Third Man of the Double Helix*, 205–6.

41. Wilkins, *The Third Man of the Double Helix*, 206.

42. Watson, *The Double Helix*, 104.

43. Wilkins, *The Third Man of the Double Helix*, 206–7.

Chapter 25: The MRC Report

1. Hedy Lamarr, *Ecstasy and Me: My Life as a Woman* (New York: Fawcett Crest, 1967), 249.

2. Francis Crick, *What Mad Pursuit: A Personal View of Scientific Discovery* (New York: Basic Books, 1988), 75.

3. Horace Freeland Judson, *The Eighth Day of Creation: Makers of the Revolution in Biology* (Cold Spring Harbor, NY: Cold Spring Harbor Laboratory Press, 2013), 139.

4. James D. Watson, *The Double Helix: A Personal Account of the Discovery of the Structure of DNA*, edited by Gunther Stent (New York: Norton, 1980), 104.

5. The Rex was originally a barnlike structure called the Rendezvous, which from 1911 to 1919 was a roller-skating rink. After the Great War, the building was converted into the Rendezvous Theatre to accommodate the silent picture craze but it burned to the ground in 1931. The theatre reopened the following year as the Rex. James D. Watson, *The Annotated and Illustrated Double Helix*, edited by Alexander Gann and Jan Witkowski (New York: Simon and Schuster, 2012), 193; "flea pit" in *The Cambridge Dictionary*, https://dictionary.cambridge.org/dictionary/english/fleapit.

6. During the Second World War, Lamarr invented a "wi-fi" radio guidance system for torpedoes which prevented them from being jammed up by enemy radio signals and thrown off course. Lamarr, a self-taught inventor, patented this technology with the musician George Antheil in 1942, but it was not installed on U.S. naval ships until 1962. Richard Rhodes, *Hedy's Folly: The Life and Breakthrough Inventions of Hedy Lamarr, the Most Beautiful Woman in the World* (New York: Doubleday, 2012).

7. The 1933 Czech erotic romance was Lamarr's first starring role. The dark-haired eighteen-year-old played the bored wife of a much older man. The most titillating scenes featured close-ups of her face in the throes of orgasm (fueled by the sadistic director, Gustav Machatý, who repeatedly jabbed her in the buttocks and elbows with a pin), her "bare bottom bounc[ing] across the screen," swimming in the nude, and a handful of other glimpses of her body which at the time were viewed as "sizzling," if not absolutely forbidden. Instantly banned in both the United States and Germany, *Ecstasy* enjoyed a comeback of sorts in the 1950s. The version Watson saw was heavily cut and its lines of "uncontrolled passion" were dubbed into stilted English. Lamarr, *Ecstasy and Me*, 21–25.

8. Watson, *The Double Helix*, 104.

9. Watson, *The Double Helix*, 104–5.

10. M. F. Perutz, M. H. F. Wilkins, and J. D. Watson, "DNA Helix," *Science* 164, no. 3887 (1969): 1537–39; report by John Randall to the Medical Research Council, December 1952, JRP, RNDL 2/2/2,; see also "Letters and Documents related to R. E. Franklin's X-ray diffraction studies at King's College, London, in my Laboratory," JRP, RNDL 3/1/6.

11. Judson, *The Eighth Day of Creation*, 142.

12. Horace Judson interview with Francis Crick, July 3, 1975, HFJP; Judson, *The Eighth Day of Creation*, 142.

13. Judson, *The Eighth Day of Creation*, 142.

14. Horace Judson interview with Francis Crick, July 3, 1975, HFJP; Judson, *The Eighth Day of Creation*, 142.

15. Horace Judson interview with Francis Crick, July 3, 1975, HFJP; Judson, *The Eighth Day of Creation*, 143; Watson, *The Double Helix*, 99.

16. Robert Olby interviews with Francis Crick, March 6, 1968, and August 7, 1972, cited in Robert Olby, *The Path to the Double Helix* (Seattle: University of Washington Press, 1974), 404.

17. This tart description appears in a letter Chargaff wrote to Maurice Wilkins, just after the famous Watson and Crick model was published; May 8, 1953, ECP.

18. Erwin Chargaff, "A Quick Chase Up Mount Olympus," review of *The Double Helix* by James D. Watson), *Science* 159, no. 3822 (1968): 1448–49.

19. Letter from Max Perutz to John Randall, February 13, 1969, JRP, RNDL 2/4.

20. Letter from Landsborough Thomson to Max Perutz, February 4, 1969, and letter from H. P. Himsworth to Max Perutz, July 26, 1968, both in JRP, RNDL 2/4 and 2/2/2.

21. Perutz, Wilkins, and Watson, "DNA Helix."

22. Perutz, Wilkins, and Watson, "DNA Helix."

23. Letter from Max Perutz to Harold Himsworth, April 6, 1953, Medical Research Council Archives, FD1, British National Archives, Richmond, UK. This remarkable letter was discovered by Georgina Ferry and appears in her *Max Perutz and the Secret of Life* (London: Chatto and Windus, 2007), 151–54.

24. Memorandum from Maurice Wilkins to John Randall, December 19, 1968, MWP, K/PP178/3/35/7.

25. Letter from John Randall to W. L. Bragg, January 13, 1969, WLBP, 12/98. In another letter to Bragg, November 5, 1968, Randall wrote, "I have always felt that an initial joint publication by Watson, Crick and Wilkins would have been the best thing, but I was unable to press this with you at the time because Wilkins himself did not appear to want it" (12/90).

26. Max F. Perutz, "How the Secret of Life was Discovered," *Daily Telegraph*, April 27, 1987 reprinted as "Discoverers of the Double Helix" in Max F. Perutz, *Is Sci-*

ence Necessary? Essays on Science and Scientists (New York: E. P. Dutton, 1989), 181–83.

27. Watson, *The Double Helix*, 105.

28. J. N. Davidson, *The Biochemistry of the Nucleic Acids* (London: Methuen, 1950).

29. Albert Neuberger, "James Norman Davidson, 1911–1972," *Biographical Memoirs of Fellows of the Royal Society* 19 (1973): 281–303.

30. Erwin Chargaff and J. N. Davidson, eds., *The Nucleic Acids: Chemistry and Biology*, 2 vols. (New York: Academic, 1955).

31. Davidson, *The Biochemistry of the Nucleic Acids*, 5–19.

32. Watson, *The Double Helix*, 105.

33. Watson, *The Double Helix*, 105.

34. Professor Gulland died in a train crash on October 26, 1947, while traveling on the London Northeastern Railway from Edinburgh to King's Cross. James D. Watson, *The Double Helix*, 106. J. M. Gulland, D. O. Jordan, and C. J. Threlfall, "212. Deoxypentose Nucleic Acids. Part I. Preparation of the Tetrasodium Salt of the Deoxypentose Nucleic Acid of Calf Thymus," *Journal of the Chemical Society* 1947: 1129–30; J. M. Gulland, D. O. Jordan, and H. F. W. Taylor, "213. Deoxypentose Nucleic Acids. Part II. Electrometric Titration of the Acidic and the Basic Groups of the Deoxypentose Nucleic Acid of Calf Thymus," *Journal of the Chemical Society* 1947: 1131–41; J. M. Creeth, J. M. Gulland, and D. O. Jordan, "214. Deoxypentose Nucleic Acids. Part III. Viscosity and Streaming Birefringence of Solutions of the Sodium Salt of the Deoxypentose Nucleic Acid Thymus," *Journal of the Chemical Society* 1947: 1141–45; J. M. Creeth, "Some Physico-Chemical Studies on Nucleic Acids and Related Substances," PhD thesis, University of London, 1948; S. E. Harding, G. Channell, and Mary K. Phillips-Jones, "The Discovery of Hydrogen Bonds in DNA and a Re-evaluation of the 1948 Creeth Two-Chain Model for its Structure," *Biochemical Society Transactions* 48 (2018): 1171–82; H. Booth and M. J. Hey, "DNA Before Watson and Crick: The Pioneering Studies of J. M. Gulland and D. O. Jordan at Nottingham," *Journal of Chemical Education* 73, no. 10 (1996): 928–31; A. Peacocke, "Titration Studies and the Structure of DNA," *Trends in Biochemical Sciences* 30, no. 3 (2005): 160–62; K. Manchester, "Did a Tragic Accident Delay the Discovery of the Double Helical Structure of DNA?," *Trends in Biochemical Sciences* 20, no. 3 (1995): 126–28. As an aside, Linus Pauling gave the prestigious Jesse Boot Lecture at Nottingham in 1948 but did not meet with Jordan or Creeth. At the time, the doctoral student Creeth was fiddling with a two-chain model of DNA but did not realize the implications because he was focusing on hydrogen bonding.

35. Watson, *The Double Helix*, 106.

36. Watson, *The Double Helix*, 106.

37. Watson, *The Double Helix*, 108.

38. Watson, *The Double Helix*, 108.

Chapter 26: Base Pairs

1. The phrase "base pairs" is used here as a pun drawn from the bonding of the nucleotide bases in DNA. "Base Pairs" was one of the original titles Watson proposed for what became *The Double Helix*. See manuscripts of *The Double Helix*, JDWP.

2. Jerry Donohue, "Honest Jim?," *Quarterly Review of Biology* 51 (June, 1976): 285–89. Donohue was a chemistry professor at the University of Southern California, 1953–66 and at the University of Pennsylvania from 1966 until his death in 1985. In the 1970s, he became a harsh critic of the Watson–Crick model of DNA. See letters from Jerry Donohue to Francis Crick, May 6, 1970, and August 10, 1970, and letter from Francis Crick to Jerry Donohue, May 20, 1970, FCP, PP/CRI/D/2/11/; see also Jerry Donohue, "Fourier Analysis and the Structure of DNA," *Science* 165, no. 3898 (September 12, 1969): 1091–96; Jerry Donohue, "Fourier Series and Difference Maps as Lack of Structure Proof: DNA Is an Example," *Science* 167, no. 3826 (March 27, 1970): 1700–2; F. H. C. Crick, "DNA: Test of Structure?," *Science* 167, no. 3926 (March 27, 1970): 1694; M. H. F. Wilkins, S. Arnott, D. A. Marvin, and L. D. Hamilton, "Some Misconceptions on Fourier Analysis and Watson–Crick Base Pairing," *Science* 167, no. 3926 (March 27, 1970:): 1693–94.

3. James D. Watson, *The Double Helix: A Personal Account of the Discovery of the Structure of DNA*, edited by Gunther Stent (New York: Norton, 1980), 110.

4. On February 25, Max Delbrück handwrote a postscript to a copy of the paper's submission letter to the editor of *PNAS*: "Jim: [Albert] Sturtevant thinks your theory is all wrong . . . Marguerite [Vogt] thinks your theory has a chance but the paper [is] written much too positively. We all think the evidence is very thin and the formulation difficult to read. However, since you don't want to change it, and since I want to do experiments rather than rewrite your paper, and since it will do you good to learn what it means to publish prematurely, I sent it off today with only a few commas and missing words amended." Letter from Max Delbrück to James D. Watson, February 25, 1953, JDWP. See also letter from Max Delbrück to E. B. Wilson, February 25, 1953, JDWP; J. D. Watson and W. Hayes, "Genetic Exchange in *Escherichia Coli* K 12: Evidence for Three Linkage Groups," *Proceedings of the National Academy of Sciences* 39, no. 5 (May, 1953): 416–26.

5. Papers sent to the *Proceedings of the National Academy of Sciences* must be "sponsored" by an elected Academy member. Watson was not yet a member and thus needed Delbrück's sponsorship. See letters from James D. Watson to Max Delbrück, September 23, 1952, October 6, 1952, October 22, 1952, November 25, 1952, and January 15, 1953, MDP, box 23, folders 21 and 22.

6. Watson, *The Double Helix*, 110. See also letter from Max Delbrück to James D. Watson, February 25, 1953, JDWP; letter from Max Delbrück to E. B. Wil-

son, February 25, 1953, JDWP; and the paper that resulted: Watson and Hayes, "Genetic Exchange in *Escherichia coli* K 12"; James D. Watson to Max Delbrück, February 20, 1953, MDP, box 23, folder 22.

7. James D. Watson to Max Delbrück, February 20, 1953, MDP, box 23, folder 22.

8. James D. Watson to Max Delbrück, February 20, 1953, MDP, box 23, folder 22.

9. Watson, *The Double Helix*, 110. Elizabeth II's accession to the throne occurred soon after the death of her father, King George VI, on February 6, 1952; her coronation, as Queen of England, was celebrated on June 2, 1953. Postboxes with her royal marker, ER II (Elizabeth Regina II) began appearing in the spring of 1952.

10. Watson, *The Double Helix*, 110.

11. Horace Freeland Judson, *The Eighth Day of Creation: Makers of the Revolution in Biology* (Cold Spring Harbor, NY: Cold Spring Harbor Laboratory Press, 2013), 129.

12. Thomas Hager, *Force of Nature: The Life of Linus Pauling* (New York: Simon and Schuster, 1995), 425–26.

13. June M. Broomhead, "The structure of pyrimidines and purines. II. A determination of the structure of adenine hydrochloride by X-ray methods," *Acta Crystallographica* 1 (1948): 324–29; June M. Broomhead, "The structures of pyrimidines and purines. IV. The crystal structure of guanine hydrochloride and its relation to that of adenine hydrochloride," *Acta Crystallographica* 4 (1951): 92–100; June M. Broomhead, "An X-ray investigation of certain sulphonates and purines," PhD thesis, Cambridge University, 1948. Her married name was Lindsay and some of her later publications reflect this.

14. Judson, *The Eighth Day of Creation*, 145.

15. Judson, *The Eighth Day of Creation*, 146–47; Jerry Donohue, "The Hydrogen Bond in Organic Crystals," *Journal of Physical Chemistry* 56 (1952): 502–10; Horace Judson interview with Jerry Donohue, October 5, 1973, HFJP; Jerry Donohue Papers, box 5, folders 20 and 21, University of Pennsylvania Archives, Philadelphia, PA.

16. Watson, *The Double Helix*, 110.

17. Watson, *The Double Helix*, 110.

18. Watson, *The Double Helix*, 112.

19. Author interview with James D. Watson (no. 4), July 26, 2018.

20. Watson, *The Double Helix*, 112.

21. L. C. Pauling and R. B. Corey, "Structure of the Nucleic Acids," *Nature* 171 (February 21, 1953): 346; L. C. Pauling and R. B. Corey, "A Proposed Structure for the Nucleic Acids," *Proceedings of the National Academy of Sciences* 39 (1953): 84–97.

22. Rosalind Franklin, notes on the Pauling–Corey triple helix paper, February 1953, RFP, FRKN 1/3 and 1/4; Judson, *The Eighth Day of Creation*, 141.

23. Brenda Maddox, *Rosalind Franklin: The Dark Lady of DNA* (New York: HarperCollins, 2002), 195, 200; R. E. Franklin and R. G. Gosling, "Molecular con-

figuration in sodium thymonucleate," *Nature* 171 (1953): 740–41; R. E. Franklin and R. G. Gosling, "The Structure of Sodium Thymonucleate Fibers. I. The Influence of Water Content," *Acta Crystallographica* 6 (1953): 673–77; R. E. Franklin and R. G. Gosling, "The Structure of Thymonucleate Fibers. II: The Cylindrically Symmetrical Patterson Function," *Acta Crystallographica* 6 (1953): 678–85; see also R. E. Franklin and R. G. Gosling, "The Structure of Sodium Thymonucleate Fibers III. The Three-Dimensional Patterson Function," *Acta Crystallographica* 8 (1955): 151–56. See also J. D. Watson and F. H. C. Crick, "A structure for deoxyribose nucleic acid," *Nature* 171 (1953): 737–38; M. H. F. Wilkins, A. R. Stokes, and H. R. Wilson, "Molecular structure of deoxypentose nucleic acids," *Nature* 171 (1953): 738–40.

24. The *Acta Crystallographica* papers were received for publication by the journal's English editor and slotted to appear in the September 1953 issue; Maddox, *Rosalind Franklin*, 199–201.

25. Robert Olby, *Francis Crick: Hunter of Life's Secrets* (Cold Spring Harbor, NY: Cold Spring Harbor Laboratory Press, 2009), 165; Robert Olby, *The Path to the Double Helix* (Seattle: University of Washington Press, 1974), 410–14.

26. Olby, *Francis Crick*, 165.

27. Watson, *The Double Helix*, 112, 114.

28. Richard Sheridan, *The Rivals*, in *The School for Scandal and Other Plays* (London: Penguin, 1988), 29–124.

29. Watson, *The Double Helix*, 114.

30. This likely apocryphal tale was told a few centuries later by Marcus Vitruvius Pollo in his *Ten Books on Architecture*. Archimedes' discovery regarding displaced volumes immediately allowed him to develop a means of testing the purity of gold. The full text of an English translation by Morris H. Morgan is available at http://www.gutenberg.org/ebooks/20239.

31. Another contribution by the Greeks, this dramatic trick of an unprobable and abrupt solution to a particular problem is credited to the playwright Euripides (480–406 BCE).

32. Watson, *The Double Helix*, 113–14. Watson claims that the formation of a third hydrogen bond between guanine and cytosine was considered, but rejected because a crystallographic study of guanine hinted that it would be very weak. Today, this conjecture is known to be incorrect; three strong hydrogen bonds exist between guanine and cytosine. Linus Pauling made this correction, but it should be noted that the original 1953 Watson–Crick model did <u>not</u> include the third hydrogen bond between cytosine and guanine. L. C. Pauling and R. B. Corey, "Specific Hydrogen-Bond Formation Between Pyrimidines and Purines in Deoxyribonucleic Acids," *Archives of Biochemistry and Biophysics* 65 (1956): 164–81.

33. Author interview with James D. Watson (no. 2), July 25, 2018.

34. Watson, *The Double Helix*, 114–15.

35. Watson, *The Double Helix*, 115.

36. Deb Amlen, "How to Solve the New York Times Crossword," *New York Times*, https://www.nytimes.com/guides/crosswords/how-to-solve-a-crossword-puzzle.

37. Watson, *The Double Helix*, 115; Olby, *Francis Crick*, 166.

38. Olby, *The Path to the Double Helix*, 412.

39. Watson, *The Double Helix*, 115.

40. Olby, *Francis Crick*, 167–68.

41. Watson, *The Double Helix*, 115.

42. Watson, *The Double Helix*, 115.

43. Watson, *The Double Helix*, 115.

44. Ivan Noble, "'Secret of Life' Discovery Turns 50," *BBC News*, February 28, 2003, http://news.bbc.co.uk/2/hi/science/nature/2804545.stm.

Chapter 27: It's So Beautiful

1. Francis Crick, *What Mad Pursuit: A Personal View of Scientific Discovery* (New York: Basic Books, 1988), 78–79. Invitation to the May 1, 1953, meeting of the Hardy Club in Kendrew's rooms at Peterhouse College, featuring James Watson's reading of a paper, "Some Comments on desoxyribonucleic acid"; letters from James D. Watson to Francis Crick, FCP, PP/CRI/H/1/42/3, box 72.

2. James D. Watson, *The Double Helix: A Personal Account of the Discovery of the Structure of DNA*, edited by Gunther Stent (New York: Norton, 1980), 116.

3. Author interview with James D. Watson (no. 1), July 23, 2018.

4. Watson, *The Double Helix*, 116; see also Robert Olby, *Francis Crick: Hunter of Life's Secrets* (Cold Spring Harbor, NY: Cold Spring Harbor Laboratory Press, 2009), 168–69; Robert Olby, *The Path to the Double Helix* (Seattle: University of Washington Press, 1974), 399–416; Horace Freeland Judson, *The Eighth Day of Creation: Makers of the Revolution in Biology* (Cold Spring Harbor, NY: Cold Spring Harbor Laboratory Press, 2013), 148–52.

5. Watson, *The Double Helix*, 116.

6. Olby, *The Path to the Double Helix*, 414; Robert Olby recorded Interviews with Francis Crick for the Royal Society, March 8, 1968 and August 7, 1972, Collections of the Royal Society, London.

7. Olby, *The Path to the Double Helix*, 414.

8. Watson, *The Double Helix*, 116–17.

9. Watson, *The Double Helix*, 117.

10. Judson, *The Eighth Day of Creation*, 627–29.

11. Watson, *The Double Helix*, 118.

12. Watson, *The Double Helix*, 117.

13. Crick, *What Mad Pursuit*, 77.

14. "The Race for the Double Helix," documentary television program, narrated by Isaac Asimov, *Nova*, PBS, March 7, 1976.

15. Watson, *The Double Helix*, 117–18.

16. Watson, *The Double Helix*, 118.

17. Watson, *The Double Helix*, 118.

18. The most prevalent viral type in England in 1953 was prosaically labeled Influenza Virus A/England/1/51, quite similar to the Liverpool strain of 1950–51; the second most prevalent type that year, A/England/1/53, probably originated in Scandinavia; A. Isaacs, R. Depoux, P. Fiset, "The Viruses of the 1952–53 Influenza Epidemic," *Bulletin of the World Health Organization* 11, no. 6 (1954): 967–79; *The Registrar General's Statistical of England and Wales for the Year 1953* (London: Her Majesty's Stationery Office, 1956), 173–88. For an exegesis on the history of influenza pandemics, see Howard Markel et al., "Nonpharmaceutical Interventions Implemented by U.S. Cities During the 1918–1919 Influenza Pandemic," *Journal of the American Medical Assocation* 298, no. 6 (2007: 644–54; Howard Markel and J. Alexander Navarro, eds., *The American Influenza Epidemic of 1918–1919: A Digital Encyclopedia*, http://www.influenzaarchive.org.

19. Letter from Rosalind Franklin to Adrienne Weill, March 10, 1953, ASP, box 2, folder 15.1; Brenda Maddox, *Rosalind Franklin: The Dark Lady of DNA* (New York: HarperCollins, 2002), 205–6.

20. WLBP, MS WLB 54A/282; MS WLB 32E/7. See also Graeme K. Hunter, *Light is a Messenger: The Life and Science of William Lawrence Bragg* (Oxford: Oxford University Press, 2004), 196, 279.

21. Watson, *The Double Helix*, 118.

22. Watson, *The Double Helix*, 118.

23. Watson, *The Double Helix*, 120.

24. Watson, *The Double Helix*, 120.

25. Watson, *The Double Helix*, 120.

26. Alexander Todd, *A Time to Remember: The Autobiography of a Chemist* (Cambridge: Cambridge University Press, 1983), 88.

27. Todd, *A Time to Remember*, 89. See also W. L. Bragg, J. C. Kendrew, and M. F. Perutz, "Polypeptide Chain Configurations in Crystalline Proteins," *Proceedings of the Royal Society of London A: Mathematical and Physical Sciences* 203, no. 1074 (October 10, 1950): 321–57; L. C. Pauling, R. B. Corey, and H. R. Branson, "The structure of proteins: Two hydrogen-bonded helical configurations of the polypeptide chain," *Proceedings of the National Academy of Sciences* 37, no. 4 (1951): 205–11; M. F. Perutz, "New X-ray Evidence on the Configuration of Polypeptide Chains: Polypeptide Chains in Poly-γ-benzyl-L-glutamate, Keratin and Hæmoglobin," *Nature* 167, no. 4261 (1951): 1053–54.

28. Todd conjectured that if only the physicists and chemists had worked more

closely together during this era, the chemists might "have enabled the physicists to make the imaginative jump a year or so earlier, but probably not much more." Todd, *A Time to Remember*, 89.

29. Todd, *A Time to Remember*, 89.

30. Watson, *The Double Helix*, 120.

31. Olby, *The Path to the Double Helix*, 416.

32. Watson, *The Double Helix*, 120.

33. Watson, *The Double Helix*, 120.

34. Crick, *What Mad Pursuit*, 75.

35. Watson, *The Double Helix*, 120.

36. Watson, *The Double Helix*, 120. This is a paraphrase of what Wilkins actually wrote in his letter to Francis Crick, March 7, 1953, FCP, PP/CRI/H/1/42/4.

Chapter 28: Defeat

1. Letter from Maurice Wilkins to Francis Crick, March 7, 1953, FCP, PP/CRI/H/1/42/4.

2. William Shakespeare, *The Life and Death of Richard II*, Act II, scene i.

3. William Shakespeare, sonnets 127–52.

4. Michael Schoenfeldt, *The Cambridge Introduction to Shakespeare's Poetry* (Cambridge: Cambridge University Press, 2010), 98–111.

5. William Shakespeare, sonnet 147.

6. Horace Judson interview with Raymond Gosling, July 21, 1975, HFJP.

7. James D. Watson, *The Double Helix: A Personal Account of the Discovery of the Structure of DNA*, edited by Gunther Stent (New York: Norton, 1980), 126.

8. Max Delbrück, undated memorandum, "The Pauling seminar on his triple helix DNA structure was held on Wednesday, March 4, 1953," MDP, box 23, file 22; Thomas Hager, *Force of Nature: The Life of Linus Pauling* (New York: Simon and Schuster, 1995), 425; Watson, *The Double Helix*, 126.

9. James D. Watson, "Succeeding in Science: Some Rules of Thumb," *Science* 261, no. 5129 (September 24, 1993): 1812–13.

10. Letter from Peter Pauling to Linus and Ava Helen Pauling, March 14, 1953, LAHPP, http://scarc.library.oregonstate.edu/coll/pauling/dna/corr/bio5.041.6-peter pauling-lp-19530301-transcript.html.

11. Letter from Peter Pauling to Linus and Ava Helen Pauling, March 14, 1953, LAHPP, http://scarc.library.oregonstate.edu/coll/pauling/dna/corr/bio5.041.6-peter pauling-lp-19530301-transcript.html.

12. Letter from Francis Crick to Linus Pauling, March 2, 1953, California Institute of Technology Archives, Pasadena, CA, cited in Hager, *Force of Nature*, 424.

13. Maurice Wilkins, *The Third Man of the Double Helix* (Oxford: Oxford University Press, 2003), 211; Horace Freeland Judson, *The Eighth Day of Creation: Makers of*

the Revolution in Biology (Cold Spring Harbor, NY: Cold Spring Harbor Laboratory Press, 2013), 152.

14. Wilkins, *The Third Man of the Double Helix*, 211.

15. Wilkins, *The Third Man of the Double Helix*, 211.

16. Wilkins, *The Third Man of the Double Helix*, 211–12.

17. Wilkins, *The Third Man of the Double Helix*, 212. "Their [the Wilkins laboratory] improved structure for the B form differed from the original in details of the backbones, most significantly by a shift in the angle of sugars, which brings the bases in more snugly to the center. The improved model took Wilkins seven years"; Judson, *The Eighth Day of Creation*, 167. See also Maurice Wilkins, "The Molecular Configuration of Nucleic Acids," in *Nobel Lectures, Physiology or Medicine 1942–1962* (Amsterdam: Elsevier, 1964), 755–82, available at https://www.nobelprize.org/prizes/medicine/1962/wilkins/lecture.

18. Watson, *The Double Helix*, 122.

19. Horace Judson interview with William Lawrence Bragg, January 28, 1971, HFJP.

20. Letter from Maurice Wilkins to Max Perutz, June 30, 1976, HFJP.

21. Wilkins, *The Third Man of the Double Helix*, 215.

22. Watson, *The Double Helix*, 122. In 2018, Watson applauded the "good British manners" of both Wilkins and Franklin when first seeing his and Crick's DNA double helix model; author interview with James D. Watson (no. 4), July 26, 2018.

23. Wilkins, *The Third Man of the Double Helix*, 213–15.

24. Wilkins, *The Third Man of the Double Helix*, 215.

25. Brenda Maddox, *Rosalind Franklin: The Dark Lady of DNA* (New York: HarperCollins, 2002), 209.

26. Anne Sayre interview with Jerry Donohue, December 19, 1975, ASP, box 2; Maddox, *Rosalind Franklin*, 209.

27. Letter from James D. Watson to Max Delbrück, March 12, 1953, MDP, box 23, folder 22.

28. Watson, *The Double Helix*, 127.

29. Watson, *The Double Helix*, 122.

30. Anne Sayre, *Rosalind Franklin and DNA* (New York: Norton, 1975), 168–69.

31. Judson, *The Eighth Day of Creation*, 628.

32. Watson, *The Double Helix*, 124.

33. "Due Credit," *Nature* 496 (April 18, 2013): 270. The last line paraphrases Isaac Newton's famous 1675 adage, "If I have seen further than others, it is by standing on the shoulders of giants." The phrase has been attributed to Bernard of Chartres, who may have first uttered it in the twelfth century.

34. Watson, *The Double Helix*, 124–26.

35. Judson, *The Eighth Day of Creation*, 148.

36. Author interview with James D. Watson (no. 4), July 26, 2018.

37. Sayre, *Rosalind Franklin and DNA*, 213–14; Francis Crick, "How to Live with a Golden Helix," *The Sciences* 19 (September 1979): 6–9.

38. Maddox, *Rosalind Franklin*, 202.

39. Mansel Davies, "W. T. Astbury, Rosie Franklin, and DNA: A Memoir," *Annals of Science* 47 (1990): 607–18, quote is on 617–18. Mansel Davies (1913–95) was a student of William Astbury's and a prominent physicist, X-ray crystallographer, and expert on molecular structure. See Sir John Meurig Thomas, "Professor Mansel Davies," obituary, *Independent*, January 17, 1995, https://www.independent.co .uk/news/people/obituariesprofessor-mansel-davies-1568365.html.

40. Watson, *The Double Helix*, 126.

41. Robert Olby, *Francis Crick: Hunter of Life's Secrets* (Cold Spring Harbor, NY: Cold Spring Harbor Laboratory Press, 2009), 169.

42. Judson, *The Eighth Day of Creation*, 151.

43. Watson, *The Double Helix*, 126.

44. Author interview with James D. Watson (no. 4), July 26, 2018.

45. Watson, *The Double Helix*, 127.

46. Letter from James D. Watson to Max Delbrück, March 22, 1953, MDP, box 23, folder 22.

47. Author interview with James D. Watson (no. 4), July 26, 2018.

48. A copy of the complete letter, Francis Crick to Michael Crick, March 19, 1953, can be found in FCP, PP/CRI/D/4/3, box 243. The original sold at auction on April 10, 2013, by Christie's of New York, for $6,059,750, the world record price for a letter at the time; Jane J. Lee, "Read Francis Crick's $6 Million Letter to Son Describing DNA," *National Geographic*, April 11, 2013, https://blog .nationalgeographic.org/2013/04/11/read-francis-cricks-6-million-letter-to-son -describing-dna/.

Chapter 29: It Has Not Escaped Our Notice

1. J. D. Watson and F. H. C. Crick, "A Structure for Deoxyribose Nucleic Acid," *Nature* 171, no. 4356 (April 25, 1953): 737–38.

2. Letter from James D. Watson and Francis Crick to Linus Pauling, March 21, 1953, LAHPP, http://scarc.library.oregonstate.edu/coll/pauling/dna/corr/sci9.001 .32-watsoncrick-lp-19530321.html.

3. Letter from James D. Watson and Francis Crick to Linus Pauling, March 21, 1953, LAHPP, http://scarc.library.oregonstate.edu/coll/pauling/dna/corr/sci9.001 .32-watsoncrick-lp-19530321.html.

4. Letter from James D. Watson to his parents, March 24, 1953, WFAT, "Cambridge Letters, to his Family, September 1953–September 1953."

5. Held since 1922, and attended by some of the brightest physicists and chemists in the world, the Solvay Conferences were funded by the wealthy Belgian chemist

and industrialist Ernest G. J. Solvay. Although Solvay did not attend university, because of pleurisy, he spent the rest of his life associating with brilliant chemistry and physics professors through his philanthropic work. He developed the Solvay process, which makes soda ash (anhydrous sodium hydroxide, a key ingredient in the manufacture of glass, paper, rayon, soaps, and detergents) from brine and limestone. See Institut International de Chimie Solvay, *Les Protéines, Rapports et Discussions: Neuvième Conseil de Chimie tenu à l'université de Bruxelles du 6 au 14 Avril 1953* (Brussels: R. Stoops, 1953).

6. Thomas Hager, *Force of Nature: The Life of Linus Pauling* (New York: Simon and Schuster, 1995), 388–89, 427.

7. James D. Watson, *The Double Helix: A Personal Account of the Discovery of the Structure of DNA*, edited by Gunther Stent (New York: Norton, 1980), 127; letter from James D. Watson to Max Delbrück, March 12, 1953, MDP, box 23, folder 22.

8. Hager, *Force of Nature*, 428.

9. Hager, *Force of Nature*, 387–89, 427–28.

10. Brenda Maddox, *Rosalind Franklin: The Dark Lady of DNA* (New York: Harper-Collins, 2002), 209.

11. Maddox, *Rosalind Franklin*, 209; "Due Credit," editorial, *Nature* 496 (April 18, 2013): 270.

12. When Macmillan sold the extant files of *Nature* to the British Museum in 1966, the historic and important materials regarding the Watson and Crick DNA paper had already been destroyed. Letter from A. J. V. Gale to Horace Judson, October 3, 1976, and letter from David Davies, editor of *Nature*, to Horace Judson, September 1, 1976, HFJP, file "A. J. V. Gale/*Nature*"; Maddox, *Rosalind Franklin*, 211.

13. Letter from Francis Crick to Maurice Wilkins, March 17, 1953; on the reverse side is a draft of a note to A. J. V. Gale, the co-editor of *Nature*, about their DNA paper, or "letter" as it was referred to in *Nature* parlance: "Both Prof Bragg & Perutz have read the letter and have approved our sending it to you. We would be grateful if you could give us a rough idea if & when you are likely to be able to publish it." See: A. Gann and J. Witkowski, "The lost correspondence of Francis Crick, *Nature* 467 (2010): 519–24.

14. Letter from Maurice Wilkins to Francis Crick, March 18, 1953, FCP, PP/CRI/H/1/42/3, box 72, quoted in Robert Olby, *The Path to the Double Helix* (Seattle: University of Washington Press, 1974), 417–18.

15. Letter from Maurice Wilkins to Erwin Chargaff, June 3, 1953, ECP.

16. The day before, April 1, Gerald Pomerat, the influential assistant director of the natural sciences program at the Rockefeller Foundation, was visiting the Cavendish Laboratory to examine Bragg's protein research, which was funded by the Foundation. Pomerat's diary makes clear that the key story in Cambridge was the excitement over DNA and the two "somewhat mad hatters who bubble over

their new structure." J. Witkowski, "Mad Hatters at the DNA Tea Party," *Nature* 415 (2001): 473–74.

17. James D. Watson, *Girls, Genes and Gamow: After the Double Helix* (New York: Knopf, 2002), 8.

18. Martin J. Tobin, "Three Papers, Three Lessons," *American Journal of Respiratory and Critical Care Medicine* 167, no. 8 (2003): 1047–49.

19. Horace Freeland Judson, *The Eighth Day of Creation: Makers of the Revolution in Biology* (Cold Spring Harbor, NY: Cold Spring Harbor Laboratory Press, 2013), 154.

20. Watson, *The Double Helix*, 129.

21. Watson and Crick, "A Structure of Deoxyribose Nucleic Acid."

22. M. H. F. Wilkins, A. R. Stokes, and H. R. Wilson, "Molecular Structure of Deoxypentose Nucleic Acids," *Nature* 171, no. 4356 (April 25, 1953): 738–40.

23. R. E. Franklin and R. G. Gosling, "Molecular Configuration in Sodium Thymonucleate," *Nature* 171, no. 4356 (April 25, 1953): 740–41. See also Roger Chartier, *The Order of Books: Authors and Libraries in Europe Between the 14th and 18th Centuries* (Palo Alto, CA: Stanford University Press, 1994); and Roger Chartier, *The Cultural Uses of Print in Early Modern France* (Princeton: Princeton University Press, 2019).

24. Judson, *The Eighth Day of Creation*, 148.

25. Franklin and Gosling, "Molecular Configuration in Sodium Thymonucleate."

26. Maddox, *Rosalind Franklin*, 211–12. In fact, Franklin accepted the Watson–Crick model only as a hypothesis. In a September 1953 paper, she wrote that "discrepancies prevent us from accepting it in detail." See R. E. Franklin and R. G. Gosling, "The structure of sodium thymonucleate fibres: The influence of water content. Part I," and "The structure of sodium thymonucleate fibres: The cylindrically symmetrical Patterson function. Part II," *Acta Crystallographica* 6 (1953): 673–77, 678–85; see also Brenda Maddox, "The Double Helix and the 'Wronged Heroine,'" *Nature* 421, no. 6291 (January 23, 2003): 407–8.

27. Watson, *The Double Helix*, 129.

28. Watson, *The Double Helix*, 130.

29. Watson, *The Double Helix*, 129.

30. Watson, *The Double Helix*, 130.

31. Letter from Linus Pauling to Ava Helen Pauling, April 6, 1953, LAHPP, http://scarc.library.oregonstate.edu/coll/pauling/dna/corr/safe1.021.3.html.

32. Linus Pauling notebook, Solvay Congress, April 1953, LAHPP, http://scarc.library.oregonstate.edu/coll/pauling/dna/notes/safe4.083-031.html. In the published version of Bragg's "Note Complémentaire" on Watson and Crick's work—a short report constituting the first formal announcement of their DNA model—Pauling gave a smoother imprimatur to the proceedings: "I feel that it is very likely that the Watson–Crick model is essentially correct." See "Discussion des rapports de

MM. L. Pauling et L. Bragg," and J. D. Watson and F. H. C. Crick, "The Stereo-chemical Structure of DNA," both in Institut International de Chimie Solvay, *Les Protéines. Rapports et Discussions*, 113–18, 110–12.

33. "Linus Pauling Diary: Trips to Germany, Sweden and Denmark, July and August, 1953," 89, LAHPP, http://scarc.library.oregonstate.edu/coll/pauling/dna/notes/safe4.082-017.html.

34. *Lifestory: Linus Pauling*, BBC, 1997, in Linus Pauling and the Nature of the Chemical Bond, website maintained by the LAHPP, http://scarc.library.oregonstate.edu/coll/pauling/bond/audio/1997v.1-pasadena.html.

35. Hager, *Force of Nature*, 429.

36. Hager, *Force of Nature*, 431; John L. Greenberg oral history interview with Linus Pauling, May 10, 1984, 23, Archives of the California Institute of Technology, Pasadena, CA, http://oralhistories.library.caltech.edu/18/1/OH_Pauling_L.pdf.

37. Jim Lake, "Why Pauling Didn't Solve the Structure of DNA," correspondence, *Nature* 409, no. 6820 (February 1, 2001): 558.

38. Hager, *Force of Nature*, 429–30.

39. *Oxford English Dictionary* (Oxford: Oxford University Press, 1989), https://www.oed.com/oed2/00048049;jsessionid=0389830C953F30EA35E2A97FD896F289.

40. Maurice Wilkins, *The Third Man of the Double Helix* (Oxford: Oxford University Press, 2003), 164–65.

41. Maddox, *Rosalind Franklin*, 209–10.

42. Watson and Crick, "A Structure for Deoxyribose Nucleic Acid."

43. Letter from Maurice Wilkins to Francis Crick, March 23, 1953, Sydney Brenner Collection, SB/11/1/77/, Cold Spring Harbor Laboratory Archives, Cold Spring Harbor, NY; Maddox, *Rosalind Franklin*, 210.

44. Maurice Wilkins to Francis Crick, March 18, 1953, cited in Olby, *The Path to the Double Helix*, 418; Maddox, *Rosalind Franklin*, 211.

45. Oddly, Wilkins spells the nickname both "Rosy" and "Rosie" in the same letter. Letter from Maurice Wilkins to Francis Crick, March 23, 1953, Sydney Brenner Collection, SB/11/1/77.

46. Eugene Garfield, "Bibliographic Negligence: A Serious Transgression," *Scientist* 5, no. 23 (November 25, 1991): 14.

47. James D. Watson and Francis Crick, "A Structure for DNA," FCP, PP/CRI/H/1/11/2, box 69.

48. Maddox, *Rosalind Franklin*, 210.

49. W. T. Astbury, "X-ray Studies of Nucleic Acids," *Symposia of the Society for Experimental Biology (I. Nucleic Acids)*, 1947: 66–76; M. H. F. Wilkins and J. T. Randall, "Crystallinity in sperm heads: molecular structure of nucleoprotein in vivo," *Acta Biochimica et Biophysica* 10, no. 1 (1953): 192–93.

50. Watson and Crick, "A Structure for Deoxyribose Nucleic Acid."

51. F. H. C. Crick and J. D. Watson, "The Complementary Structure of Deoxyri-

bonucleic Acid," *Proceedings of the Royal Society A: Mathematical, Physical and Engineering Sciences* 223 (1954): 80–96.

52. Francis Crick, *What Mad Pursuit: A Personal View of Scientific Discovery* (New York: Basic Books, 1988), 66.

53. Watson and Crick, "A Structure for Deoxyribose Nucleic Acid." In *What Mad Pursuit*, Crick noted that some critics have called the closing sentence of their famous paper as coy. In his memoir, he recalled his efforts to include a line or two about the genetic implications of DNA: "I was keen that the paper should discuss genetic implications. Jim was against it. He suffered from periodic fears that the structure might be wrong and that he had made an ass of himself. I yielded to his point of view but insisted that something be put in the paper, otherwise someone would certainly write to make the suggestion, assuming we had been too blind to see it. In short, it was a claim to priority" (66).

54. Watson and Crick have often complained that they gained no immediate fame or credit for the discovery of the double helix. This is true regarding the popular media, and a few years passed before their DNA model was incorporated into college and medical school coursework and textbooks. Watson explained the absence of cheering as the result of "the feeling that we didn't deserve it—because we hadn't done any experiments and it was other people's data"; see Victor K. McElheny, *Watson and DNA: Making a Scientific Revolution* (New York: Perseus, 2003), 65. At the fortieth anniversary of the double helix, Crick told an audience that, as far as instant acclaim went, there was "Not a bit of it, not a bit of it"; see Stephen S. Hall, "Old School Ties: Watson, Crick, and 40 Years of DNA," *Science* 259, no. 5101 (March 12, 1993): 1532–33. There was some newspaper coverage of the Solvay announcement, and only a smattering of newspaper articles on the *Nature* paper. The *New York Times* did not cover the DNA story until May 16, 1953, under the title, "Form of 'Life Unit' in Cell Is Scanned"; a longer piece appeared on June 12, 1953. An article by Ritchie Calder, "Why You Are Nearer to the Secret of Life," appeared in the May 15, 1953, edition of the *London News Chronicle* and there was a short article in the Cambridge undergraduate school paper, *The Varsity*, on May 30, 1953. Nonetheless, one hundred or more scientists who really counted at the time understood the importance of their discovery rather quickly and redirected their own research agendas accordingly.

55. Maddox, *Rosalind Franklin*, 206.

56. Letter from John Randall to Rosalind Franklin, April 17, 1953, JRP, RNDL 3/1/6.

57. Franklin asked Sayre this question in 1953; Anne Sayre, *Rosalind Franklin and DNA* (New York: Norton, 1975), 168, 214.

58. Maddox, *Rosalind Franklin*, 221–22.

59. Steven Rose interview with Maurice Wilkins, "National Life Stories. Leaders of National Life. Professor Maurice Wilkins, FRS," C408/017 (London: British Library, 1990), 60, 116.

60. Steven Rose interview with Maurice Wilkins, 60, 104.

61. Letter from Rosalind Franklin to John Randall, April 23, 1953, JRP, RNDL 3/1/6.

62. The Franklin–Wilkins Building is located at 150 Stamford Street, in the Water-loo campus of the University of London. It now houses the Dental Education Centre and the Franklin–Wilkins Library, which "supports the needs of nursing and midwifery students [and] contains extensive management, bioscience and educational holdings," as well as a small collection of law books for law students taking courses in the building. See https://www.kcl.ac.uk/visit/franklin-wilkins -building.

63. Jenifer Glynn, *My Sister Rosalind Franklin: A Family Memoir* (Oxford: Oxford University Press, 2012), 127.

64. Author interview with Jenifer Glynn, May 7, 2018; Glynn, *My Sister Rosalind Franklin*, 127. Brenda Maddox reported an interview with Dr. Simon Altmann in 1999, and subsequent letters in 2000 and 2001, where he claimed that Frank-lin told him how she came to her lab one day and "found her notebooks being read." She also was worried that her supervisors were not protecting her and, instead, conferring with Watson and Crick. Unfortunately, with the passage of years, Altman could not place the time of this discussion and was working in Argentina between early 1952 and the spring of 1953. Maddox, *Rosalind Franklin*, 194, 210, 343.

65. Author interview with James D. Watson (no. 2), July 24, 2019.

66. Maddox, *Rosalind Franklin*, 212.

67. Letter from John Randall to Rosalind Franklin, April 17, 1953, JRP, RNDL 3/1/6.

68. James Boswell, *The Life of Samuel Johnson* (London: Penguin, 1986), 116. John-son uttered this chauvinistic quip on July 31, 1763, after Boswell told Johnson he had attended a Quaker meeting; see Howard Markel, "The Death of Dr. Samuel Johnson: A Historical Spoof on the Clinicopathologic Conference," in Howard Markel, *Literatim: Essays at the Intersections of Medicine and Culture* (New York: Oxford University Press, 2020), 15–24.

69. The "holy fool" refers to a cadre of "prophets" who followed Christ under the guise of acting mad, or "foolish," and who insisted that they were able to reveal the truth of the Gospel. In Watson's case, his god and gospel were contained within DNA. Horace Judson interview with Maurice Wilkins, June 26, 1971, HFJP; Judson, *The Eighth Day of Creation*, 156–57.

70. Author interview with James D. Watson (no. 2), July 24, 2019.

71. Watson, *The Double Helix*, 132–33.

72. Maddox, *Rosalind Franklin*, 254.

73. Maddox, *Rosalind Franklin*, 240–41, 246, 262–63, 268–69, 295.

74. Watson, *The Double Helix*, 133.

75. Letter from C. P. Snow to W. L. Bragg, March 14, 1968, after reading the man-uscript of *The Double Helix*, MFP, 4/2/1. In another communication, John Mad-

dox, the editor of *Nature*, wrote that "discomfited scientists had a duty to do more than strike Jim Watson off their Christmas card lists. They had a duty to come forward with their own accounts; Victor McElheny, review of *The Path to the Double Helix* by Robert Olby, *New York Times Book Review*, March 16, 1975, BR19.

76. Franklin and Gosling, "Molecular Configuration in Sodium Thymonucleate."

77. J. D. Watson and F. H. C. Crick, "Genetical Implications of the Structure of Deoxyribonucleic Acid," *Nature* 171, no. 4361 (May 30, 1953): 964–67.

78. Judson, *The Eighth Day of Creation*, 156.

79. Watson, *The Double Helix*, 130–31.

80. Watson, *The Double Helix*, 131.

81. Maurice Goldsmith, *Sage: A Life of J. D. Bernal* (London: Hutchinson, 1980), 166.

82. Letter from J. D. Bernal to Birkbeck Administration, January 6, 1955; letter from Rosalind Franklin to J. D. Bernal, undated (May 26, 1955?), both in RFP, "Rosalind Franklin File Kept by Professor J. D. Bernal," FRKN 2/31.

83. Letter from Rosalind Franklin to J. D. Bernal, July 25, 1955, RFP, "Rosalind Franklin File Kept by Professor J. D. Bernal," FRKN 2/31; Maddox, *Rosalind Franklin*, 256, 262–65.

84. Maddox, *Rosalind Franklin*, 254.

85. Letter from W. L. Bragg to Francis Crick, November 23, 1956 (83P/20); letter from Francis Crick to W. L. Bragg, December 8, 1956 (83P/37); invitation letter to Rosalind Franklin from W. L. Bragg, June 26, 1956 (85B/164); letter from Rosalind Franklin to W. L. Bragg, July 23, 1956 (85B/165), all in WLBP.

86. Muriel Franklin came tantalizingly close to identifying Caspar as a potential beau during her correspondence with Anne Sayre, and Brenda Maddox has suggested a love relationship between Franklin and Caspar which may have begun in 1955 or 1956, while he was a fellow under her at Birkbeck College, London (*Rosalind Franklin*, 258, 274–75, 280–81, 283, 295–96, 304). After Franklin's death, Caspar kept a photograph of her on his desk, and he named his first daughter after her. Some have even commented that the woman he married looked like her. Anna Ziegler dramatized an imagined romance in the play *Photograph No. 51*, in Anna Ziegler, *Plays One* (London: Oberon, 2016), 199–274. Jenifer Glynn holds that the stories about Caspar are "pure fantasy" (author interview, May 7, 2018). Caspar later worked with both James Watson and Aaron Klug.

87. Some have asserted that Franklin's comment "I wish I were" suggests a physical relationship with Donald Caspar, but her words could also represent the obvious wish that she was healthy and pregnant rather than riddled with malignant tumors. Maddox, *Rosalind Franklin*, 284, see also 279.

88. Maddox, *Rosalind Franklin*, 144.

89. Maddox, *Rosalind Franklin*, 285, quotes from her medical chart at University

College Hospital: "Prof. Nixon, UCH notes for Miss Rosalind Franklyn [sic] Right oophorectomy and left ovarian cystectomy, Case No. AD 1651, September 4, 1956;" See also K. A. Metcalfe, A. Eisen, J. Lerner-Ellis, and S. A. Narod, "Is it time to offer BRCA1 and BRCA2 testing to all Jewish women?," *Current Oncology* 22, no. 4 (2015): e233–36; F. Guo, J. M. Hirth, Y. Lin, G. Richardson, L. Levine, A. B. Berenson, and Y. Kuo, "Use of BRCA Mutation Test in the U.S., 2004–2014," *American Journal of Preventive Medicine* 52, no. 6 (2017): 702–9.

90. Glynn, *My Sister Rosalind Franklin*, 149–50.

91. Maddox, *Rosalind Franklin*, 315. Watson made a similar statement about Franklin's so-called "poor relations" with her parents to the author during lunch at the Cold Spring Harbor Laboratory on July 23, 2018.

92. Glynn, *My Sister Rosalind Franklin*, 142; author interview with Jenifer Glynn, May 7, 2018.

93. Maddox, *Rosalind Franklin*, 304–5. In his biography of Aaron Klug, Kenneth C. Holmes claims it was Crick who came to Birkbeck to invite Franklin and Klug to work at the new Laboratory for Molecular Biology in Cambridge; Kenneth C. Holmes, *Aaron Klug: A Long Way from Durban* (Cambridge: Cambridge University Press, 2017), 103–4.

94. Maddox, *Rosalind Franklin*, 305.

95. Anne Sayre interview with Gertrude "Peggy" Clark Dyche, May 31, 1977, ASP, box 7, "Post Publication Correspondence A–E."

96. Letter from Muriel Franklin to Anne Sayre, November 23, 1969, ASP, box 2, folder 15.2.

97. The term "spinster" was less derogatory than it would be considered today; it was often used as a legal term for a woman who had never married. Rosalind Franklin's death certificate, April 15, 1958, cited in Maddox, *Rosalind Franklin*, 307.

98. J. D. Bernal, "Dr. Rosalind E. Franklin," *Nature* 182, no. 4629 (1958): 154.

99. The third line on her tombstone, in Hebrew, says, "Rochel, daughter of Yehuda" (her Hebrew name and her father's Hebrew name); the final line represents the Hebrew initials for תהיה נשמתה צרורה בצרור החיים ("Let her soul shall be bound in the bundle of life"), from 1 Samuel 25:29. Willesden is one of the most prominent Anglo-Jewish cemeteries in Great Britain. See "Tomb of Rosalind Franklin, Non-Civil Parish-1444176," Historic England, https://historicengland.org.uk/listing/the-list/list-entry/1444176.

100. Author interview with Jenifer Glynn, May 7, 2018.

101. Glynn, *My Sister Rosalind Franklin*, 160. The *Oxford English Dictionary* defines "swot" as a slang term meaning "to work hard at one's studies." The term originated at the Royal Military College, Sandhurst, c. 1850, when the assignments of a mathematics professor named William Wallace were said to make one *swot* (from "sweat"). On her deathbed, Rosalind told her brother Colin of her ambition to become a member of the Royal Society. See also "Rosalind Franklin was

so much more than the 'wronged heroine' of DNA," editorial, *Nature* 583 (July 21, 2020): 492.

Part VI: The Nobel Prize

1. John Steinbeck, Nobel Prize Banquet speech, December 10, 1962, https://www .nobelprize.org/prizes/literature/1962/steinbeck/25229-john-steinbeck-banquet -speech-1962/.

Chapter 30: Stockholm

1. James D. Watson, Nobel Prize Banquet toast, December 10, 1962. https://www .nobelprize.org/prizes/medicine/1962/watson/speech/.
2. Adam Smith interview with James D. Watson, December 10, 2012, Nobel Media AB 2019, https://www.nobelprize.org/prizes/medicine/1962/watson/interview/.
3. Ragnar Sohlman, *The Legacy of Alfred Nobel* (London: Bodley Head, 1983).
4. Howard Markel, "The Story Behind Alfred Nobel's Spirit of Discovery," *PBS News-Hour,* https://www.pbs.org/newshour/health/the-story-behind-alfred-nobela-spirit -of-discovery.
5. James D. Watson, *Avoid Boring People: Lessons from a Life in Science* (New York: Knopf, 2007), 179.
6. Watson, *Avoid Boring People*, 179.
7. Maurice Wilkins, *The Third Man of the Double Helix* (Oxford: Oxford University Press, 2003), 241.
8. The Nobel Prize in Physiology or Medicine, 1962, https://www.nobelprize.org/ prizes/medicine/1962/summary/.
9. John Steinbeck's *The Grapes of Wrath* (New York: Viking, 1939) won the 1939 National Book Award and the 1940 Pulitzer Prize. See also William Souder, *Mad at the World: A Life of John Steinbeck* (New York: Norton, 2020).
10. Wilkins, *The Third Man of the Double Helix*, 241; Paul Douglas, "An Interview with James D. Watson," *Steinbeck Review* 4, no. 1 (February, 2007): 115–18.
11. The Nobel Prize in Physics, 1962, https://www.nobelprize.org/prizes/physics/ 1962/summary/.
12. The Nobel Medal for Physiology or Medicine, https://www.nobelprize.org/prizes/ facts/the-nobel-medal-for-physiology-or-medicine. Virgil, *Aeneid*, book 6, line 663; Aeneas is in the underworld and looking upon the spirits of past human beings who made great contributions to the betterment of humankind by their unique creations and discoveries in what we now call *artes et scientiae*, the arts and sciences. The original line is *"Inventas aut qui vitam excoluere per artes"* (in William Morris's 1876 translation it reads, "and they who bettered life on earth by new-found mastery"). Since 1980, the medals have been made of "18 carat recycled gold."
13. A Unique Gold Medal, https://www.nobelprize.org/prizes/about/the-nobel

-medals-and-the-medal-for-the-prize-in-economic-sciences; *Dr. James D. Watson's Nobel Medal and Related Papers*, auction catalogue, Christies: New York, December 4, 2014. The medal sold for $4.1 million to the Russian billionaire Alisher Usmanov, who promptly returned the medal to Watson. Watson said he would use some of the proceeds to support research at Cold Spring Harbor Laboratory and Trinity College, Dublin; see Brendan Borrell, "Watson's Nobel medal sells for U.S. $4.1 million," *Nature*, December 4, 2014, https://www.nature.com/news/watson-s-nobel-medal-sells-for-us-4-1-million-1.16500.

14. The morning after the awards ceremony, December 11, 1962, Jim Watson went to the Enskilda Bank to change his one-third share of the prize into American dollars; Watson, *Avoid Boring People*, 189. The amount of money for each of the Nobel Prizes awarded in 2020 was set at 10 million Swedish krona, or about $1,145,000; press release from the Nobel Foundation, September 24, 2020, https://www.nobelprize.org/press/?referringSource=articleShare#/publication/5 f6c4a7438241500049eca4a/552bd85dccc8e20c00e7f979?&sh=false.

15. James D. Watson, *Genes, Girls and Gamow: After the Double Helix* (New York: Knopf, 2002), 252.

16. Since 1974, the banquet has been held in the Blue Hall, to accommodate more guests. See also Philip Hench, "Reminiscences of the Nobel Festival, 1950," *Proceedings of the Staff Meetings of the Mayo Clinic* 26 (November 7, 1951): 417–37, available at https://www.nobelprize.org/ceremonies/reminiscences-of-the-nobel-festival-1950/.

17. Ulrica Söderlind, *The Nobel Banquets: A Century of Culinary History, 1901–2001* (Singapore: World Scientific, 2005), 148–52; menu available at https://www.nobelprize.org/ceremonies/nobel-banquet-menu-1962/.

18. Steinbeck, Nobel Prize Banquet speech. The axiom Steinbeck uses to conclude his speech ("In the end is the Word, and the Word is Man—and the Word is with Men") is adapted from John 1:1 (King James Version): "In the beginning was the Word, and the Word was with God, and the Word was God." The Steinbeck Nobel Prize files are "Utlånde av Svenska Akademiens Nobelkommitté, 1962; Förslag till utdelning av nobelpriset i litteratur år 1962" [Lent by the Swedish Academy's Nobel Committee, 1962; Proposal for the Awarding of the Nobel Prize in Literature in 1962]; Per Hallström, "John Steinbeck, 1943," Archives of the Swedish Academy, Stockholm.

19. Douglas, "An interview with James D. Watson."

20. Erling Norrby, *Nobel Prizes and Nature's Surprises* (Singapore: World Scientific, 2013), 348–50; see also Wilkins, *The Third Man of the Double Helix*, 242–43.

21. Watson, *Avoid Boring People*, 183, 192.

22. Horace Freeland Judson, *The Eighth Day of Creation: Makers of the Revolution in Biology* (Cold Spring Harbor, NY: Cold Spring Harbor Laboratory Press, 2013), 556.

23. Watson's toast did not match the soaring orations uttered by the thirty-fifth American president, or that of his superb speechwriter, Theodore Sorenson. Watson, *Avoid Boring People*, 187.

24. Watson, Nobel Prize Banquet toast.

25. Watson, *Avoid Boring People*, 187.

26. Maurice Wilkins, "The Molecular Configuration of Nucleic Acids," in N*obel Lectures, Physiology or Medicine 1942–1962* (Amsterdam: Elsevier, 1964), 754–82; see also James D. Watson, "The Involvement of RNA in the Synthesis of Proteins," ibid., 785–808, and Francis H. C. Crick, "On the Genetic Code," https://www.nobelprize.org/prizes/medicine/1962/crick/lecture/.

27. Norrby, *Nobel Prizes and Nature's Surprises*, 373–74.

28. Klug received the 1982 prize in Chemistry for his "development of crystallographic electron microscopy and his structural elucidation of biologically important nucleic acid-protein complexes." Aaron Klug, "From Macromolecules to Biological Assemblies," in *Nobel Lectures, Chemistry 1981–1990* (Singapore: World Scientific, 1992), available at https://www.nobelprize.org/prizes/chemistry/1982/klug/lecture/.

29. Watson, *Avoid Boring People*, 189.

30. Watson, *Avoid Boring People*, 193.

31. Watson, *Avoid Boring People*, 187, 189.

32. Judson, *The Eighth Day of Creation*, 556–57.

Chapter 31: Closing Credits

1. *The Man Who Shot Liberty Valance*, directed by John Ford, screenplay by James Warner Bellah and Willis Goldbeck based on a short story by Dorothy M. Johnson, Paramount Pictures, 1962.

2. Email from Ann-Mari Dumanski, Karolinska Institutet, to the author, August 6, 2018.

3. Email from Ann-Mari Dumanski, Karolinska Institutet, to the author, August 21, 2020.

4. The biochemist Erwin Chargaff remained intensely bitter about being overlooked in 1962 for the prize awarded to Watson, Crick, and Wilkins. Adding a hefty dose of salt to his wound were 1988 and 2001 invitations from the Nobel Committee for him to nominate "one or more candidates for the Nobel Prize in Physiology or Medicine." Dr. Chargaff knew he could not nominate himself. He died in 2002, Nobel-less. ECP, B: C37, Series IIC. See also Horace Freeland Judson, "No Nobel Prize for Whining," op-ed, *New York Times*, October 20, 2003, A17; David Kroll, "This Year's Nobel Prize in Chemistry Sparks Questions About How Winners Are Selected," *Chemical and Engineering News* 93, no. 45 (November 11, 2015): 35–36.

5. The Physics prize is a close second. As of 2020, out of the 114 prizes awarded to 216 laureates since 1901, 47 went to one laureate, 32 were shared by two, and 35 were shared by three; in Chemistry, out of 112 prizes awarded to 186 laureates, 63 were awarded to a single laureate, 24 were shared by two, and 25 were shared by three. Of the 100 Peace Prizes, 68 went to individual laureates, 30 to two, and 2 to three; of the 52 Economics prizes, which were awarded to 84 laureates, 25 went to individuals, 19 to two, and 7 to three; and of the 113 Literature prizes, only 4 awards were shared by two writers and the rest, 109, were awarded individually. The Peace prize has often been awarded to organizations (27); the 2020 award, for example, was awarded to the UN-based World Food Programme. See https://www.nobelprize.org/prizes/facts/nobel-prize-facts/.

6. The Swedish economist, diplomat, and second UN Secretary-General, Dag Hammerskjöld, died in a plane crash on September 18, 1961, at age fifty-six; he was nominated for the 1961 Peace Prize before his death. The Swedish poet and permanent secretary of the Swedish Academy, Karl Karlfeldt, died on April 8, 1931, at age sixty-six; he, too, was nominated before his death. Following the announcement of the 2011 Nobel Prize in Physiology or Medicine, it was discovered that one of the laureates, Ralph Steinman, had passed away three days earlier. An interpretation of the purpose of the rule above led the board of the Nobel Foundation to conclude that Dr. Steinman should remain a laureate because the Nobel Assembly at Karolinska Institute had made the announcement without knowing of Steinman's death. See Nobel Prize Facts, https://www.nobelprize.org/prizes/facts/nobel-prize-facts/.

7. Email from Dr. Karl Grandin to the author, July 22, 2019; see also Nobel Prize Facts, https://www.nobelprize.org/prizes/facts/nobel-prize-facts.

8. Horace Judson interview with William Lawrence Bragg, January 28, 1971, HFJP; letter from W. L. Bragg to Arne Westgren, Nobel Committee in Chemistry, January 9, 1960, Archives of the Center for the History of Science, Royal Swedish Academy of Sciences, Stockholm.

9. Letter from Linus Pauling to the Nobel Committee in Chemistry, March 15, 1960, LAHPP, http://scarc.library.oregonstate.edu/coll/pauling/dna/corr/sci9.001.47-lp-nobelcommittee-19600315.html.

10. The 1960 Chemistry nominators were W. L. Bragg, D. H. Campbell, W. H. Stein, H. C. Urey, J. Cockcroft, S. Moore, L. C. Pauling, and J. Monod, and the nominators in Physiology or Medicine were M. Stoker, E. J. King (he nominated only Crick and Perutz); in 1961, the Chemistry nominators were A. Szent-Gyorgi, G. Beadle, and R. M. Herriott; in 1962, G. H. Mudge, G. Beadle, C. H. Stuard-Harris, P. J. Gaillard, and F. H. Sobels. Archives of the Center for the History of Science, Royal Swedish Academy of Sciences, Stockholm. See also Nobel Prize nominations for Medicine or Physiology: *Karol. Inst.*

Nobelk. 1960. P.M. Försändelser Och Betänkanden; Sekret Handling, 1961. Betänkande angående F.H.C. Crick, J.D. Watson och M.H.F. Wilkins av Arne Engström; (Shipments and Reports; Secret Action, 1961. Report on F.H.C. Crick, J.D. Watson, and M.H.F. Wilkins by Arne Engström) Nobel Prize Nominations for Medicine or Physiology. *Karol. Inst. Nobelk. 1961; P.M. Försändelser Och Betänkanden; Sekret Handling, 1962. Betänkande angående F.H.C. Crick, J.D. Watson och M.H.F. Wilkins av Arne Engström* (Shipments and Reports; Secret Action, 1962. Report on F.H.C. Crick, J.D. Watson, and M.H.F. Wilkins by Arne Engström), Nobel Prize Nominations for Medicine or Physiology. *Karol. Inst. Nobelk. 1962.* Nobel Prize Committee in Physiology or Medicine, Nobel Forum, Karolinska Institute, Stockholm, Sweden. For secondary accounts, see Erling Norrby, *Nobel Prizes and Nature's Surprises* (Singapore: World Scientific, 2013), 333, 370; A. Gann and J. Witkowski, "DNA: Archives Reveal Nobel Nominations," correspondence, *Nature* 496 (2013): 434.

11. Arthur Conan Doyle, "The Gloria Scott," *The Adventures and Memoirs of Sherlock Holmes* (New York: Modern Library, 1946), 427. This 1893 story originally appeared in *The Strand* magazine and then as part of the collection of stories *The Memoirs of Sherlock Holmes* (London: George Newnes, 1893). In the story, Conan Doyle portrays the vicious murder of a prison ship captain by a "sham chaplin . . . [who stood over a dead captain] with his brains smeared over the chart of the Atlantic . . . [and] with a smoking pistol in his hand at his elbow." See also William Safire, "The Way We Live Now: On Language, Smoking Gun," *New York Times Magazine*, January 26, 2003, 18, https://www.nytimes.com/2003/01/26/magazine/the-way-we-live-now-1-26-03-on-language-smoking-gun.html.

12. Gunnar Hägg, "Arne Westgren, 1889–1975," *Acta Crystallographica* 32, no. 1 (1976): 172–73.

13. Arne Westgren, "Bilaga 8: Yttrande rörande förslag att belöna J. D. Watson, F. H. C. Crick och M. H. F. Wilkins med nobelpris," ("Appendix 8: Opinion on Proposals to award the Nobel Prize to J. D. Watson, F. H. C. Crick, and M. H. F. Wilkins") in *Protokoll vid Kungl: Vetenskapsakademiens Sammankomster för Behandling av Ärenden Rörande Nobelstiftelsen, År 1960* (Minutes at the Royal Swedish Academy of Sciences' Meetings for Processing Matters Concerning the Nobel Foundation, 1960), Center for the History of Science, Royal Swedish Academy of Sciences, Stockholm. Translated by Erling Norrby in *Nobel Prizes and Nature's Surprises*, 337–38. I am indebted to Professor Erling Norrby of the Royal Swedish Academy of Sciences, a former Nobel Prize committee member, for his excellent translation and elucidation of this critical report and his generosity in sharing it with me and permission to quote from it.

14. W. L. Bragg, nomination for the Nobel Prize in Chemistry, January 9, 1960, *Ärenden Rörande Nobelstiftelsen. År 1960* (Matters Concerning the Nobel Foundation, 1960), Center for the History of Science, Royal Swedish Academy of Sciences, Stockholm.

15. The Farewell Symphony was written in 1772 by Haydn, who served as *Kappelmeister* of the orchestra that played for Prince Esterházy. As summer segued into fall, the orchestra members appealed to Haydn to do something to get the prince to leave his summer home in rural Hungary so that they might return to their families. The musical ploy apparently worked and the court musicians returned to their homes the next day. Daniel Coit Gilman, Harry Thurston Peck, and Frank Moore Colby, eds., *The New International Encyclopedia* (New York: Dodd, Mead, 1905), 43; James Webster, *Haydn's "Farewell" Symphony and the Idea of Classical Style* (Cambridge: Cambridge University Press, 1991).

16. "Decoding Watson," *American Masters*, PBS, January 2, 2019, http://www.pbs.org/wnet/americanmasters/american-masters-decoding-watson-full-film/10923/?button=fullepisode.

17. Amy Harmon, "For James Watson, the Price Was Exile," *New York Times*, January 1, 2019, D1; Amy Harmon, "Lab Severs Ties with James Watson, Citing 'Unsubstantiated and Reckless' Remarks," *New York Times*, January 11, 2019, https://www.nytimes.com/2019/01/11/science/watson-dna-genetics.html.

18. Author interview with James D. Watson (no. 1), July 23, 2018.

19. James D. Watson, *Genes, Girls and Gamow: After the Double Helix* (New York: Knopf, 2002), 250.

20. Author interview with James D. Watson (no. 1), July 23, 2018. In 1970, Maurice Wilkins told Anne Sayre that had Franklin been alive, the Nobel would have been awarded only to Watson and Crick. Sayre recounts that "this thought haunts and depresses him"; Anne Sayre interview with Maurice Wilkins, June 15, 1970, ASP, box 4, folder 32.

21. James D. Watson, "Striving for Excellence," *A Passion for DNA: Genes, Genomes and Society* (Cold Spring Harbor, NY: Cold Spring Harbor Laboratory Press, 2001), 117–21, quote is on 120. In his 2010 book *Avoid Boring People*, Watson credits the eminent Columbia University historian Jacques Barzun for encouraging him to tell "the story of our discovery as a very human drama" (213).

22. Author interview with James D. Watson (no. 2), July 24, 2018.

23. James D. Watson, *The Double Helix: A Personal Account of the Discovery of the Structure of DNA*, edited by Gunther Stent (New York: Norton, 1980), 7. At the time, one of the best-selling books in Britain was the novel *Lucky Jim* by Kingsley Amis (London: Victor Gollancz, 1954), which depicted the life of a lecturer, James Dixon, at a provincial university. Brenda Maddox has speculated that Watson liked the novel so much he based the format of *The Double Helix*

on it, with himself playing the "bumbling honest Jim Dixon" and Franklin "the neurotic female lecturer Margaret Peel ('quite horribly well done', said the *New Statesman*), with her tasteless clothes and utter ignorance of how to appeal to a man." Brenda Maddox, *Rosalind Franklin: The Dark Lady of DNA* (New York: HarperCollins, 2002), 315.

24. Maddox, *Rosalind Franklin*, 314.

Illustration Credits

permission of the Franklin Trust and the Churchill Archives of Cambridge University)

85   Rosalind Franklin in Cabane des Evettes taking a break from a mountain climb, c. 1950 (reproduced with permission of the Franklin Trust and the Churchill Archives of Cambridge University)

93   Linus Pauling and Ava Helen Miller, 1922 (Linus and Ava Helen Pauling Papers, Oregon State University)

95   Linus Pauling in his Caltech laboratory, c. 1930s (Linus and Ava Helen Pauling Papers, Oregon State University)

98   Linus and Ava Helen Pauling and their children, c. 1941 (Linus and Ava Helen Pauling Papers, Oregon State University)

100 Robert Corey and Linus Pauling, 1951 (California Institute of Technology Archives)

108 James D. Watson, age ten, 1938 (James D. Watson Collection, Cold Spring Harbor Laboratory Archives)

109 Jim Watson as a Quiz Kid in 1942 (James D. Watson Collection, Cold Spring Harbor Laboratory Archives)

113 Jim Watson, graduation photo, 1947, University of Chicago (James D. Watson Collection, Cold Spring Harbor Laboratory Archives)

115 Max Delbrück and Salvador Luria at Cold Spring Harbor, summer 1952 (California Institute of Technology Archives)

117 Max Delbrück and the Phage Group, 1949 (California Institute of Technology Archives)

121 Jim and Betty Watson in Copenhagen, 1951 (James D. Watson Collection, Cold Spring Harbor Laboratory Archives)

123 Copenhagen, the city of bicycles (University of Michigan Center for the History of Medicine)

132 The Naples Zoological Station (University of Michigan Center for the History of Medicine)

134 The library of the Naples Zoological Station (Stazione Zoologica Anton Dohrn Archives)

140 Paestum: the Second Temple of Hera (University of Michigan Center for the History of Medicine)

144 The University of Michigan, c. 1950 (Bentley Historical Library at the University of Michigan)

148 John Kendrew, c. 1962 (Science Photo Library)

156 Max Perutz, c. 1962 (Associated Press Images)

163 Watson and Crick walking along the backs of King's and Clare colleges, 1952 (James D. Watson Collection, Cold Spring Harbor Laboratory Archives)

174 King's College Biophysics Unit interdepartmental cricket match, c. 1951 (King's College, London/Science Photo Library)

180 Punting on the Cam River, alongside King's and Clare colleges, Cambridge (University of Michigan Center for the History of Medicine)

187 The main lobby and stairwell of King's College, London (University of Michigan Center for the History of Medicine)

189 The lecture theatre, King's College, London (University of Michigan Center for the History of Medicine)

203 Dorothy Hodgkin and Linus Pauling, 1957 (Linus and Ava Helen Pauling Papers, Oregon State University)

223 King's Cross station (University of Michigan Center for the History of Medicine)

231 Watson and Crick. Detail from a group portrait of the staff of the Cavendish Laboratory, 1952 (Cavendish Physics Laboratory, Cambridge University)

239 Pauling testifying on the denial of his passport, c. 1953–55 (Linus and Ava Helen Pauling Papers, Oregon State University)

242 Ruth Shipley at the beginning of her career at the U.S. State Department, 1920 (University of Michigan Center for the History of Medicine)

248 Erwin Chargaff (National Library of Medicine)

259 The Abbaye de Royaumont (University of Michigan Center for the History of Medicine)

260 Alfred D. Hershey (National Library of Medicine)

268 Rosalind Franklin on holiday in Tuscany, Italy, spring 1950 (reproduced with permission of the Franklin Family Trust and the Churchill Archives of Cambridge University)

283 Louis Budenz (left) testifying before the Canwell Fact-Finding Committee on Un-American Activities in the State of Washington, January 27, 1948 (Museum of History and Industry, Seattle, Washington)

288 Clare Memorial Hall (author's photo)

289 Bridge across the Cam River to Clare College (University of Michigan Center for the History of Medicine)

296 Peter Pauling, 1954 (Linus and Ava Helen Pauling Papers, Oregon State University)

303 "Photograph No. 51" (Churchill Archives of Cambridge University and the Franklin Family Trust; Papers of Rosalind Franklin, FRKN 1/3)

318 Sir William Lawrence Bragg, c. 1953 (Cavendish Physics Laboratory, Cambridge University)

343 Tautomeric forms of guanine and thymine (James D. Watson Collection, Cold Spring Harbor Laboratory Archives)

349 Eureka! The correct adenine–thymine and guanine–cytosine (*keto* form) base pairs used to construct the double helix (James D. Watson Collection, Cold Spring Harbor Laboratory Archives)

356 Watson and Crick's double helix DNA model, 1953 (A. Barrington Brown/Science Source Images)

362   Crick and Watson having a cup of tea in Room 103 of the Austin Wing, Cavendish Laboratory (A. Barrington Brown/Science Source Images)

367   The third man: Maurice Wilkins (National Library of Medicine)

372   Rosalind Franklin at the microscope (reproduced with permission of the Franklin Family Trust and the Churchill College Archives Centre, University of Cambridge)

384   Neuvième Conseil de Chimie Solvay, April 1953 (courtesy of Instituts Internationaux de Physique et de Chimie Solvay and the Linus and Ava Helen Pauling Papers, Oregon State University)

407   James Watson with his sister, Elizabeth, and his father, James, Sr., arriving in Copenhagen for the Nobel Prize ceremony, 1962 (James D. Watson Collection, Cold Spring Harbor Laboratory Archives)

409   The Nobel laureates, 1962 (James D. Watson Collection, Cold Spring Harbor Laboratory Archives)

415   Francis Crick with his wife, Odile, and daughters Gabrielle, age twelve, and Jacqueline, age eight, at the Nobel Prize dinner, 1962 (James D. Watson Collection, Cold Spring Harbor Laboratory Archives)

Index

Page references for photos and illustrations are in *italic*.